Novel Approach to Sustainable Temperate Horticulture

In the current scenario, marked by a continual improvement in living standards, it becomes imperative to boost the productivity as well as the efficiency of agriculture, especially horticulture, which holds the potential for significant economic prosperity aligning with Sustainable Development Goal (SDG) number 8, "Decent Work and Economic Growth". Modern technological interventions, such as geospatial technology and geographic information system (GIS) technology, can be harnessed to yield effective results in addressing challenges and providing enhanced decision support, particularly in the planning of horticultural resource management. Cultivation and production of fresh produce faces several challenges, including prolonged juvenile phases and reproductive cycles with extended breeding periods, creating bottlenecks in the process. The evolving trends in biotechnology offer promising solutions for improving the selection of desirable traits. Biotechnological techniques aimed at improving fruit efficiency encompass tissue culture, induction of genetic variability, germplasm conservation, and molecular breeding/genomics. These methods involve the study of genetic diversity, DNA fingerprinting, and Quantitative Trait Loci (QTL) analysis for marker-assisted selection. Over the past few decades, the global population has consistently risen, raising concerns about the ability of the current food system to adequately feed the anticipated 10 billion people in the next 30 years. While this challenge is deemed achievable, certain changes in both food production and consumption systems are essential to ensure sustainability, reduce food loss and waste, and contribute to a global shift toward healthier and more sustainable diets. Implementing sustainable models of crop production represents a significant undertaking. To address the growing food demand amid deteriorating production environments, there is a need for promising technologies and effective management options to enhance productivity.

This book is poised to be a valuable resource for horticultural scientists operating in universities, government agencies, and research centers, offering insights into achieving sustainable cultivation practices for fruits. It stands out as the first of its kind, providing in-depth knowledge on environmentally friendly methods for cultivating temperate fruit crops, to reduce harmful emissions and pollution. This book will delve into the application of geographic information system (GIS) technology for estimating horticulture area expansion and crop yield. Additionally, it will encompass recent biotechnological interventions in horticulture, circular agriculture models, and emerging non-thermal food preservation techniques as significant components.

Features:

- Aims to provide a comprehensive and integrated overview of current techno-statistical techniques employed in horticulture, delving into the associated livelihood benefits derived from the practice.
- Explores the novel geographical trends to identify the site suitability indices of several temperate fruits.
- Offers a comprehensive and integrated exploration of recent trends in biotechnological approaches aimed at enhancing food production, quality, and safety.

Novel Approach to Sustainable Temperate Horticulture

Edited by
Gulzar Ahmad Nayik, Sajad H. Wani,
Vajahat Khursheed, and Mohammad Javed Ansari

CRC Press
Taylor & Francis Group
Boca Raton London New York

CRC Press is an imprint of the
Taylor & Francis Group, an **informa** business

Designed cover image: Shutterstock

First edition published 2025
by CRC Press
2385 NW Executive Center Drive, Suite 320, Boca Raton FL 33431

and by CRC Press
4 Park Square, Milton Park, Abingdon, Oxon, OX14 4RN

CRC Press is an imprint of Taylor & Francis Group, LLC

Library of Congress Cataloging-in-Publication Data
Names: Nayik, Gulzar Ahmad, editor. | Wani, Sajad H., editor. |
Khursheed, Vajahat, editor. | Ansari, Mohammad Javed, editor.
Title: Novel approach to sustainable temperate horticulture / edited by
Gulzar Ahmad Nayik, Sajad H Wani, Vajahat Khursheed, and Mohammad Javed Ansari
Description: First edition | Boca Raton, FL : CRC Press, 2024 |
Includes bibliographical references and index
Identifiers: LCCN 2024006544 (print) | LCCN 2024006545 (ebook) |
ISBN 9781032535241 (hardback) | ISBN 9781032535258 (paperback) |
ISBN 9781003412489 (ebook)
Subjects: LCSH: Horticultural crops. | Sustainable agriculture.
Classification: LCC SB318 .N68 2024 (print) | LCC SB318 (ebook) |
DDC 635—dc23/eng/20240516
LC record available at https://lccn.loc.gov/2024006544
LC ebook record available at https://lccn.loc.gov/2024006545

ISBN: 978-1-032-53524-1 (hbk)
ISBN: 978-1-032-53525-8 (pbk)
ISBN: 978-1-003-41248-9 (ebk)

DOI: 10.1201/9781003412489

Typeset in Times
by codeMantra

G.A. Nayik dedicates this book to his better half whose love and encouragement has been an inspiration, while the rest of the editors dedicate this to their beloved families.

Contents

Preface

Horticultural crops are high-valued crops generating higher profits than staple food crops per unit of land, and the income thus generated can be used for different purposes in terms of eradicating hunger through meeting the food requirements and other necessities. Despite facing numerous challenges, remarkable progress was made in horticulture in recent years from the point of view of the expansion of area and production to its large-scale improvement in productivity. Above all, now the horticulture is not merely a means of diversification but forms an integral part of food, nutritional security, and poverty alleviation and also an essential ingredient of economic security which in turn is in compliance of the United Nations Sustainable Development Goals (SDGs). In the present scenario when there is ever-increasing improvement in the standards of living, it becomes the need of the hour to enhance the productivity or we can say the efficiency of agriculture particularly that of the horticulture which has the potential of maximum economic prosperity Sustainable Development Goal (SDG) number 8 "Decent Work and Economic Growth". Modern technical interventions like geospatial technology and geographic information system (GIS) technology could be leveraged to obtain suitable results to meet the challenges and facilitate enhanced decision support, including planning for horticultural resources management. Remote sensing is a proven tool that can play an important role to achieve the goal. Methods like site-specific crop management and site suitability analysis for different horticulture crops, including measuring and analyzing the edaphic characteristics horticulture, have been widely used and acknowledged.

Horticulture fruits such as apple pear, nectarine, peach, plum, apricot, and strawberry are of immense importance due to their nutraceutical values. But various problems put bottlenecks in their cultivation and production which include long juvenile phase and reproductive cycles with lengthy breeding cycles. The emerging trends of biotechnology hold promise for enhancing the selection of desirable traits. Beginning with plant tissue culture, biotechnology encompasses the role of various advanced and emerging trends in promoting the production of horticulture fruits. To a large extent, some of the techniques have proven to be a boon to the perennial horticulture fruit species. The biotechnology techniques for enhancing the efficiency of fruits are tissue culture, induction of genetic variability, germplasm conservation, and molecular breeding/genomics studying genetic diversity, DNA fingerprinting, and QTL analysis for marker-assisted selection. There is a need to exploit this method using plant hormones-related genes for dwarfing/canopy and modifying other desirable traits through genetic manipulation, which need to be resolved. Biotechnology also needs to be promoted to advance early disease detection and development of abiotic and biotic resistance.

During the last decades, the world's population has been continuously increasing; consequently, so has the concern about the capability of the current food system for providing sufficient food for the 10 billion expected people in 30 years. Even though this challenge is considered feasible, some changes in the food production and consumption system are needed to achieve sustainability, reduce food waste, or contribute to a shift toward global healthier and more sustainable diets. Implementing sustainable models of crop production is really an enormous endeavor. Promising technologies and management options are needed to increase productivity to meet the growing food demand despite deteriorating production environments. Increasing global competition has put the focus on improving sensory quality and shelf-life of temperate fruits. This book will help in achieving sustainable cultivation of fruits for horticultural scientists working in universities, government, and other research centers. This book is first of its kind that will give in-depth knowledge about eco-friendly cultivation of temperate fruit crops to reduce environmentally damaging emissions and pollution. This book will focus on how we can use GIS in estimating horticulture area expansion and crop yield. Recent biotechnological interventions in horticulture, circular agriculture models, and emerging non-thermal food preservation techniques will also be major part of this book.

Dr. Gulzar Ahmad Nayik, Dr. Sajad H. Wani,
Dr. Vajahat Khursheed, and Dr. Mohammad Javed Ansari

Editors Biography

Gulzar Ahmad Nayik completed his Master's Degree in Food Technology from Islamic University of Science & Technology, Awantipora, Jammu and Kashmir, India, and his PhD from Sant Longowal Institute of Engineering & Technology, Longowal, Sangrur, Punjab, India. He has published over 100 peer-reviewed research and review papers, authored/co-authored over 60 book chapters, and edited 20 books with Springer, Elsevier, Royal Society of Chemistry, and Taylor & Francis. He has also delivered a number of presentations at various national and international conferences, seminars, workshops, and webinars. He was shortlisted twice, in 2017 and 2018, for the prestigious Inspire-Faculty Award from the Indian National Science Academy, New Delhi, India. He was nominated for India's prestigious National Award (Indian National Science Academy Medal for Young Scientists-2019–20). He also fulfills the roles of editor, associate editor, assistant editor, and reviewer for many food science and technology journals. He has received many awards, appreciations, and recognitions, and holds membership of various international societies and organizations. He is currently editing several book projects with Elsevier, Taylor & Francis, Springer Nature, etc.

Sajad H. Wani is currently working as a lecturer in the Department of Biotechnology, Government Degree College Shopian, affiliated with University of Kashmir Srinagar, India. Dr. Sajad is a former National Postdoctoral Fellow of Division of Biotechnology, Central Institute of Temperate Horticulture, Srinagar, J&K, India. He has a Doctor of Philosophy (Ph.D.) in Biotechnology from the Department of Biotechnology, University of Kashmir, J&K, India, and has also qualified J&K SET-2010 in Biotechnology. He has been awarded international travel support award by DST to attend third international conference on agriculture and biological science at Qingdao University China. He has been awarded Young Scientist Award and Best Thesis Award for his contribution toward the biological sciences in several reputed journals. Furthermore, he is the author of five books and 18 book chapters. In addition, Dr. Sajad has contributed in more than 50 research papers in reputed journals. He is well versed with several biological techniques: PCR, RT-PCR, Tissue Culture, Cell Culture, Bioevaluation Studies, HPLC, Column Chromatography, ELISA, and Spectroscopic Techniques and has completed several training courses related to applied biotechnology from prominent universities and institutes like NBAIR-ICAR, SKUAST-K, CITH-ICAR, etc. Dr. Sajad has presented his research at more than 20 conferences at national as well as international and participated in around six workshops.

Vajahat Khursheed completed his Bachelor's, Master's, and Doctorate degrees in Geography from Aligarh Muslim University, Aligarh, UP, India. Dr. Vajahat is currently working as a Lecturer in the Department of Geography, Government Degree College Shopian, J&K affiliated with the University of Kashmir Srinagar, India. Vajahat Khursheed has also qualified JRF in Geography. He has published many peer-reviewed research and review papers. He has also published few book chapters. He has also attended few national and international conferences, seminars, and workshops.

Dr. Mohammad Javed Ansari is currently working as an Assistant Professor in the Department of Botany, Hindu College Moradabad, UP, India. Dr. Ansari has also worked as an Assistant Professor in the Department of Plant Protection, College of Food & Agricultural Sciences, King Saud University Riyadh, Saudi Arabia, from 2010 to 2017. He did his Ph.D. in Biotechnology in the Indian Institute of Technology Roorkee, India, and M.Sc. in Botany from Ch. Charan Singh University Meerut, India. Dr. Ansari has published over 250 research/review articles in high-impact-factor journals. Dr. Ansari has worked as Principal Investigator in many funded projects in King Saud University Riyadh, Saudi Arabia. Dr. Ansari has participated in many national and international conferences (Abu Dhabi, Belgium, Indonesia, Turkey, USA, Italy, Romania, Germany, Spain, Cuba, Australia, Egypt, Ukraine, Ethiopia, Portugal, etc.). Dr. Ansari is editorial board member of many renowned journals and also member of many international scientific associations and societies. Dr. Ansari has supervised many students at master's and doctorate levels.

List of Contributors

Fareed Afzal
Department of Food Science
Government College University Faisalabad
Faisalabad, Pakistan

Muhammad Sajid Arshad
Department of Food Science
Government College University Faisalabad
Faisalabad, Pakistan

Yogesh Kumar Awasthi
Department of Food Technology
School of Applied and Life Science, Uttaranchal University
Dehradun, Uttarakhand, India

M. Cuneyt Bagdatli
Faculty of Architecture
Department of City and Regional Planning
Nigde Ömer Halisdemir University
Nigde, Türkiye

Jahangir A. Bhat
Department of School Education
J&K, India

Nayana Borah
Department of Life Sciences
School of Sciences, Jain (Deemed-to-be) University
Bangalore, India

Tridip Boruah
PG Department of Botany
Madhab Choudhury College
Barpeta, Assam, India

Shivali Chauhan
Department of Fruit Science
Dr YSP University of Horticulture and Forestry Nauni
Solan, HP, India

Kaberi Saikia Das
PG Department of Botany
Madhab Choudhury College
Barpeta, Assam, India

Kanika Dulta
Department of Food Technology
School of Applied and Life Science
Uttaranchal University
Dehradun, Uttarakhand, India

Wadah Elsheikh
Faculty of Agricultural Sciences and Technologies
Nigde Ömer Halisdemir University
Nigde, Türkiye

Umar Farooq
Department of Food Science and Technology
MNS-University of Agriculture
Multan, Punjab, Pakistan

Zannatul Ferdoush
Department of Nutrition & Food Engineering
Daffodil International University
Daffodil Smart City, Birulia, Savar, Dhaka, Bangladesh

Showkat A. Ganaie
Department of Geography
Govt. Degree College Shopian
J&K, India

Iram Hafiz
Institute of Chemistry
University of Sargodha
Sargodha, Pakistan

Hadia Hanif
Department of Food Science
Government College University Faisalabad
Faisalabad, Pakistan

Sneha Hasnu
PG Department of Botany
Madhab Choudhury College
Barpeta, Assam, India

Jayashree, V.H.
Department of Life Sciences, School of Sciences
Jain (Deemed-to-be) University
Bangalore, India

Meghali Kalita
Assam State Biodiversity Board
Aranya Bhawan
Guwahati, Assam, India

Bhawna Kaushal
Division of Fruit Science
SKUAST-K
Shalimar, Srinagar, J&K, India

Shakeela Khalid
Department of Zoology
University of Sargodha
Sargodha, 40100, Pakistan

Waseem Khalid
Faculty of Chemical Sciences and Technologies
Department of Organic Chemistry
University of Castilla-La Mancha
Ciudad Real, Spain
University Institute of Food Science and Technology
The University of Lahore
Lahore, Pakistan

Gözde Koşarsoy Ağçeli
Department of Biology
Hacettepe University
Ankara, Turkey

M.A. Kuchay
Department of Fruit Science
Central Institute of Temperate Horticulture
Rangreth, Srinagar, J&K, India

Keshav Kumar
Department of Biotechnology
Dr. Yashwant Singh Parmar University of Horticulture and Forestry
Solan, Himachal Pradesh, India

Sandeep Kumar, C.
Department of Life Sciences
School of Sciences
Jain (Deemed-to-be) University
Bangalore, India

Sharath Kumar, N.
ICAR-Central Institute of Temperate Horticulture
Old Airfield
Rangreth, Srinagar, J&K, India

Yong Kun Ma
School of Food and Biological Engineering
Jiangsu University
Jiangsu, Zhenjiang, China

Mahum
Faculty of Food and Home Sciences
Department of Food Science and Technology
Muhammad Nawaz Shareef University of Agriculture
Multan, Pakistan

Zahra Maqbool
Department of Food Science
Government College University Faisalabad
Faisalabad, Pakistan

Maryam Maqsood
Faculty of Food and Home Sciences
Department of Food Science and Technology
Muhammad Nawaz Shareef University of Agriculture
Multan, Pakistan

Mohd Mazhar
Department of Bioengineering
Integral University
Lucknow, India

Md. Anisur Rahman Mazumder
Department of Nutrition & Food Engineering
Daffodil International University
Daffodil Smart City, Birulia, Savar, Dhaka, Bangladesh

Javid Iqbal Mir
Department of Biotechnology
Central Institute of Temperate Horticulture
Rangreth, Srinagar, J&K, India

Andres Moreno
Faculty of Chemical Sciences and Technologies
Department of Organic Chemistry
University of Castilla-La Mancha
Ciudad Real, Spain

Bismillah Mubeen
School of Food and Biological Engineering
Jiangsu University
Jiangsu, Zhenjiang, China

Shamas Murtaza
School of Food and Biological Engineering
Jiangsu University
Jiangsu, Zhenjiang, China
and
Department of Food Science and Technology
MNS-University of Agriculture
Multan, Punjab, Pakistan

Hammad Naeem
Department of Food Science and Technology
MNS-University of Agriculture
Multan, Punjab, Pakistan

Gulzar Ahmad Nayik
Department of Food Science & Technology
Govt. Degree College Shopian, J&K, India

Deepika Negi
Department of Biotechnology
Dr. Yashwant Singh Parmar University of Horticulture and Forestry
Solan, Himachal Pradesh, India

Tripti Negi
Dev Bhoomi Uttarakhand University
Dehradun

Samara Qaisar alvi
Department of Food Science and Technology
MNS-University of Agriculture
Multan, Punjab, Pakistan

Saima Rafique
Faculty of Agriculture
Department of Food Science and Technology
University of Poonch
Rawalakot, AJ&K

Afzal Rahman
Department of Food Technology and Rural Industries
Bangladesh Agricultural University
Mymensingh, Bangladesh

Md. Shamsur Rahman
Department of Food Technology and Rural Industries
Bangladesh Agricultural University
Mymensingh, Bangladesh

Juwel Rana
Department of Nutrition & Food Engineering
Daffodil International University
Daffodil Smart City, Birulia, Savar, Dhaka, Bangladesh

Izza Faiz ul Rasool
Department of Food Science
Government College University Faisalabad
Faisalabad, Pakistan

Hadiqa Faiz ul Rasul
Center of Agricultural Biochemistry and Biotechnology (CABB)
University of Agriculture Faisalabad
Faisalabad, Pakistan

Saroat Rawdkuen
Food Science and Technology Program
School of Agro-Industry
Mae Fah Luang University
Chiang Rai, Thailand

Nighat Raza
Department of Food Science and Technology
MNS-University of Agriculture
Multan, Punjab, Pakistan

Sanabil
School of Food and Biological Engineering
Jiangsu University
Jiangsu, Zhenjiang, China
and
Faculty of Science and Technology
Department of Food Science and Technology
University of Central Punjab
Lahore, Punjab, Pakistan

Uzma Noor Shah
Department of Life Sciences
School of Sciences, Jain (Deemed-to-be) University
Bangalore, India

Renu Bala Sharma
Department of Food Technology
Dev Bhoomi Uttarakhand University
Dehradun

Abdul Mannan Shakir
Department of Food Science and Technology
MNS-University of Agriculture
Multan, Punjab, Pakistan

Muhammad Shehbaz
Department of Food Science and Technology
MNS-University of Agriculture
Multan, Punjab, Pakistan

Rattan Singh
Department of Food Technology
Guru Jambeshwar University of Science & Technology
Hisar, Haryana, India

Priyanka Tangariya
Department of Food Technology
School of Applied and Life Science
Uttaranchal University
Dehradun, Uttarakhand, India

Mohammad Tauseef Ali
Division of Fruit Science
SKUAST-K
Shalimar, Srinagar, J&K, India

Thottiam Vasudevan Ranganathan
Department of Food Processing Technology
School of Agriculture and Biosciences
Karunya Institute of Technology and Sciences
Coimbatore, Tamil Nadu, India

1 Estimation of Horticulture Area Expansion and Crop Yield Quantification by Using Remotely Sensed Data and Geographic Information Systems

M. Cuneyt Bagdatli and Wadah Elsheikh
Nigde Ömer Halisdemir University

1.1 INTRODUCTION

Every nation's economy depends heavily on agriculture. For a nation with a robust economy, it is a big trading sector. Food production at a low cost is the primary goal of every farmer, large-scale farm management, and regional agricultural agency (Kumar et al., 2015). Stopping the trend of declining overall crop production, reducing the rate of resource degradation, and raising farm incomes are required to solve the future issues of feeding the world's 9 billion people. Aside from these challenges, agricultural growth and development have also been seriously threatened by the fragmentation of land ownership, the liberalization of agricultural commerce, and climate change. Future agricultural productivity may be significantly boosted by the adoption of recently created technology (Hakkim et al., 2016). Since well-documented spatial maps are obtained by GISs and information acquired from satellite RS data, the efficacy of large-scale farming depends on this convergence (Brisco et al., 1998). It is now used for a variety of reasons in agriculture and associated industries.

Horticultural crops play a significant role in improving the economic conditions of farmers and business owners, boosting exports, expanding employment, and ensuring the population's access to food and nutrition. A current and reliable database is necessary for systematic planning and decision-making in order to enhance crop management and plant more area with horticulture crops. With the help of advanced technology called remote sensing (RS), management plans for scientifically based management can be generated by collecting and updating information. To identify crops, calculate area, pinpoint diseases and pests, and do other horticultural tasks,

DOI: 10.1201/9781003412489-1

some studies employed satellite data (Usha and Singh, 2013). Recent developments in these technologies give us the chance to collect and use digital spatial data and high-resolution satellite pictures (Merchant and Narumalani, 2009). Additionally, this technology offers users a range of geospatial information management tools and techniques for gathering, storing, integrating, querying, displaying, and analyzing geospatial data at different sizes (Goodchild, 2011), while RS technology uses sensors placed on various platforms, such as satellites, airborne RS, and ground-based equipment, to gather images and other data about crops and soil, which are then processed by computers to improve agricultural decision-making processes (Weiss et al., 2020). In many instances, satellite RS is the only practical source of data that may be utilized to supplement in situ monitoring networks. The ability to measure practically all elements of the hydrological cycle directly and indirectly is now possible thanks to satellite-based sensors (Zhang et al., 2016). Agricultural monitoring and planning have long used RS retrievals of the fluctuation in vegetation status and, consequently, plant yield and health. The extensive coverage (up to global) of satellite data also makes it possible to evaluate risk in relation to local food production, trade, and water security (Dalin et al., 2017). Despite the fact that several of these satellite RS products are still in their infancy and have significant restrictions, difficulties, and caveats in their usage, the broad spatial coverage and high temporal resolution enable them to offer nearly real-time access to information about the whole planet (Sheffield et al., 2018).

Because agricultural production systems are becoming more complex, policymakers are interested in learning how to use geography as a "dimension" of agriculture and modern technology to increase output and develop better production techniques (Thakur et al., 2017). The adoption of geographic information system (GIS) technology in agriculture has increased the likelihood of creating spatial explicit frameworks that are even better, making it easier to create dynamic agriculture databases and interactive systems (Nie et al., 2010). However, the availability of hydro-meteorological and agricultural monitoring networks is typically low, and there is often a large amount of latency in many locations, making them useless for real-time decision-making. The ability to monitor and forecast hazards is typically low in developing regions, where there is arguably the greatest need for information, as a result of the decline in hydro-meteorological and agricultural monitoring networks over the past 30 years, as well as the persistent underinvestment in infrastructure and training (Fay et al., 2017). Unprecedented opportunities to support and enhance crop yield estimation and the estimation of horticulture area expansion are made possible by the exponential growth of satellite-based information over the past 10 years, as evidenced by the increased number of satellites launched with continuously improving resolution and the multiple derived products. A wider audience can now handle and receive this abundance of information thanks to technical advancements, which are also eliminating conventional barriers to access and exploitation of satellite data (Sheffield et al., 2018). Nonetheless, in order to fully realize the potential of satellite-based information for enhancing agriculture, it is still necessary to address persistent issues with accuracy, sampling, and continuity. Moreover, this richness of new data presents additional issues that must be resolved. To clarify and assemble the many GIS and RS applications in agriculture, some thorough literature investigations have recently been conducted. In this chapter, we will highlight some studies on estimating horticulture area expansion and crop yield quantification using remotely sensed data and GISs.

1.2 REMOTE SENSING (RS)

In order to gather and update information to create scientific management plans, sophisticated technologies like RS and geomatics science are used. In order to produce precise, large-scale data about the earth's surface and atmosphere, a variety of different sensors, including microwave radiometers, cameras, laser meters, magnetic sensors, and sensors, collect electromagnetic data. Recent improvements in radiometric, geographical, and temporal resolutions, as well as the accessibility of airborne sensors or unmanned aerial vehicles (UAVs), have changed the effectiveness and methodology of horticulture RS. These data and images are digitized, making it simple to quantify and change them using computers, robust computational processing, and, in some cases, free software (Marinelli et al., 2019).

RS is also referred to as the process of gathering data and information about an object or event without coming into physical contact with it. RS is a phenomenon with numerous applications in geology, surveying, forestry, photography, and many others. Nonetheless, RS has found widespread application in agriculture. There are numerous agricultural applications for RS (Meraj et al., 2021).

Recent years have seen a tremendous expansion in the applications and performance of RS as a result of extraordinary technological advancements (Gruen, 2008). The observation space has expanded beyond the conventional topographic mapping and land cover classification, making it more challenging to define the boundaries of RS. Furthermore, it is challenging to organize RS technologies from the perspective of standard data collection since sensors have developed, platforms have increased, and many systems are so tightly linked that they do not match traditional classifications (Toth and Jóźków, 2016).

Technology for RS is constantly improving thanks to improvements in imaging sensors and the processing, storage, and communication capabilities of the information infrastructure. Additionally, new platforms are introduced, which have recently shown impressive growth (Colomina and Molina, 2014). In reality, during the last 5 years, the rate of change has steadily increased as new image sensors and advanced data and information processing technologies have been launched, greatly widening the applications (Pajeres, 2015).

There are numerous critical criteria that are unique to certain applications. For instance, bathymetric mapping has recently drawn greater interest because of requirements connected to climate change (Brock and Purkis, 2009). Direct or proximate sensors based on the ground collect real-time measurements of the water status at a specific place. Data are typically gathered from several locations throughout a field for decision-making reasons, enabling the use of geospatial interpolation techniques like Kriging (Ha et al., 2013). However, this type of data gathering is laborious, ineffective, and prone to mistakes, particularly when measuring the water quality of a vast area. The launching of various earth observation satellite systems, which normally fly between 180 and 2000 km in height, was motivated by the need to monitor and observe farms on a greater spatial scale (Belward et al., 2015). High-altitude manned aircraft and unmanned aerial systems (UAS) were used to fill the spatial division among high-resolution ground observations and comparatively low-resolution satellite data (McCabe et al., 2017). All of the aforementioned RS platforms are utilized for determining the water status for horticulture crops, depending on the requirements of the client (Mancini et al., 2019).

Each RS platform has a unique mix of advantages and disadvantages. Depending on the necessary geographic and temporal resolution for the RS crop water status data, one of those systems will be chosen above the others. UAS might prove more helpful for mapping the intra-field changes, whereas manned aircraft and satellites could be effective for describing an area on a broader scale. Ground systems mounted on vehicles provide some of the same measurement capabilities as RS, albeit on a smaller scale (Gutiérrez et al., 2018). These devices are able to cover a significantly larger area as compared to ground-based direct measurements since they can travel inside horticultural lines and gauge the water condition of neighboring plants while the vehicle is in motion (Aquino et al., 2018).

1.2.1 Utilizing Remote Sensing for Crop Yield Quantification and Estimating Horticultural Area Expansion

Early crop yield projections over agricultural lands are necessary for planning for food security and estimating agricultural economic return. The procedure and methods for tracking the growth of crops and predicting their yields have been enhanced because of the ongoing development of RS and GIS technology (Mann and Warner, 2017). Many global strategies have been put to the test over time in an effort to find the best agricultural yield prediction methods. The first estimation method to be used was one based on field surveys. At first, the yield for the following year was predicted using information from the previous year's yield, in-depth field surveys, and farmer-level interviews. The two primary approaches for yield forecasting up to 1929 worldwide were crop reporters' judgments and meteorological occurrences (Meena et al., 2018). A crucial component of production estimation is crop yield information. Every crop genotype has a specific yield potential that can be met in an experimental field under ideal circumstances. Yet, in the actual world, factors like soil, weather, and cultivation techniques, including planting date, irrigation, and fertilizer, affect crop output. Moreover, biotic stressors like disease and pests have an impact on crop output. Given that it delivers a rapid, precise, synoptic, and objective estimation of numerous crop characteristics, satellite-based RS is an appropriate substitute for assessing and forecasting crop condition and yield. One key resource for yield modeling is data from RS (Dadhwalet al., 2003).

Crop yield can be predicted by crop vigor. Using vegetation indices collected at various wavelengths, it can be evaluated. Models for simulating plant growth have been used to monitor crop growth, health, and production. The use of plant growth systems across wide areas has been constrained because most of them were developed at field scales. Synthetic aperture radar technology and crop modeling are combined to assess and predict yield (Setiyono et al., 2014). By measuring the loss of leaf area, RS data can be used to identify and assess the health of a crop. Chlorophyll normally breaks down as a result of pest infestations, and RS can be used to detect the decrease in chlorophyll levels in plants (Gebeyehu, 2019).

Horticultural crops frequently experience significant fluctuations in their production and consumption, which leads to an extremely unstable market and price. For market planning and export of goods, accurate statistics on the area and production of horticultural items are crucial. Hence, RS is crucial in evaluating the supply scenario. The size and productivity of crops like potatoes, which are cultivated in large fields, may be estimated with over 90% accuracy (Nageswara Rao et al. 2004). Along with other horticultural crops, vegetables are an essential component of diet and a source of income for growers. Both the supply and demand of certain commodities fluctuate significantly, which makes the market and pricing unpredictable. Accurate statistics on the region in which a particular crop is grown are helpful for market planning and export. Among other techniques, the supply scenario has been assessed using satellite-based RS. The evaluation of land cover area is one of the most evident uses of RS. Since it is frequently incorrect to predict area 4–5 months before the harvest using historical trend, economic analysis, surveys, and other instruments, the significance of RS in area estimations of horticultural crops is currently being researched (Usha and Singh, 2013). Nageswara Rao et al. (2004) claimed that it is possible to estimate the area and productivity of crops such as potatoes that are grown over huge contiguous fields with an accuracy rating of over 90%. Due to overlapping spectral signatures, it was possible to estimate the area beneath mango plantations older than 5 years with ease but not the younger ones. Early in the season, mulberry was discovered to be spectrally identical to several vegetable crops, but afterward showed considerable distinction. Due to overlapping spectral signatures, it is easier to estimate the area under mango orchards with trees older than 5 years than that with younger mango trees (Usha and Singh, 2013).

Gupta and Sharma (1990) visually analyzed Landsat-5 TM data from 1986 on a 1:50,000 scale to locate the abundant mango plantations in the Malihabad Tehsil of Lucknow district, Uttar Pradesh, India. High-resolution RS data offer a lot of potential in this field. Mango-growing areas might be distinguished from other orchards using time series moderate resolution imaging spectral radiometers (MODIS) Normalized Difference Vegetation Index (NDVI) profile indicators of phenotypic changes brought on by mango flowering. Because many mango orchards were small in size, the results showed that finer-resolution RS data were more reliable than coarse resolution in terms of the statistical accuracy of estimated area (Usha and Singh, 2013). Using SPOT 20 m, N'Doume et al. (2001) have verified their estimation of coffee and cocoa orchards in the Ivory Coast, West Africa. Citrus, grapes, plantation crops, and apple area estimation all utilized multispectral Indian RS satellite Linear Imaging Self-scanning Sensor (IRS LISS-III) data with a spatial resolution of 23 m (Sharma and Panigrahy, 2007). Shrivastava and Gebelein (2007) carried out a study in Florida to locate citrus groves for economic evaluation using Landsat Enhanced Thematic Mapper Plus (ETM+) (30 m) images. Their research showed a strong correlation between citrus production, income, and citrus area coverage determined by remotely sensed data. Using image data from the IRS LISS-III and Indian RS Advanced Wide

Field Sensor (IRS AWiFS) (23 and 55 m, respectively), Sharma and Panigrahy (2007) created an apple orchard database for the entire state of Himachal Pradesh in India. Through the use of relatively low-resolution pictures, they were effective in delineating. The hilly districts of India's northeastern states were identified and prioritized for passion fruit plantations using the same low-quality IRS imagery.

Nagaraja (2009) made an effort to determine the acreage and map the spatial distribution of mango-growing areas using linear imaging self-scanning (LISS) and moderate resolution imaging spectral radiometers (MODIS) data for the Saharanpur district, Uttar Pradesh (UP), in India. Spectral angle mapping, categorization, decision trees, supervised and unsupervised learning, and decision trees were all applied to the acreage estimation. The study thoroughly demonstrated the value of IRS IC LISS-III data in identifying mango orchard locations and calculating acreage. The SPOT 5 multispectral imaging data can be used to identify crop kinds and estimate crop acreage (Yang et al., 2011).

Moreover, RS has been used to estimate crop yields for a variety of annual crops, but little study has been done on estimating crop yields for vegetables and fruit trees (Mann and Warner, 2010). Growers can receive final crop production estimates via RS techniques, which can also accurately depict yield changes between fields or orchards. In-depth yield data may be acquired from a field with the expanded usage of harvester-mounted yield monitors, and RS imaging enables data to be examined more precisely. Only a few crops, including citrus, pistachio, and tomato crops, now have commercial yield monitors available in the research and development stages (Usha and Singh, 2013). Citrus trees were mapped using an automated ultrasonic system and a sensor-based automatic yield monitoring system. They discovered a linear relationship between fruit yield and the sizes of the trees detected by ultrasonography. Using sensors, color photography, and ultrasonic sensors, citrus yield monitoring was carried out (Schumann and Hostler, 2008). Use of image processing allowed for the measurement of wild blueberry yield (Zaman et al., 2008). Swain et al. (2010) created a computer vision system for mapping the production of wild blueberries.

Ye et al. (2008) also investigated the relationships between specific canopy characteristics found in airborne multispectral images and citrus tree fruit yields, and they discovered that mature canopy leaves were more significantly correlated with fruit yield for the current growing season, while younger leaves were more significantly correlated with fruit yield for the following growing season. Multispectral imageries with visible and Near Infrared (NIR) bands would be sufficient for the investigation of fruit and nut plants since these bands primarily contain the distinguishable spectral characteristics, whereas some fruit crops have a very wide range for the highest spectral reflectance (800–1100 nm band). Due to the high spectral resolution and ability to detect subtle spectral features that are diagnostic of specific minerals and plant characteristics, it is therefore possible to use airborne visible/infrared imaging spectrometer (AVIRIS) imagery from a National Aeronautics and Space Administration (NASA) high-altitude aircraft that records reflected light in 224 visible and NIR wavelengths. Airborne hyperspectral photography, IKONOS, and Quick Bird photos have all been used to analyze crop yields (Yang et al., 2007). In order to forecast productivity and determine crop water requirements, O'Connell and Goodwin (2004) used remotely sensed high-resolution aerial photographs to identify the tree canopy of a peach orchard.

Ye et al. (2007) used partial least squares regression models as opposed to Vegetation Indices VIs and multiple linear regression models to predict citrus tree yields from their canopy features obtained from airborne hyperspectral pictures. Numerous studies have shown that airborne hyperspectrum photography has a significant potential for determining fruit output and detecting plant stress (Panda et al., 2010).

Thermal imaging has also been used to determine how many fruits are present in groves and orchards. The assumption that the fruit has a temperature that is noticeably different from its surroundings is the basis for its detection. Despite the fact that both fruits and biomass are essentially made up of water, fruits often have a higher mass than other parts of biomass, such as leaves, which causes a drastically different transient thermodynamic reaction and time constant. Fruits have different thermodynamic properties than the items around them; therefore, this theory is generally accurate. However, in addition to ambient temperature, relative humidity, wind speed, and incident radiation intensity, the temperature of the fruits is also influenced by these factors. The temperature temporal variation in the citrus canopy was examined in an effort to determine the ideal period for fruit identification. Throughout the late afternoon, about 16:00, till midnight, there was a comparatively big temperature difference between the fruit and canopy. This made the fruit seem better in the thermal photos and made it easier to recognize fruit (Bulanon et al., 2008). Because the success rate for robotic harvesting is so low, thermal imaging was accordingly combined with additional vision systems to boost efficiency. The fruits were identified by Stajnko et al. (2004) using pseudo-color thermal imaging and color image processing technologies. The pseudo-color image's color coding may then affect the image processing results. The number of fruits humanly counted in the photos, and the number of apples automatically detected had a strong correlation. Also, it was possible to estimate the fruit diameter from the thermal images, with riper fruits yielding better accuracy. Fruit identification accuracy can be improved by combining thermal and Red-Green-Blue RGB multi-modal pictures (Bulanon et al., 2008). The co-registration of the images is the initial stage in fusing RGB and thermal images. Combining many similarity measures for automatic registration has proven to be more effective than utilizing each similarity measure separately (Wachs et al., 2009).

Airborne multispectral and hyper spectral images can be used to find yield trends and within-field control zones, which might be very useful to use for within-season and after-season management even though precise yield projections aren't always achievable throughout the growing season. There are times during the growing season when spectral measurements from the crop canopy can be used to estimate yield, but there are many other variables to consider, such as weather, image availability, and the lack of user-friendly and trustworthy models (Lee et al., 2010).

Numerous studies demonstrate how crop production assessment uses GIS and RS technologies. Memon et al. (2019) demonstrated how combining multiple RS-based spectral indices and multispectral Landsat satellite data may be used to estimate the proportion of wheat straw cover and its effects on rice crop yields. Planning for long-term agricultural sustainability in rice–wheat cropping systems can be aided by the knowledge. The results of Hassan and Goheer's study (2021) show that crop yield data, vegetation indices produced from moderate resolution imaging spectroradiometer satellite imagery, and a GIS modeling technique can all be used to accurately estimate wheat crop production before harvest. A useful tool for simulating rice crop yield was supplied by a GIS-based environment policy integrated climate model employed in another work by Muslim et al. (2015). To estimate crop yield fluctuations regionally, the model incorporated crop-level data at the regional level with soil data, farm management data, and climate data. Similarly, Al-Gaadi et al. (2016) used soil-adjusted vegetation index and NDVI collected from Landsat satellite pictures taken during the potato growth stages to forecast potato tuber crop production. Crop yield forecasting models that are based on GIS and RS technology may be more widely used to guide spatially oriented agricultural strategies. For instance, policy action could be developed to change the specific elements that affect agricultural yields based on the findings of these models (Meng et al., 2014). It is essential to predict crop yields far in advance of harvest, particularly in an area with unpredictable environmental conditions. In order to plan and make policies that will ensure food security and predict agricultural economic return, it is crucial to monitor the growth circumstances of agricultural crops and make predictions about their potential output (Muslim et al., 2015). In order to feed a growing population in low- and middle-income countries (LMICs), agricultural production systems must work to close the yield gap between the yields currently attained by farmers and those that may be attained under rain-fed subsistence farming systems. Hochman et al. (2013) developed a methodology to assess and map wheat yield gaps by combining statistical yield and cropping area data, RS data, cropping system simulation, and GIS mapping in order to resolve this disparity. From the foregoing, it is clear that RS offers many advantages for crop monitoring and calculating the extension of horticultural areas.

1.2.2 Unmanned Aircraft Systems for Crop Yield Quantification and Estimating Horticultural Area Expansion

Unmanned aerial systems (UASs), also referred to as drones and unmanned aerial vehicles, have advanced significantly across the globe as legal frameworks have gradually made room for this expanding industry (Wargo et al., 2016). Agricultural RS makes use of both rotary- and fixed-wing unmanned aerial systems (UASs). Regarding sensor payload, adaptability, and coverage, each variation has advantages and disadvantages. In this regard, the literature provides a list of state-of-the-art UAS, their classification, an overview of their structural characteristics, and flight parameters, in the context of agricultural use (Zecha et al., 2013).

A rotary-wing UAS may have 4, 6, 8, or more rotors, making it resemble a helicopter (Urbahs and Jonaite, 2013). While rotary-wing UASs are more agile and have a wider range of motions, fixed-wing UAS must move forward at a certain speed to sustain power (Gautam and Ha, 2013). Due to its flexibility and unique capabilities, rotary-wing UAS can hover, perform vertical takeoffs and landings, move vertically (up and down), or return to the previous location. In contrast, fixed-wing UAS can travel farther in a single trip because of their longer flight times, greater payload capacities, and faster flight speeds (Shi et al., 2016). The advantages of both fixed-wing and rotary-wing aircraft are combined in recently developed fixed-wing UASs, making them a suitable platform for agricultural use (Gautam and Pagay, 2020).

UAS offers flexibility in terms of spatial resolution, observation size, spectral bands, and temporal resolution to enable data collection on any day with good weather. Similar to satellites and human-powered aircraft, the UAS is inoperable in the presence of precipitation, severe winds, and extreme temperatures. By simply adjusting the flying altitude, the UAS provides extra flexibility to analyze a larger area with lower spatial resolution or a smaller area with significantly better detail (Candiago et al., 2015). Temporarily, the UAS can be scheduled with little notice at a user-defined time, enabling time-sensitive applications like recording significant phonological stages of crop growth. Due to the interchangeability and spectral flexibility that UAS provide, any desired combination of sensors and spectral bands can be utilized to target specific features (Gautam and Pagay, 2020).

Before being used in the grower decision-making process, UAS-acquired image data need to be post-processed. The turnaround time for mosaicking UAS pictures is now between 1 day and 1 week, depending on the size of the dataset, the processing power, and the spatial and spectral quality of the finished result (Turner et al., 2013). For established horticultural crops in particular, the spatial quality of the data may not be as important as the spectral quality. For better spectral quality, the spectrum sensors must be calibrated and atmospheric effects must be adjusted. After processing aerial images, it was shown that the spectral data from UAS were significantly correlated with ground-based data (Berni et al., 2009).

Thermal, multispectral, and RGB sensors are the most often utilized UAS-based sensor types to research crop water status, whereas LiDAR and hyperspectral sensors are employed less frequently (Pajares, 2015). Broader physiological characteristics of the crop can be recorded using spectral sensors, for instance, biomass and greenness, which are typically correlated with crop water status (Di Gennaro et al., 2017). The biophysical and chemical characteristics of crops can be directly observed using narrower band spectral sensors. For example, the solar-induced chlorophyll fluorescence (SIF) and the photochemical reflectance index (PRI), which gauge a plant's photosynthetic effectiveness (Zarco-Tejada et al., 2012; Cendrero-Mateo et al., 2019). Thermal-based sensors detect the crop's surface temperature, which provides information about the plant's biotic and abiotic stresses (Khanal et al., 2017). In general, 3D metrics like plant size and shape can be precisely measured utilizing 3D point clouds produced by a digital RGB camera and LiDAR (Matese and Filippo, 2015).

RS, a branch of general sensing, computing science, and information technology, is significantly impacted by the general growth trends. Sensors can generate a lot more data than can currently be managed with the tools and techniques available. New methods—often referred to as "data analytics"—are needed to solve the issues that huge data in a cloud computing environment bring about. Because of the ongoing paradigm shift in navigation, a process of convergence between the RS and navigation domains has been initiated. While navigation requires imaging to travel in regions without GPS and/or other Radio Frequency (RF) signal-based navigation solutions, RS demands higher georeferencing capability as sensors get more sophisticated (Toth and Jóźków, 2016).

One of the several RS applications that has profited from the development and implementation of several commercial and customized airborne imaging systems during the past 20 years is precision agriculture. In recent years, UASs have developed into a versatile and cost-effective platform for airborne RS. These innovations fill the gap in spatial resolution between ground-based platforms and platforms used in piloted aircraft for RS. Conventional manned aircraft provide an accessible and flexible platform for airborne RS, including the tens of thousands of agricultural aircraft in use in the United States (Yang, 2016). With its numerous techniques and agricultural uses, RS has transformed crop monitoring and interventions for raising farm output (Weiss et al., 2020). RS is essential for achieving the objectives of precision agriculture, together with GPS, GIS, and other instruments. This combination is essential for a number of applications (Ghosh and Kumpatla, 2022). In order to better monitor and characterize crops, it is necessary to combine RS observations with varying spatial and spectral capabilities with auxiliary data from other sources (such as ground measurement networks, national statistics, and historical time series). The development of solid observatory infrastructures, ground measurement networks, and open source data and codes, which provide long-term records, as well as their promotion, makes such techniques particularly promising (Delgado et al., 2018).

1.3 GEOGRAPHIC INFORMATION SYSTEMS (GIS)

The GIS is described as a strong system made up of tools for data collection, storage, and retrieval at will as well as for the analysis, transformation, and display of the spatial data for a particular purpose (Delgado et al., 2019). It serves a critical purpose by providing the spatial context and information on a variety of properties, each of which is available as a data layer. Users also have access to tools for modifying both spatial and non-spatial data, which is displayed in illustrative and approachable map forms. A variety of industries, including geography, environmental sciences, natural resources, forestry, agriculture, food, manufacturing, banking, and health services, are being significantly impacted by GISs (Soomro, 2015). Whereas the RS involves learning about the earth from a distant object, usually an aircraft or a satellite orbiting the globe, the RS involves learning about the earth from a close-up object. These craft are outfitted with sensors that can snap images of the earth's outermost layer in the radar, visible, and other spectral range. GIS, on the other hand, uses computer-based tools to collect, store, analyze, and display spatial data. These two techniques are frequently used in the social and natural sciences to evaluate natural resources, track environmental changes, and comprehend a variety of social phenomena as they relate to ecological and environmental issues (Manson et al., 2015).

The introduction of digital agriculture, which is recognized as the fourth agricultural revolution, has fundamentally altered how farming is done thanks to advancements in geospatial technology, sensors, artificial intelligence, robotics, and other tools and technologies. For the precise identification of farming issue areas and to keep track of all activities along the whole agriculture value chain, image and non-image data as well as spatial context are required. The use of GIS offers clear and understandable information visualization for data-driven decision-making to boost crop productivity. It does this through the use of its analytical modules and component tools, as well as the data generated by its partner technologies like RS and GPS. Although GIS has long been used for agricultural reasons, current technological improvements have caused the diversity of uses to rapidly increase. Over the past few decades, there has been a tremendous increase in the use of GIS technologies for diverse agricultural applications at the local, regional, national, and worldwide levels. These applications most typically involve the use of GIS together with partner technologies like RS, GPS, and data analytics in order to gain a full picture of a farm or region and to provide intervention or corrective actions for the crops and/or soils. Another advantage of GIS is the flexibility it provides for using the same data for different purposes or applications because it is linked to a common referencing system. Additionally, we can add more data and combine it with already-existing data to conduct a collaborative analysis to produce unique findings (Ghosh and Kumpatla, 2022).

Due to its role in easing the collection, storage, retrieval, and analysis of data on features and locations as well as its utility for data-driven solutions, particularly in site-specific management, GIS is known as the brain of precision agriculture (Kumar et al., 2017). Digital GIS maps differ from traditional maps in that they have numerous layers of data, each containing information or a map on a particular attribute, such as soil survey, precipitation, nutrient status, pest infestation, and yield. Additionally, GIS provides analytical capability through the application of geospatial analytics and statistical tools, enabling the identification of correlations between features. The conclusions from these assessments can be used to inform management practice decisions (Ghosh and Kumpatla, 2022).

No matter the approach taken to describe agricultural systems, continual validation is still essential, and much more work needs to be done to assess the abilities of RS products, both qualitatively and quantitatively. This aids in detecting the products' shortcomings so that improvements can be made to the algorithms for acquiring data as well as the models themselves, as well as to the training data sets for machine learning techniques (Weiss et al., 2020). In fact, the utilization of RS data has become more democratic, bypassing the scientific community to benefit everyone in society (such as policy makers, farmers, and breeders). In light of this, one of the current challenges is figuring out how to qualify the data so that users can understand what product they are using and how reliable it can be in light of the specifics of their application. New validation techniques are also being researched as part of crowdsourcing investigations (See et al., 2016), with examples including crop map confirmation via farmer surveys or mobile applications (Fritz et al., 2019).

Many techniques for RS are well suited to vast, uniform expanses. This can be attributed to both the homogeneity assumption, which reduces the interactions between input and output variables, and the historical technical limitations of sensors. This also explains why there are so few studies that concentrate on agricultural systems made up of smallholder farms and fragmented landscapes. Given that they maintain the food chain while being incredibly sensitive to natural disasters and climate change, research into these agricultural systems has to be prioritized, especially in less developed countries (Morton, 2007). It is now necessary to adapt RS algorithms to these types of systems, which exhibit significant horizontal heterogeneities. Additionally, as a result of the emergence of novel agricultural systems like agroforestry, which combine different species in one area to better address climatic and sustainable development limits, both horizontal and vertical heterogeneities are increasingly included by RS algorithms (Weiss et al., 2020). RS can be viewed as a useful tool for agricultural actors when it comes to implementing coordinated activities to achieve agricultural sustainability.

1.3.1 Application of GIS for Estimation of Horticulture Area Expansion and Crop Yield Quantification

Planning land management for both sustainable agriculture and the sustainable use of natural resources is based on agricultural productivity. Thus, a precise and balanced structure between nature and agricultural ecosystems is essential for defining land suitability in terms of economically feasible, sustainable, and resilient production systems (Kocjančič et al., 2018). Many agricultural products have been modeled using GIS. Although these studies are helpful in understanding the regional variance of agricultural products, they often only considered a small number of variables. As a result, the assessment of land suitability for agricultural output was ignored (Balezentis et al., 2020).

Monitoring crop growth, health, and an accurate or nearly exact yield projection are essential for managing food security since they not only help to estimate economic return but also the production of food. Several research revealed that conventional techniques for estimating agricultural production could result in subpar evaluation and inaccurate crop area appraisal (Dadhwal and Ray, 2000). Furthermore, gathering crop and yield data using these approaches is time-consuming, labor-intensive, and expensive. The capacity to assess the temporal and spatial variability of crop dynamics and yield output using RS, GPS, and GIS technologies is quite helpful in this case (Thomas, 1997).

The nation needs a current understanding of agricultural yields in order to establish rules and costs for the import and export of food as well as to precisely estimate and track crop yields. The Food and Agricultural Organization of the United Nations (FAO) has given it top priority, as have the majority of countries. Crop yield estimation mainly concentrates on two areas: first, estimating the crop planting area, and second, building the yield model by utilizing both the yield model and long-distance RS monitoring. GIS allows for the swift and accurate completion of this operation. The production process begins with the creation of RS photos of crop development using a RS technique, followed by the interpretation of the information contained in the RS images, evaluation of various spatial data in GIS, determination of crop types, and computation of planting area using statistics. After examining how the environment and condition have changed during the crop growth process, we utilize the model feature of the GIS system to construct a crop growth model and multiple estimating models for various scenarios. The model is then updated to integrate the aforementioned variables, projecting the yield and development of large-area crops (Zhang and Cao, 2019). Using a GIS tool and the SALUS (System Approach to Land Use Sustainability) model, Thanh Van Hoang and Tien Yin Chou conducted a case study of Haiphong Province in northern Vietnam. Haiphong Province is used as a case study in research that demonstrates the effects of climate change on agriculture in northern Vietnam. Second, the SALUS model is incorporated into the GIS to forecast the yield of rice in case studies as well as other environmental parameters. This comparison shows a confidence level of more than 80% when the model is adequately modeled using the root mean square error. This model has accurately predicted rice yield in a changing climate (Van Hoang and Chou, 2015).

With the help of the accompanying analysis techniques and geospatial data obtained in a spatiotemporal manner, it is possible to anticipate predicted yields based on crop growth and health and monitor changes in crop health. A common method for assessing crop health is the development of vegetation indices based on surface reflectance from crop canopies at two or more wavelengths. Numerous vegetation indicators can be employed to examine crop development dynamics, stress from biotic sources, the quantity and quality of vegetation, and other crucial evaluations (Ghosh and Kumpatla, 2022). For instance, Adhav et al. (2021) used a variety of vegetation indices to assess crop health and also fluctuations in health conditions. With the use of the ArcMap 10.5 program, they aggregated all vegetative indices, reclassified the merged indices, and used the reclassified indices to describe health scenarios categorically. Farmers can find regions that require rapid management intervention using this type of depiction (Adhav et al., 2021).

Combining the two key technologies of GIS and RS with the required input from others allows for the creation of models for estimating agricultural yields at different geographical scales and an efficient approach for assessing crop health. While crop health may be monitored and yield predictions can be made using remotely sensed photos and related analytics, GIS technology makes it possible to gather, store, retrieve, and display data that are geographically linked. The utilization of remotely sensed geospatial data gathered by satellites, aircraft, or UAVs enables the analysis of a crop's health. The characteristics of the soils that support the crops as well as their numerous crop qualities can be discovered using these data. With the help of the accompanying analysis techniques and geospatial data obtained in a spatiotemporal manner, it is possible to anticipate predicted yields based on crop growth and health and monitor changes in crop health. A common method for assessing crop health is the development of vegetation indices based on surface reflectance from crop canopies at two or more wavelengths. There are several vegetation indices available for measuring the quantity and health of vegetation, crop development dynamics, stress brought on by biotic or abiotic sources, and other useful assessments (Xue and Su, 2017).

Particularly in terms of revisit frequency, the versatility of UAV platforms appears to be better than that of satellite data. However, the use of such platforms is severely constrained by meteorological factors (such as the need for low wind speeds and a lack of precipitation); by current regulations; by the spatial coverage, which can be reduced by autonomy or battery; or by the maximum payload, which limits the use of multiple sensors at once. Although in-flight and vicarious calibration techniques for satellite platforms are well characterized, a consistent procedure is not yet available for the calibration of UAV-borne sensors (Barsi et al., 2014). In fact, at particular points during the flight, the latter take pictures of calibration targets (Herrero-Huerta et al., 2014), while the key environmental parameters (such optical lighting and thermal infrared temperature) may change dramatically throughout the flight due to things like cloud shadows, wind patterns, and solar courses. Due to this, the calibration may not be successful, which will result in inaccurate signal quantification (Wang et al., 2019). Ultimately, proximal sensing often yields sufficient data accuracy but is constrained to a low spatial and temporal coverage due to expense (e.g., Internet of Things) or labor (Weiss et al., 2020).

The use of UAVs for data collecting and analysis has shown to be crucial for farmers in order to implement timely and suitable corrective actions in order to maintain or increase productivity. A recent study evaluated the effectiveness of using multispectral UAV imagery and a random forest machine learning (ML) algorithm to estimate the chlorophyll content of maize at different growth stages. In order to help farmers manage their crops, the study also generated a map of chlorophyll variation that reflected the spatial heterogeneity of chlorophyll in the field. Such RS- and GIS-based diagnostics and interventions are crucial for smallholder farmers (Brewer et al., 2022). Another study assessed crop health using several chlorophyll indices and adjusted vegetation indexes using data from two separate satellites and ArcMap (of ArcGIS) for geospatial analytics. The results provided information that might be utilized to manage nutrient applications to increase agricultural yield (Singhal et al., 2018). The accessibility of an accurate acreage map of the crop of interest, the cropping systems in a particular area, and technologies for predicting yield before the reproductive phase or harvesting of the crop are two essential prerequisites for implementing location-specific management practices and interventions. Both objectives can be accomplished with the aid of RS and GIS technology. While GIS tools can offer the spatial context, NDVI, a widely utilized vegetation index, serves the dual purposes of monitoring crop health and forecasting crop yield. There are numerous studies that used NDVI and GIS to predict yield (Ghosh and Kumpatla, 2022). In a field study to create an effective model for predicting potato tuber yield using RS and GIS techniques, two vegetation indices, the NDVI and soil-adjusted vegetation index (SAVI), created from images acquired by Landsat-8 and Sentinel-2 satellites, were found to be extremely effective in yield prediction (Al-Gaadi et al., 2016). Also, using the indices, they were able to produce maps of the research region that showed the productivity differences between different zones. Both authorities and farmers can use this information to estimate production accurately. Farmers can use it to adopt the necessary management procedures. While most researchers have utilized GIS- and RS-based vegetation indicators to predict crop yields, a few of researchers have combined GIS with crop modeling or physiological models and have shown that these combinations perform well in this area (Kadiyala et al., 2015). Due to their usefulness in establishing management techniques, evaluating the impact of climatic fluctuations on crop performance, and forecasting yields, crop simulation models have gained popularity (Jones et al., 2003). In a study that brought together RS data, a crop growth model, and GIS, it was shown that yield estimates from RS pictures were more precise than estimates from an alternative method that linked the Oryza 2000 rice crop model with GIS climatic layers and soil properties. This proved how effective combining RS, GIS, and crop models was for estimating crop yield (Sailaja et al., 2013). The researchers were able to capture the spatial variety of input variables and how it affected yield projections by fusing GIS and RS with a soybean growth model (Carbone et al., 1996). The findings showed geographical variation in the predicted yield estimations, which was mostly linked to soil properties and rainfall. The availability of these regional patterns from the simulated yield estimates is very helpful for estimating productivity in areas susceptible to abiotic stresses, such as droughts, as well as for identifying factors that affect yield. Additionally, efforts have been undertaken to create web-based decision support systems based on simulation models and GIS to aid in the making of agronomic decisions (Kadiyala et al., 2015). Another method entailed combining GIS and an inference engine (IE) with the erosion productivity impact calculator (EPIC), a model for analyzing the relationship between soil erosion and agricultural output at the field level (Tan and Shibasaki, 2003). Applications for assessing damage to some high-value crops have also been made with RS and GIS. One crop that has extreme yield variations is cranberries, which are caused by the peculiarities of the soil, which also affect the availability of nutrients and water. A spatial variation map for the crop was produced by Oudemans et al. (2002) using GIS, GPS, and RS. This map allows for the examination of crop losses inside individual fields or across the entire field.

As a result of developments in digital technologies, which have been using GIS as a key partner technology for assessing crops, soils, and their habitats, the number of applications and significance of GIS have expanded. Over the past couple decades, the use of GIS in agriculture has expanded significantly. GIS is used throughout the entire agricultural value chain. In addition to the historical, current, and well-liked uses of GIS in land suitability/use planning and management of water, soil, and biotic and abiotic resources, the development of digital agricultural tools and technologies has increasingly taken advantage of the capabilities of GIS in new and emerging applications in high-fidelity crop monitoring, yield prediction, precision farming, and supply chain management for both primary produce and biomass utilization toward energy production (Ghosh and Kumpatla, 2022). Table 1.1 shows some of the previous studies for applications of RS and GIS in the estimation of horticulture area expansion and crop yield quantification.

GIS and RS technology utilization generally has drawbacks and is not a magic bullet for successful evidence-based policy and practice. The effective use of geospatial technology, high-quality data, and significant management resources are all necessary for its implementation to be successful. In low-resource countries like LMICs, the technology's acceptance and wider use are constrained by its high cost and lack of qualified workers (Naik et al., 2013).

Accurate agricultural yield simulation is challenging due to the range of cropping methods and levels of technology employed. To effectively measure the crop yield gap, the input data's quality would need to be enhanced. This would include using more precise meteorological parameters, better soil characterization, and data on land use that is distributed across multiple regions (Muslim et al., 2015). Additionally, it would call for the creation of instrumented, georeferenced validation

TABLE 1.1
Applications of RS and GIS in the Estimation of Horticulture Area Expansion and Crop Yield Quantification

The Location	Models/Data Used	Main Results	References
Saharanpur district, India	Linear imaging self-scanning (LISS) of IRS IC LISS-III data	IRS IC LISS-III data are useful for mango orchard identification and acreage estimation.	Nagaraja (2009)
Himachal Pradesh, India	IRS LISS-III and Indian RS Advanced Wide Field Sensor (IRS AWiFS) imagery data	RS data were used to create apple orchard maps and statistics for area and production estimation.	Sharma and Panigrahy (2007)
Florida, USA	Landsat ETM Plus (30 m) imagery	The findings revealed an important relationship between citrus production/income and citrus area coverage derived from remotely sensed imagery.	Shrivastava and Gebelein (2007)
south Texas, USA	SPOT 5 satellite image data	High-resolution SPOT 5 imagery is useful for crop estimation.	Yang et al. (2011)
Burkina Faso	Linear regression models were developed using the in situ crop production estimates and temporal metrics derived from Sentinel-2 time series.	The Sentinel-2 data can be an important resource for upscaling field measurements of crop production.	Karlson et al. (2020)
United States at county and field scale	Field-level analysis was carried out using NASA's Harmonized Landsat Sentinel-2 (HLS) product. 3D convolutional neural network (CNN) and CNN followed by long short-term memory (LSTM) were used.	The results highlight the potential of using satellite data for yield estimation at different management scales.	Ghazaryan et al. (2020)
Hungary	The RS data	The established robust technique for yield estimation proven to be reliable and consistent for operational application. The technique is easy to use and reasonably priced.	Ferencz et al. (2004)
Guanzhong, China	Meteorological data and two remotely sensed indices, Vegetation Temperature Condition Index (VTCI), and Leaf Area Index (LAI).	An accurate wheat yield prediction will increase the productivity and sustainability of agricultural operations in addition to assisting with educated crop management decisions.	Tian et al. (2021)
Hunan Province, China	Data from the HJ-1A/B Chinese Environmental Satellite.	The possibility of estimating the rice area at the county level by using RS data efficiently has been demonstrated. The method can also be used in mountainous regions and areas of fragmented cultivation.	Li et al. (2014)

sites that provide comprehensive survey data to direct a cycle of continuous improvement around yield gap measurement (Hochman et al., 2013). Therefore, more precise inputs for yield gap assessment would be produced by further development of present RS technologies and the development of more integrated cropping system models. Another impediment to the use and acceptance of GIS and RS in agriculture is the absence of generally agreed-upon standards for data interchange. Due to the increased accessibility of spatial data consumption in LMICs, many of these data are routinely collected and stored with different spatial units, formats, metadata, timings, and space intervals. Because of this, some data are made unusable, geographic data integration is impeded, and thorough analysis of data—especially data acquired from numerous sensors and platforms—is impeded. For agricultural spatial data, it is vital to develop uniform standards (Mathenge et al., 2022). However, GIS plays a key and important role in obtaining maximum agricultural output and high efficiency through current and new applications, in conjunction with existing and cutting-edge associate technologies.

1.4 CONCLUSION

Nowadays, RS and other geoinformation technology are used to perform high-tech farming with accuracy. RS has the potential to be applied in a number of domains, including crop yield prediction and crop biomass identification, thanks to its quick development. To identify, categorize, map, track, and get ready for the mitigation and management of disasters as a whole, RS offers a trustworthy source of information. The information may be used for precise horticulture, market forecasting, site-specific management, and export. Modern techniques like RS help with data gathering and updating so that scientific management strategies can be created. Microwave radiometers, laser meters, magnetic sensors, and cameras, among other kinds of sensors, gather electromagnetic data to offer precise, thorough data about the earth's surface and atmosphere. Computers can quickly measure and alter these data and these pictures because they are digital. RS has mostly solved our problems with data gathering, storage, processing, and other challenges. Also, it makes tracking all-season crops and preparing statistical data simpler. RS and GIS have shown to be useful tools for horticulture and crop yield monitoring in a number of situations.

REFERENCES

Adhav, V., Das, S., & Patkar, V. (2021). Crop health monitoring using geospatial technologies for nashik district, Maharashtra. International Journal of Modern Agriculture, 10(2), 1395–1409.

Al-Gaadi, K. A., Hassaballa, A. A., Tola, E., Kayad, A. G., Madugundu, R., Alblewi, B., & Assiri, F. (2016). Prediction of potato crop yield using precision agriculture techniques. PLoS One, 11(9), e0162219.

Aquino, A., Millan, B., Diago, M. P., & Tardaguila, J. (2018). Automated early yield prediction in vineyards from on-the-go image acquisition. *Computers and Electronics in Agriculture, 144*, 26–36.

Balezentis, T., Chen, X., Galnaityte, A., & Namiotko, V. (2020). Optimizing crop mix with respect to economic and environmental constraints: An integrated MCDM approach. *Science of the Total Environment*, 705, 135896.

Barsi, J. A., Schott, J. R., Hook, S. J., Raqueno, N. G., Markham, B. L., & Radocinski, R. G. (2014). Landsat-8 thermal infrared sensor (TIRS) vicarious radiometric calibration. *Remote Sensing*, *6*(11), 11607–11626.

Belward, A. S., & Skøien, J. O. (2015). Who launched what, when and why; trends in global land-cover observation capacity from civilian earth observation satellites. *ISPRS Journal of Photogrammetry and Remote Sensing*, *103*, 115–128.

Berni, J. A., Zarco-Tejada, P. J., Suárez, L., & Fereres, E. (2009). Thermal and narrowband multispectral remote sensing for vegetation monitoring from an unmanned aerial vehicle. *IEEE* Transactions on Geoscience and Remote Sensing, *47*(3), 722–738.

Brewer, K., Clulow, A., Sibanda, M., Gokool, S., Naiken, V., & Mabhaudhi, T. (2022). Predicting the chlorophyll content of maize over phenotyping as a proxy for crop health in smallholder farming systems. *Remote Sensing*, 14(3), 518.

Brisco, B., Brown, R. J., Hirose, T., McNairn, H., & Staenz, K. (1998). Precision agriculture and the role of remote sensing: a review. Canadian Journal of Remote Sensing, *24*(3), 315–327. Doi: 10.1080/07038992.1998.10855254

Brock, J. C., & Purkis, S. J. (2009). The emerging role of lidar remote sensing in coastal research and resource management. Journal of Coastal Research, (10053), 1–5.

Bulanon, D. M., Burks, T. F., & Alchanatis, V. (2008). Study on temporal variation in citrus canopy using thermal imaging for citrus fruit detection. *Biosystems Engineering*, *101*(2), 161–171. Doi: 10.1016/j.biosystemseng.2008.08.002

Candiago, S., Remondino, F., De Giglio, M., Dubbini, M., & Gattelli, M. (2015). Evaluating multispectral images and vegetation indices for precision farming applications from UAV images. *Remote Sensing*, *7*(4), 4026–4047.

Carbone, G. J., Narurnalani, S., & King, M. (1996). Application of remote sensing and technologies with physiological crop models. *Photogrammetric Engineering & Remote Sensing*, *62*(2), 171–179.

Cendrero-Mateo, M. P., Wieneke, S., Damm, A., Alonso, L., Pinto, F., Moreno, J., ... & Mac Arthur, A. (2019). Sun-induced chlorophyll fluorescence III: Benchmarking retrieval methods and sensor characteristics for proximal sensing. Remote Sensing, *11*(8), 962.

Colomina, I., & Molina, P. (2014). Unmanned aerial systems for photogrammetry and remote sensing: A review. ISPRS Journal of photogrammetry and remote sensing, 92, 79–97.

Dadhwal, V. K., & Ray, S. S. (2000). Crop assessment using remote sensing-Part II: Crop condition and yield assessment. *Indian Journal of Agricultural Economics*, *55*, 55–67.

Dadhwal, V. K., Sehgal, V. K., Singh, R.P., & Rajak, D.R. (2003). Wheat yield modelling using satellite remote sensing with weather data: Recent Indian experience. Mausam, 54(1), 253–262.

Dalin, C., Wada, Y., Kastner, T., & Puma, M. J. (2017). Groundwater depletion embedded in international food trade. *Nature*, *543*(7647), 700–704.

Delgado, J. A., Vandenberg, B., Kaplan, N., Neer, D., Wilson, G., D'Adamo, R., ... & Derner, J. D. (2018). Agricultural Collaborative Research Outcomes System (AgCROS): A network of networks connecting food security, the environment, and human health. *Journal of Soil and Water Conservation*, *73*(6), 158A–164A.

Delgado, J. A., Vandenberg, B., Neer, D., & D'Adamo, R. (2019). Emerging nutrient management databases and networks of networks will have broad applicability in future machine learning and artificial intelligence applications in soil and water conservation. *Journal of Soil and Water Conservation*, *74*(6), 113A–118A. Doi: 10.2489/jswc.74.6.113A

Di Gennaro, S. F., Matese, A., Gioli, B., Toscano, P., Zaldei, A., Palliotti, A., & Genesio, L. (2017). Multisensor approach to assess vineyard thermal dynamics combining high-resolution unmanned aerial vehicle (UAV) remote sensing and wireless sensor network (WSN) proximal sensing. *Scientia Horticulturae*, *221*, 83–87.

Fay, M., Andres, L. A., Fox, C., Narloch, U., & Slawson, M. (2017). *Rethinking Infrastructure in Latin America and the Caribbean: Spending Better to Achieve More*. Washington, DC: World Bank Publications.

Ferencz, C., Bognar, P., Lichtenberger, J., Hamar, D., Tarcsai, G., Timár, G., ... & Ferencz-Árkos, I. (2004). Crop yield estimation by satellite remote sensing. *International Journal of Remote Sensing*, *25*(20), 4113–4149. Doi: 10.1080/01431160410001698870

Fritz, S., See, L., Bayas, J. C. L., Waldner, F., Jacques, D., Becker-Reshef, I., ... & McCallum, I. (2019). A comparison of global agricultural monitoring systems and current gaps. *Agricultural Systems*, *168*, 258–272.

Gautam, D., & Ha, C. (2013). Control of a quadrotor using a smart self-tuning fuzzy PID controller. *International Journal of Advanced Robotic Systems*, *10*(11), 380.

Gautam, D., & Pagay, V. (2020). A review of current and potential applications of remote sensing to study the water status of horticultural crops. *Agronomy*, *10*(1), 140.

Gebeyehu, M. N. (2019). Remote sensing and GIS application in agriculture and natural resource management. *International Journal of Environmental Sciences & Natural Resources*, *19*(2), 45–49.

Ghazaryan, G., Skakun, S., König, S., Rezaei, E. E., Siebert, S., & Dubovyk, O. (2020). Crop yield estimation using multi-source satellite image series and deep learning. In *IGARSS 2020-2020 IEEE International Geoscience and Remote Sensing Symposium* (pp. 5163–5166). Waikoloa, HI: IEEE. Doi: 10.1109/IGARSS39084.2020.9324027

Ghosh, P., & Kumpatla, S. P. (2022). GIS Applications in Agriculture. *Geographic Information Systems and Applications in Coastal Studies*. (pp. 89-114). IntechOpen Doi: 10.5772/intechopen.104786.

Goodchild, M. F. (2011). Spatial thinking and the GIS user interface. *Procedia-Social and Behavioral Sciences*, *21*, 3–9. Doi: 10.1016/j.sbspro.2011.07.002

Gruen, A. (2008). Scientific-technological developments in photogrammetry and remote sensing between 2004 and 2008. In *Advances in Photogrammetry, Remote Sensing and Spatial Information Sciences: 2008 ISPRS Congress Book* (pp. 39–44). CRC Press. London

Gupta, P. C., & Sharma, R. (1990). Application of satellite remote sensing technique in delineation and hectarge estimation of mango orchards in parts of Uttar Pradesh, India. *Proceedings of the GIS Development* (pp. 12–14). Uttar pradesh, India.

Gutiérrez, S., Diago, M. P., Fernández-Novales, J., & Tardaguila, J. (2018). Vineyard water status assessment using on-the-go thermal imaging and machine learning. PLoS One, 13(2), e0192037.

Ha, W., Gowda, P. H., & Howell, T. A. (2013). A review of downscaling methods for remote sensing-based irrigation management: Part I. Irrigation Science, 31, 831–850.

Hakkim, V. A., Joseph, E. A., Gokul, A. A., & Mufeedha, K. (2016). Precision farming: the future of Indian agriculture. *Journal of Applied Biology and Biotechnology*, *4*(6), 68–72. Doi: 10.7324/JABB.2016.40609

Hassan, S. S., & Goheer, M. A. (2021). Modeling and monitoring wheat crop yield using geospatial techniques: a case study of potohar region, Pakistan. *Journal of the Indian Society of Remote Sensing*, *49*(6), 1331–1342.

Herrero-Huerta, M., Hernández-López, D., Rodriguez-Gonzalvez, P., González-Aguilera, D., & González-Piqueras, J. (2014). Vicarious radiometric calibration of a multispectral sensor from an aerial trike applied to precision agriculture. *Computers and Electronics in Agriculture*, *108*, 28–38.

Hochman, Z., Gobbett, D., Holzworth, D., McClelland, T., van Rees, H., Marinoni, O., ... & Horan, H. (2013). Reprint of "Quantifying yield gaps in rainfed cropping systems: A case study of wheat in Australia". *Field Crops Research*, *143*, 65–75.

Jones, J. W., Hoogenboom, G., Porter, C. H., Boote, K. J., Batchelor, W. D., Hunt, L. A., ... & Ritchie, J. T. (2003). The DSSAT cropping system model. *European Journal of Agronomy*, *18*(3–4), 235–265.

Kadiyala, M. D. M., Nedumaran, S., Singh, P., Chukka, S., Irshad, M. A., & Bantilan, M. C. S. (2015). An integrated crop model and GIS decision support system for assisting agronomic decision making under climate change. *Science of the Total Environment*, *521*, 123–134.

Karlson, M., Ostwald, M., Bayala, J., Bazié, H. R., Ouedraogo, A. S., Soro, B., ... & Reese, H. (2020). The potential of Sentinel-2 for crop production estimation in a smallholder agroforestry landscape, Burkina Faso. *Frontiers in Environmental Science*, *8*, 85.

Khanal, S., Fulton, J., & Shearer, S. (2017). An overview of current and potential applications of thermal remote sensing in precision agriculture. *Computers and Electronics in Agriculture*, *139*, 22–32.

Kocjančič, T., Debeljak, M., Žgajnar, J., & Juvančič, L. (2018). Incorporation of emergy into multiple-criteria decision analysis for sustainable and resilient structure of dairy farms in Slovenia. *Agricultural Systems*, *164*, 71–83. Doi: 10.1016/j.agsy.2018.03.005

Kumar, N., Yamaç, S. S., & Velmurugan, A. (2015). Applications of remote sensing and GIS in natural resource management. *Journal of the Andaman Science Association*, *20*(1), 1–6.

Kumar, S., Karaliya, S. K., & Chaudhary, S. (2017). Precision farming technologies towards enhancing productivity and sustainability of rice-wheat cropping system. *International Journal of Current Microbiology and Applied Science*, *6*(3), 142–151.

Lee, W. S., Alchanatis, V., Yang, C., Hirafuji, M., Moshou, D., & Li, C. (2010). Sensing technologies for precision specialty crop production. *Computers and Electronics in Agriculture*, *74*(1), 2–33.

Li, Q., Zhang, H., Du, X., Wen, N., & Tao, Q. (2014). County-level rice area estimation in southern China using remote sensing data. Journal of Applied Remote Sensing, *8*(1), 083657.

Mann, M. L., & Warner, J. M. (2017). Ethiopian wheat yield and yield gap estimation: A spatially explicit small area integrated data approach. *Field Crops Research*, *201*, 60–74.

Manson, S. M., Bonsal, D. B., Kernik, M., & Lambin, E. F. (2015). Geographic information systems and remote sensing. In *International Encyclopedia of the Social & Behavioral Sciences: Second Edition* (pp. 64–68). Elsevier Inc. Oxford.

Mancini, T., Sibilla, F., Argiropoulos, D., Rossi, M., & Everri, M. (2019). The opportunities and risks of mobile phones for refugees' experience: A scoping review. PloS one, 14(12), e0225684.

Marinelli, M. V., Scavuzzo, M., Giobellina, B., & Scavuzzo, M. (2019). Geoscience and remote sensing on horticulture as support for management and planning, 2(2):43–54.

Matese, A., & Filippo Di Gennaro, S. (2015). Technology in precision viticulture: A state of the art review. *International Journal of Wine Research*, 7(1), 69–81.

Mathenge, M., Sonneveld, B. G., & Broerse, J. E. (2022). Application of GIS in agriculture in promoting evidence-informed decision making for improving agriculture sustainability: A systematic review. *Sustainability*, *14*(16), 9974.

McCabe, M. F., Rodell, M., Alsdorf, D. E., Miralles, D. G., Uijlenhoet, R., Wagner, W., ... & Wood, E. F. (2017). The future of Earth observation in hydrology. Hydrology and earth system sciences, 21(7), 3879–3914.

Meena, R. S., Mitran, T., Kumar, S., Yadav, G. S., Bohra, J. S., & Datta, R. (2018). Application of remote sensing for sustainable agriculture and forest management. *Information Processing in Agriculture*, *5*, 295–297.

Memon, M. S., Jun, Z., Sun, C., Jiang, C., Xu, W., Hu, Q., ... & Ji, C. (2019). Assessment of wheat straw cover and yield performance in a rice-wheat cropping system by using Landsat satellite data. *Sustainability*, *11*(19), 5369.

Meng, W., Tao, F. L., & Shi, W. J. (2014). Corn yield forecasting in northeast china using remotely sensed spectral indices and crop phenology metrics. *Journal of Integrative Agriculture, 13*(7), 1538–1545.

Meraj, G., Singh, S. K., Kanga, S., & Islam, M. N. (2021). Modeling on comparison of ecosystem services concepts, tools, methods and their ecological-economic implications: A review. *Modeling Earth Systems and Environment, 8*, 1–20. Doi: 10.1007/s40808-021-01131-6

Merchant, J. W., & Narumalani, S. (2009). Integrating remote sensing and geographic information systems. *The SAGE Handbook of Remote Sensing*, 257–268. Doi: 10.1016/j.inpa.2018.03.003

Morton, J. F. (2007). The impact of climate change on smallholder and subsistence agriculture. *Proceedings of the National Academy of Sciences, 104*(50), 19680–19685.

Muslim, M., Romshoo, S. A., & Rather, A. Q. (2015). Paddy crop yield estimation in Kashmir Himalayan rice bowl using remote sensing and simulation model. *Environmental Monitoring and Assessment, 187*, 1–12.

N'Doume, C., Lachenaud, P., Hussard, A., Nguyen, H., & Flori, A. (2001). Etude de faisabilité pour l'élaboration d'une cartographie statistique d'inventaire des vergers café et cacao en Côte d'Ivoire par télédétection satellitale. Cocoa Producers' Alliance.

Nagaraja, A. (2009). *Predicting susceptible areas of mango malformation through remote sensing and GIS* (Doctoral dissertation, IARI, Division of Fruits Horticultural Technology, New Delhi).

Nageswara Rao, P. P., Ravishankar, H. M., Raj, U., & Nagajothi, K. (2004). Production estimation of horticultural crops using irs-1d liss-iii data. *Journal of the Indian Society of Remote Sensing, 32*, 393–398.

Naik, G., Basavaraj, K. P., Hegde, V. R., Paidi, V., & Subramanian, A. (2013). Using geospatial technology to strengthen data systems in developing countries: The case of agricultural statistics in India. *Applied Geography, 43*, 99–112.

Nie, P. C., Wu, D., Zhang, W., Yang, Y., & He, Y. (2010). Hybrid combination of GIS, GPS, WSN and GPRS technology in modern digital agriculture application. *Advanced Materials Research, 108*, 1158–1163. Doi: 10.4028/www.scientific.net/AMR.108-111.1158

O´Connell, M. G., & Goodwin, I. (2004, September). Spatial variation of tree cover in peach orchards. In *International Symposium on Harnessing the Potential of Horticulture in the Asian-Pacific Region 694* (pp. 203–205).

Oudemans, P. V., Pozdnyakova, L., Hughes, M. G., & Rahman, F. (2002). GIS and remote sensing for detecting yield loss in cranberry culture. *Journal of Nematology, 34*(3), 207.

Pajares, G. (2015). Overview and current status of remote sensing applications based on unmanned aerial vehicles (UAVs). *Photogrammetric Engineering & Remote Sensing, 81*(4), 281–329.

Panda, S. S., Hoogenboom, G., & Paz, J. O. (2010). Remote sensing and geospatial technological applications for site-specific management of fruit and nut crops: A review. *Remote Sensing, 2*(8), 1973–1997.

Sailaja, B., Voleti, S. R., Subrahmanyam, D., Nathawat, M. S., & Rao, N. H. (2013). Regional rice yield estimation by integration of spatial technologies and crop model, 4(2).56–66.

Schumann, A. W., & Hostler, H. K. (2008). Yield monitoring of citrus using color photography and ultrasonic sensors. *ISHS Application of Precision Agriculture for Fruits and Vegetables*, Orlando, FL.

See, L., Mooney, P., Foody, G., Bastin, L., Comber, A., Estima, J., ... & Rutzinger, M. (2016). Crowdsourcing, citizen science or volunteered geographic information? The current state of crowdsourced geographic information. *ISPRS International Journal of Geo-Information, 5*(5), 55.

Setiyono, T., Nelson, A., & Holecz, F. (2014). Remote sensing based crop yield monitoring and forecasting. *Crop Monitoring for Improved Food Security*. Expert Meeting on Crop Monitoring for Improved Food Security, Vientiane, Laos, pp. 123–132.

Sharma, A., & Panigrahy, S. (2007). Apple orchard characterization using remote sensing and GIS in Shimla district of Himachal Pradesh. In *Proceedings of Remote Sensing and Photogrammetry Annual Conference 2007*, Newcastle, UK (pp. 11–14).

Sheffield, J., Wood, E. F., Pan, M., Beck, H., Coccia, G., Serrat-Capdevila, A., & Verbist, K. (2018). Satellite remote sensing for water resources management: Potential for supporting sustainable development in data-poor regions. *Water Resources Research*, *54*(12), 9724–9758. Doi: 10.1029/2017WR022437

Shi, Y., Thomasson, J. A., Murray, S. C., Pugh, N. A., Rooney, W. L., Shafian, S., ... & Yang, C. (2016). Unmanned aerial vehicles for high-throughput phenotyping and agronomic research. *PloS One*, *11*(7), e0159781.

Shrivastava, R. J., & Gebelein, J. L. (2007). Land cover classification and economic assessment of citrus groves using remote sensing. *ISPRS Journal of Photogrammetry and Remote Sensing*, *61*(5), 341–353.

Singhal, G., Bansod, B., & Mathew, L. (2018). Real time Crop health monitoring using Remote Sensing and ancillary information using GIS. In *19th Esri India User Conference*. Gurugram, India.

Soomro, T. R. (2015). GIS enabling smart agriculture. *Smart Agriculture: An Approach towards Better Agriculture Managemen*t, (8), 1–6.

Stajnko, D., Lakota, M., & Hočevar, M. (2004). Estimation of number and diameter of apple fruits in an orchard during the growing season by thermal imaging. *Computers and Electronics in Agriculture*, *42*(1), 31–42.

Swain, K. C., Zaman, Q. U., Schumann, A. W., Percival, D. C., & Bochtis, D. D. (2010). Computer vision system for wild blueberry fruit yield mapping. *Biosystems Engineering*, *106*(4), 389–394.

Tan, G., & Shibasaki, R. (2003). Global estimation of crop productivity and the impacts of global warming by GIS and EPIC integration. *Ecological Modelling*, *168*(3), 357–370.

Thakur, J. K., Singh, S. K., & Ekanthalu, V. S. (2017). Integrating remote sensing, geographic information systems and global positioning system techniques with hydrological modeling. *Applied Water Science*, *7*(4), 1595–1608. Doi: 10.1007/s13201-016-0384-5

Thomas, G., Taylor, J., & Wood, G. (1997). Mapping yield potential with remote sensing.

Tian, H., Wang, P., Tansey, K., Han, D., Zhang, J., Zhang, S., & Li, H. (2021). A deep learning framework under attention mechanism for wheat yield estimation using remotely sensed indices in the Guanzhong Plain, PR China. *International Journal of Applied Earth Observation and Geoinformation*, *102*, 102375.

Toth, C., & Jóźków, G. (2016). Remote sensing platforms and sensors: A survey. *ISPRS Journal of Photogrammetry and Remote Sensing*, *115*, 22–36.

Turner, D., Lucieer, A., & Wallace, L. (2013). Direct georeferencing of ultrahigh-resolution UAV imagery. *IEEE Transactions on Geoscience and Remote Sensing*, *52*(5), 2738–2745.

Urbahs, A., & Jonaite, I. (2013). Features of the use of unmanned aerial vehicles for agriculture applications. *Aviation*, *17*(4), 170–175.

Usha, K., & Singh, B. (2013). Potential applications of remote sensing in horticulture-A review. *Scientia Horticulturae*, *153*, 71–83. Doi: 10.1016/j.scienta.2013.01.008

Van Hoang, T., & Chou, T. Y. (2015). GIS Applications and Agriculture Model: A Case Study in Hai Phong Vietnam. (pp. 651–658). In 36th Asian Conference on Remote Sensing (ACRS 2015) Quezon City, Philippines.

Wachs, J., Stern, H., Burks, T., & Alchanatis, V. (2009). Multi-modal registration using a combined similarity measure. *Applications of Soft Computing*, *52*, 159–168.

Wang, S., Baum, A., Zarco-Tejada, P. J., Dam-Hansen, C., Thorseth, A., Bauer-Gottwein, P., ... & Garcia, M. (2019). Unmanned Aerial System multispectral mapping for low and variable solar irradiance conditions: Potential of tensor decomposition. *ISPRS Journal of Photogrammetry and Remote Sensing*, *155*, 58–71.

Wargo, C., Snipes, C., Roy, A., & Kerczewski, R. (2016). UAS industry growth: Forecasting impact on regional infrastructure, environment, and economy. In *2016 IEEE/AIAA 35th Digital Avionics Systems Conference (DASC)* (pp. 1–5). IEEE. California, USA.

Weiss, M., Jacob, F., & Duveiller, G. (2020). Remote sensing for agricultural applications: A meta-review. *Remote Sensing of Environment*, *236*, 111402. Doi: 10.1016/j.rse.2019.111402

Xue, J., & Su, B. (2017). Significant remote sensing vegetation indices: A review of developments and applications. *Journal of Sensors*, *2017*. 1–17.

Yang, C. (2016). Aerial imaging with manned aircraft for precision agriculture. *Resource Magazine*, *23*(4), 25–27.

Yang, C., Everitt, J. H., & Bradford, J. M. (2007). Airborne hyperspectral imagery and linear spectral unmixing for mapping variation in crop yield. *Precision Agriculture*, *8*(6), 279–296.

Yang, C., Everitt, J. H., & Murden, D. (2011). Evaluating high resolution SPOT 5 satellite imagery for crop identification. *Computers and Electronics in Agriculture*, *75*(2), 347–354.

Ye, X., Sakai, K., Asada, S. I., & Sasao, A. (2008). Inter-relationships between canopy features and fruit yield in citrus as detected by airborne multispectral imagery. *Transactions of the ASABE*, *51*(2), 739–751.

Ye, X., Sakai, K., Manago, M., Asada, S. I., & Sasao, A. (2007). Prediction of citrus yield from airborne hyperspectral imagery. *Precision Agriculture*, *8*, 111–125.

Zaman, Q. U., Schumann, A. W., & Percival, D. C. (2008). Development of an automated slope measurement and mapping system. In *2008 Providence, Rhode Island, June 29-July 2, 2008* (p. 1). American Society of Agricultural and Biological Engineers. Saint Joseph, MI, USA.

Zarco-Tejada, P. J., González-Dugo, V., & Berni, J. A. (2012). Fluorescence, temperature and narrow-band indices acquired from a UAV platform for water stress detection using a micro-hyperspectral imager and a thermal camera. *Remote Sensing of Environment*, *117*, 322–337.

Zecha, C. W., Link, J., & Claupein, W. (2013). Mobile sensor platforms: Categorisation and research applications in precision farming. *Journal of Sensors and Sensor Systems*, *2*(1), 51–72.

Zhang, F., & Cao, N. (2019). Application and research progress of Geographic Information System (GIS) in agriculture. In 2019 8th *International Conference on Agro-Geoinformatics (Agro-Geoinformatics)* (pp. 1–5). IEEE. Istanbul, Turkey.

Zhang, Y., Pan, M. & Wood, E. F. (2016). On creating global gridded terrestrial water budget estimates from satellite remote sensing. *Surveys in Geophysics*, *37*, 249–268. Doi: 10.1007/s10712-015-9354-y

2 Cost-Benefit Analysis and Measuring Livelihood Development Index of Horticulturalists in Kashmir Valley

Showkat A. Ganaie
Government Degree College Shopian, J&K, India

Jahangir H. Bhat
Department of School Education, J&K, India

2.1 INTRODUCTION

Agriculture has a pivotal role in the Indian economy (Khullar, 2014). More than 68% of the rural households rely on this sector of economy for livelihood. Horticulture has attained commercial tone in the recent times, providing significant contribution to the economy of the country. India possesses varied agro-climatic conditions which makes it able to produce a diverse bunch of horticultural crops throughout the year. The horticulture crops provide ways for diversification of country's agriculture in context of higher return (Jadhav, 2013). The contribution of this sector in agricultural Gross Domestic Production (GDP) is around 28% from about 13.08% of the area. Horticulture products have attained the primacy in world agriculture trade (Lone and Sen, 2014; Mittal, 2007).

The concept of horticulture in Kashmir Valley as a separate field culture has its history in medieval times. The history of fruit production begins with the cultivation of apples that have been grown even before 1000 BC. Apple cultivation in Kashmir has a mention in *Tuzki-Jahangiri*, whereas Sir Walter Lawrence called the valley as "*fruit industry*." The fruits grown were indigenous and for local consumption only (Bamzai, 1987). The Jammu and Kashmir has diversity in geographical conditions and climate, feasible for the cultivation of temperate, tropical and sub-tropical fruits. The disparity in rainfall distribution, lacking of irrigation facilities and improper water harvesting measures badly affect the fruit industry (Ahmad et al. (2021)). Kashmir Valley is famous throughout country and abroad in horticulture produce, the geographical conditions offer great scope for cultivation of numerous horticulture crops, and besides, the globally famous spices of *Saffron* and *Black Zeera* are

 DOI: 10.1201/9781003412489-2

grown (Masoodi, 2003). The fruit industry has a pronounced impact on GDP of the Union Territory (Akhter, 2015). Horticulture development results in higher income generation in the rural areas, hence improves the quality of life. Lack of cold storage facilities, planning, plant protection measures, unavailability of high-quality varieties and expensive transport have badly affected this sector of economy in Jammu and Kashmir (Darzi, 2016). The fruit industry provides economic security to females as they are involved in picking, grading and carriage of fruits and assist in peak season. There is a paucity of correct data related to production costs, input requirements and export of fruits. Thus, the results of this study would be of immense help to the horticultural growers in general and Kashmir Valley in particular for making decisions linked to horticulture enterprise. The scope of the study is limited to the particular area selected for study, i.e., Kashmir Valley. However, the findings of the study can be generalized. Moreover, the findings of the study with respect to investment in horticulture and its profitability at different stages of production will be useful to the Government. The findings of the research related to costs, returns and profitability will motivate the farmers to further enhance the area under horticulture cultivation. The inferences of the study will be of great help to extension agencies to improve the scientific knowledge of the farmers by guiding for better cultivation and farm management practices. The findings of the study will also facilitate research in the study area in future.

Cost-benefit analysis (CBA) is evaluating the net economic impact of a public project. The purpose of CBA is to ascertain the feasibility of the project from welfare perspective by estimating the equivalent money of the benefits and costs to the community. CBA is now widely applied to agriculture and horticulture sectors of economy to quantify and determine profitability of farmers to diversify their land to horticulture or agricultural production from conventional practices of cultivation.

2.2 LITERATURE REVIEW

The world and likewise India experienced a tremendous rise in the production and productivity of horticultural crops on account of rapid growth of urbanization, changing trend in the food habits of the people, global market integration, etc. The increasing demand for highly nutritious food has increased the demand of fruit crops, thereby led to increase in the prices of the high-value horticultural crops by manifold. The massive rise in the area and production of horticultural crops in Indian context is also due to the successful implementation of the various horticulture development programmes initiated by the Government in the various corners of the country. The present section reviews some important studies on the area, production and productivity of important horticultural crops.

Patil et al. (1997) evaluated trends in area, production and productivity of banana in Western Maharashtra and reported that area increased pronouncedly from 6600 in 1950–1951 to 644,000 hectares in 1979–1980. The banana production rose sharply by 6.89% during the same period.

Rauf et al. (2010) studied growth of apple production across districts of J&K and Himachal Pradesh. The study depicted that the apple area depicted a steady increase in all the districts of both the states except in Solan district of Himachal Pradesh.

Malik (2013) concluded that apple cultivation is the basic occupation of Kashmir Valley consisting major portion of the total fruit area of the valley. The apple production increased over the period of time.

Lone and Sen (2014) highlighted that the favourable agro-climatic condition and topographical advantage of Jammu and Kashmir offers a wide scope for the cultivation of a variety of horticultural crops like apple, pears, plum, apricot, medicinal and aromatic plants. Apple, pear, plum and walnut are the most vital fruit crop of the state, and the state has also the monopoly of producing saffron, black zeera, delicious apple, almond, etc. The study suggested development of better transportation, adequate cold storage facilities, efficient marketing system, etc., to increase the productivity and the profitability of the farmers.

2.2.1 Cost-Return Ratio

Mehta (1966) conducted a study on input–output relationship in Kullu district in Himachal Pradesh. He concluded that the output of apple in terms of money value increased with increase in number of trees per unit of area in Upper Kullu due to steep topography of area.

Wani et al. (1994) revealed in their study that the payback period of the fruit is 14 years. The net present value was Rs. 53,417.14, cost-benefit ratio 2.29, and the internal rate of return (IRR) was 26%. These indices reflect that establishment cost of apples in Kashmir is economically viable.

Wagale et al. (2007) in their study reported that cultivation cost per hectare of mango was Rs. 43,198, Rs. 44,310 and Rs. 48,103 in small, medium and large orchards, respectively. The net return was Rs. 13,594, Rs. 16,747 and Rs. 18,879 in small, medium and large orchards, respectively. Furthermore, the cost-benefit ratio was 1.26, 1.31 and 1.33 in small, medium and large farms, respectively, at total cost of production.

2.2.2 Economic Impact of Horticulture

Varied agro-climatic conditions of Jammu and Kashmir provide enough opportunities for the farmers to grow a variety of horticultural crops. Due to more productivity and income, garden culture has a key role in making better the socio-economic milieu of rural population. A few important studies on the socio-economic impact of horticulture are being presented below:

Thakur (1971a,b) concluded that economic growth could be festered in the backward hilly areas of the country by growing fruit and vegetable crops. Apple production could provide a gross income of Rs. 15,000–Rs. 20,000 per acre which was more than the income from the cereals.

Wani et al. (1993) highlighted that apple industry is a major sector of the economy of Kashmir Valley and this industry has indeed turned the backbone of rural economy.

Rather et al. (2013) opined that J&K is the major producer of apple and walnuts in India. Seventy-seven percent of apple and 90% of walnut production in

India comes from Jammu and Kashmir and the state is known as the "*Agri. Export zone for Apples and Walnuts.*"

Malik (2013) in his study opines that agriculture is backbone of Indian economy wherein horticulture is an important component. Horticulture in general and apple production in particular is the prime occupation of people in the Kashmir Valley (India) and it makes 90% of total fruit production in the valley. This sector spanning over50 lakh households' plays a pivotal part in rural economy with an annual turnover of Rs. 750 cores.

Hanan (2015) opined that fruit cultivation sector has been playing a very dominant role in trade and economy of J&K. The state horticulture sector famous for its diverse fruits supports more than 5 lac families, providing employment to nearly 25 lac people annually.

2.3 STUDY AREA

The Valley of Kashmir is a distinct geographical entity. It is one of the three regions of the erstwhile state of Jammu and Kashmir, divided by the Himalayan mountain ranges. These divisions are symbolic to a three-storey building in the middle of which lies Kashmir Valley having semi-closed ecosystem (Hussain, 2006). The oval-shaped valley having latitudes 33° 30′ N to 34°40′ N and longitudes 73°45′ E to 75°35′E spreads over an area of 15,853 km^2. The Kashmir Valley has ten districts with north to south extension of 140 and 45 km from east to west (Raza and Mohammad, 1978).

2.4 SIGNIFICANCE OF THE STUDY

The unique geographical situation, hilly topography and absence of infrastructural facilities in J&K have resulted in dependence of people on agriculture. Horticulture is the main part of agriculture system of Kashmir Valley which contributes immensely to the economy of the valley. Horticulture sector proved as viable option due to comparative advantage enjoyed by the union territory the production of certain globally popular horticulture products owing to feasible climate and land suitability (Darzi, 2016). Thus, horticulture sector plays accelerative mechanism role for general economic growth and upliftment of rural poor by elevating society's standard of living through their income levels (Govt. of J&K, 2010). Though the scope of the study is limited to the particular area selected for study, *i.e.*, Kashmir Valley, however, the findings can be generalized and will be useful to the Government. The findings will also be useful to individual farmers for making appropriate decisions in the management of horticulture to maximize production and ultimately the profit. This study would help to understand the productivity pattern, costs, gross and net returns from horticulture and facilitate the implementation of the programme of horticulture on extensive scale. The result of the study will preferably help the producers as well as farm financing institutions. The findings of the research related to costs, returns and profitability will motivate the farmers to further enhance the area under horticulture cultivation. The results of the study will also be helpful to extension agencies to improve the scientific knowledge of the farmers by guiding for better cultivation and farm management practices. The findings of the study will also facilitate research in the study area in future.

2.5 MATERIALS AND METHODS

The data layers were generated from various sources with varied nature. The Survey of India topo sheets of the Year 1971 on 1:50,000 scale were employed to delineate the study area and generate base map and various other data layers, *viz.*, physiography map, geology map, drainage map and other map layers. Sample survey was done to generate the data pertaining to socio-economic and demographic variables, influencing the horticulture system in the Kashmir Valley. Simple random sampling technique has been carried out for this purpose. The district-wise sampling size has been found on the basis of proportion of 10 years average area under apple to the total apple area in the whole Kashmir Valley. The district-wise weightage factor was obtained by dividing the average apple to the combined average apple area. Following this procedure, among the total 1200 households to be surveyed, the district-wise sample size was obtained. A total of 220 villages including some wards of Srinagar city were surveyed, and in each village/ward, at least four to seven households were surveyed.

The data pertaining to the human population have been obtained from census of India, J&K Series, 2011. The data related to area under orchards especially area under apple and walnut have been collected from Directorate of Horticulture Kashmir, Srinagar. Data related to spray schedule timings has been acquired from the Sheri-Kashmir University of Agricultural Sciences and Technology Kashmir (SKUAST-K). Software's ERDAS IMAGINE 9.0, *ArcView*GIS, etc. were used for mapping purpose.

The data sets generated were tabulated and analysed using appropriate methodologies and techniques. The flowchart of the methodology adopted is given in Figure 2.1. A brief description of methodology is given below:

2.5.1 Input–Output or Cost-Benefit Analysis (CBA)

In order to ascertain the inputs required, primary survey was conducted and the total inputs were simply calculated by using the formula below:

$$\text{Total Input or Cost}, f(I) = \sum_{I=1}^{n} (I_1 + I_2 + I_3 + \cdots + I_n)$$

where $f(I)$ is total input function, and I_1, $I_2, I_3 \ldots$ are various inputs.

Similarly, the total output or benefit has been calculated by the formula presented below:

$$\text{Total Output or Benefit}, f(O) = \sum_{I=1}^{n} (O_1 + O_2 + O_3 + \cdots + O_n)$$

where $f(O)$ is total output function, and O_1, O_2, $O_3 \ldots$ are numerous output.

The ratio between the input function and output function resulted in the net benefit. The formula used is given below:

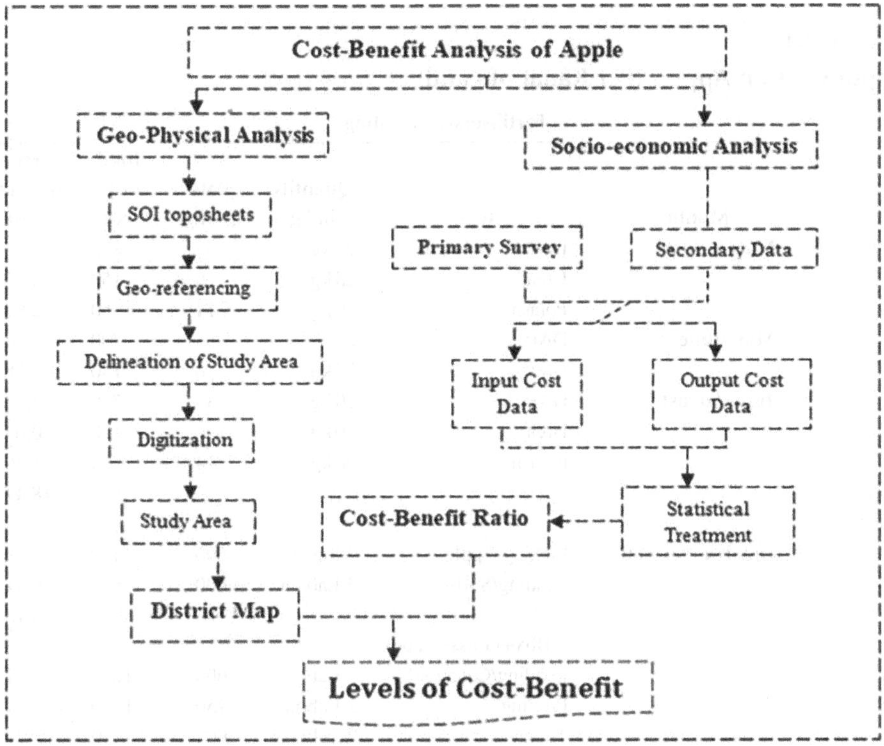

FIGURE 2.1 Flow chart of methodology.

$$N_p \text{ or } G_m = f\left[\frac{f(I)=\sum_{I=1}^{n}(I_1+I_2+I_3+\cdots+I_n)}{f(O)=\sum_{I=1}^{n}(O_1+O_2+O_3+\cdots+O_n)}\right]$$

where N_p or G_m is Net Profit or Gross margin.

2.6 RESULTS AND DISCUSSION

The main input cost on apple orchards is categorized under three major heads, fertilizers (chemical and green manure), spray cost and harvesting charges. Table 2.1 highlights the two input heads, *viz.*, fertilizers and harvesting charges.

Table2.1 depicts that total fertilizer cost per kanal including various chemical fertilizers and manure amount to 6035 INR. The harvesting charges which includes the picking of apples, grading and sorting of apples and packing and loading amount to 13,100 INR. The total input cost per kanal amounts to approximately 19,200 INR. The input cost on type of fertilizer and various processes linked to harvesting process are shown in Table 2.1. Among the fertilizers, manure cost is on top followed by Di-Ammonium Phosphate (DAP), potash and urea. In harvesting process, packing is the expensive item followed by picking, grading and pruning, etc. Fertilizers account

TABLE 2.1
Input Cost on Apples (Per Kanal of Land)

Fertilisers/Harvesting

Item	Month	Type	Quantity (in kg)	Rate (perkg)	Total Cost (INR)	Percent of Total Cost
Fertilizer	**March–April**	DAP	25 kg	35	875	4.57
		Urea	25 kg	6	150	0.78
		Potash	40 kg	14	560	2.93
	May–June	DAP	20 kg	35	700	3.66
		Urea	25 kg	6	150	0.78
	July–August	DAP	20 kg	35	700	3.66
		Urea	20 kg	6	120	0.63
		Potash	20 kg	14	280	1.46
Total (A)					**3535**	**18.47**
Harvesting		**Work**				
	October–November	Picking Apples	4 Labour	600	2400	12.54
		Grading/Sorting	3 Labour	600	1800	9.41
		Packing (Box+Grass+Tape)			5000	26.13
		Loading/Carriage	2 Labour	600	1200	6.27
		Pruning	2 Labour	750	1500	7.84
		Tillage	2 Labour	600	1200	6.27
	December	Manure (cow dung)	1 Tractor	2500	2500	13.07
Total (B)					**15,600**	**81.53**
TOTAL (A+B)					**19,135**	

Source: Primary Survey/Field Survey(2019). Survey performed by Showkat Ahmad Ganie.

for almost one-third (31.54%) of total cost followed by packing (26.13%), picking (12.54%), etc. The input cost on various sprays for orchards needed round the year is shown in Table 2.2. Table 2.2 shows that at least 10–14 sprays are recommended for the apple orchards in a year. The first spray needed is horticulture mineral oil and is used when the tree stage is "*delayed dormancy*." The second spray is recommended in the end of March when green tip appears. Any one among the four fungicide types shown against this spray (Table 2.2) can be used, and usually, *Captan 50 WP* is used. The most commonly used sprays among the basket of sprays are highlighted in the spray schedule (Table 2.2). The quantity of each fungicide, insecticide, acaricide and mineral oil is calculated as per the standard norms written on the packing of respective spray type. The third spray is scheduled after 15days of second spray, when the tree stage is "*pink bud*." In the fourth and fifth sprays, besides, fungicide, insecticide and acaricide are also needed to control crawlers of SanJose scale and mites on apples. Similarly, the subsequent sprays (sixth, seventh and so on) are done as per need. The quantity needed and rate of each unit of spray are highlighted in Table 2.2.

TABLE 2.2
Input Cost on Spray (Insecticide/Fungicide) on Apples per Kanal of Land

Spray Schedule							
			Spray Type				
Spray	**Month**	**Tree/Fruit Stage**	**Fungicide**	**Insecticide/Acaricide/Oil**	**Quantity**	**Rate**	**Total Cost**
I	March	Delayed dormancy		Horticulture mineral oil	8 L	140/L	**1120**
II	Ending March	Green tip	Mancozeb 75 WP or Propineb 70 WP or Zineb 75 WP or ***Captan 50 WP***		1 kg	550/kg	**550**
III	Mid-April	Pink Bud	Mancozeb 75 WP or Propineb 70 WP or Ziram 80 WP or Captan 50 WP or ***Dodine 65 WP*** or Dithianon 75 WP or Zineb 68%+Hexaconazole 4% 72 WP		0.25 kg	2100/kg	**550**
IV	Early May	Petal Fall	***Difenoconazole25 EC***or Flusilazole 40 EC or Trifloxystrobin 25%+ tebuconazole 50% 75WG *Apply when 60%–70% petal fall occurs	**Need Based for Insects:** ***Dimethoate 30 EC*** or Quinalphos 25 EC * Apply when 6–12 crawlers/cm^2 of San Jose scale are observed **Need Based for Mites:** Hexythiazox 5.45 EC or Spiromesifen 22.9 SC or ***Fenazaquin 10 EC*** * Apply acaricide when 4–5 mites per leaf are observed	1/7th L (fungicide) 0.75 L (Insecticide) 1/6thL (acaricide)	3500/L 500/L 2520/L	**500** **375** **420**

(*Continued*)

TABLE 2.2 (*Continued*)
Input Cost on Spray (Insecticide/Fungicide) on Apples per Kanal of Land

			Spray Schedule				
			Spray Type				Total
Spray	Month	Tree/Fruit Stage	Fungicide	Insecticide/Acaricide/Oil	Quantity	Rate	Cost
V	Mid-May	Fruit let (Pea size)	Mancozeb 75 WP or Propineb 70 or WP Zineb 75 WP or ***Captan 50 WP*** or Ziram 80 WP	**Need Based for Insects:** ***Dimethoate 30 EC*** or Quinalphos 25 EC * Apply when 6–12 crawlers/cm^2 of San Jose scale are observed **Need Based for Mites:** Hexythiazox 5.45 EC or Spiromesifen 22.9 SC or ***Fenazaquin 10 EC***	1 kg (fungicide) 0.75 L (Insecticide) 1/6thL (acaricide)	550/kg 500/L 2520/L	**550** **375** **420**
VI	Ending May to Early June	Fruit Development-I	***Dodine 65 WP*** or Difenoconazole 25 EC or Zineb 68%+Hexaconazole 4% 72 WP		0.25 kg	2100/kg	**550**
VII	Mid-June	Fruit Development-II	Flusilazole 40 EC or ***Metiram 55%+Pyraclostrobin 5% 60 WG*** or Myclobutanil 10WP		1 kg	350/kg	**350**
VIII	Ending June to Early July	Fruit Development-III	***Mancozeb 75 WP*** or Zineb 75 WP or Ziram 27 SC or Ziram 80 WP or Chlorothalonil 75 WP	**Need Based for Insects:** Dimethoate 30 EC or ***Chlorpyriphos 20 EC*** * Apply when more than 13 crawlers/cm^2 of San Jose scale are observed **Need Based for Mites:** or Spiromesifen 22.9 SC or ***Fenazaquin 10 EC*** * Apply when population of mites is more than 10 per leaf	1.25 L (fungicide) 0.75 L (insecticide) 1/6thL (acaricide)	360/L 400/L 2520/L	**450** **300** **420**

(*Continued*)

TABLE 2.2 (*Continued*)

Input Cost on Spray (Insecticide/Fungicide) on Apples per Kanal of Land

			Spray Schedule				
			Spray Type				
Spray	Month	Tree/Fruit Stage	Fungicide	Insecticide/Acaricide/Oil	Quantity	Rate	Total Cost
		Incase rainy weather hampers	***Difenoconazole 25 EC*** or Trifloxystrobin 25%+Tebuconazole 50% 75 WG		1/7th L	3500/L	**500**
IX	Ending July	Fruit DevelopmentIV	*Metiram 55%+Pyraclostrobin 5% 60 WG* or Zineb 68%+Hexaconazole 4% 72 WP or Hexaconazole 5EC or Myclobutanil 10 WP		1 kg	350/kg	**350**
			For Marssonina/Sooty blotch/Flyspeck: ***Mancozeb 75 WP*** or Ziram 27 SC or Propineb 70 WP or Ziram 80 WP		1.25 L (fungicide)	360/L	**450**
X		Pre-harvest	**Need Based:** For long term storage, 25 days before harvest Mancozeb 75 WP or Zineb 75 WP or ***Captan 50 WP*** or Ziram 80 WP		1 kg	550/kg	**550**
	Total						**8780**

*The significance lies in the fact that application of the mentioned pesticides/acaricides is subjected to the level of pest disease the orchard are facing.

Source: Spray Schedule as per the Sheri-Kashmir University of Agricultural Sciences and Technology Kashmir (SKUAST-K) and Consumption Pattern as per Primary Survey/Field Survey, 2019.

TABLE 2.3
Output Cost from Apples per Kanal of Land

	Output Analysis			
Item	**Grade**	**Average Quantity**	**Rate per Box**	**Total Output**
Fruit (Apple)	A	95 Box	600	57,000
	B	15 Box	350	5250
Firewood	-	4 Quintal	400	1600
TOTAL				**63,850**

Source: Primary Survey/Field Survey (2019). Survey performed by Showkat Ahmad Ganie.

The average total cost on sprays per kanal of land comes out to 8780 INR. The cost share of different spray types indicates that fungicides account for more cost (5360 INR) followed by acaricides (1260 INR), horticulture mineral oil (1120 INR) and insecticides (1050 INR). The Table 2.2 depicts that share of fungicides to total sprays needed is highest (61%), followed by acaricide (14%). The lowest expenses come out for insecticides (12%).Thus, from the above discussion, it is inferred that the total input cost per kanal of apple orchards is 28,000INR approximately (27,915 INR). The output from apple orchards is fresh fruit and firewood. Firewood is obtained in the pruning process usually after harvesting season. The fruits obtained are usually grouped under two categories, viz., A-Grade and B-Grade. A-Grade is high-quality apples, and B-Grade is low quality. The B-Grade apples have either mite or SanJose or less red colour, whereas the A-Grade apples are free from such issues. Thus, A-Grade apples fetch more to growers than B-Grade apples. The average output quantity per kanal of land is presented below in Table 2.3. From Table 2.3, it is evident that the average quantity per kanal of orchard land is 95 box A-Grade apples and 15 box B-Grade apples.

The average firewood quantity per kanal is four quintals. The rate of apples and firewood is shown in Table 2.3. The combined output per kanal of apple orchards is 64,000 INR approximately (63,850 INR). Thus, the absolute gain is 35,935 INR per kanal. The cost-benefit ratio comes out to 2.29 which indicate that this economic activity is beneficial and economically viable. The payback period of apples in Kashmir Valley is 6–10 years depends upon the variety/kind of trees. Thus, in the first 5–8 years, the output is almost nil, and therefore, in the CBA, the average benefit is actually lower than what comes after statistical analysis of a particular year.

2.6.1 Socio-Economic Impact of Horticulture

In the Himalayan region, agriculture plays a dominant role in employment and income generation as large-scale industrialization is not desirable ecologically, and the infrastructural facilities are too poor to attract industries. These regions do not have raw material for medium and major industries. Therefore, emphasis has to be placed on the agriculture sector for socio-economic development of the people. The horticultural activities are more remunerative than agriculture which has an impact on the economy in particular and other parameters of society in general. The analysis relies on primary data, collected through field work with the help of a detailed schedule. On account of higher productivity and income per unit area from these crops, horticulture in the Kashmir Valley plays a vital role in improving the socio-economic conditions of the rural population. In Kashmir Valley, horticulture has become a vital source of income and livelihood among the farmers. The income generated through horticulture is playing a significant role in the social development of the area under study. The people living in the plains also shifted their cropping land use in favour of horticultural crops as these crops are more remunerative (Showkat et al., 2017). This shift contributed very significantly in raising the standard of living of these people. Farmers/growers of the all the districts of Kashmir Valley felt the positive impact on their living standard due to growing of horticultural crops. Table 2.4 clearly depicts that almost 97% (96.92%) surveyed households have built *pucca* house which indicates that horticultural output enables people to build houses, while a meagre percentage of farmers (3.08%) have still *kachha* house. There are district-level variations in respect of *pucca* and *kachha* houses' category; Bandipora has maximum *kachha* houses (6.98%), while Srinagar has none (0%). Fifty-seven percent (56.75%) surveyed households have built double-storey houses, while 43% have a single-storey house. Thus, almost 60% population engaged with horticultural activities is able to build double-storey homes. The average income of surveyed households is 31,000 INR, in which Kupwara district is at bottom (25,000 INR) and Shopian tops the list (35,000 INR). The impact of horticultural income can be seen from the fact that in terms of domestic assets on an average, 56% surveyed households possess refrigerator; 44% possess inverter; 44% possess washing machine; and 60% possess a geyser.

In terms of other assets among the surveyed households, 77% own a twowheeler and 40% own a car/fourwheeler. The district-wise variations in these assets are shown in Table 2.4. Thus, we can conclude that horticultural income in particular has raised the status of people in the Kashmir Valley. Besides, the horticulture occupies a prominent place in terms of employment share or occupational structure of Kashmir Valley as it is a labour-intensive activity which requires lot of man powers for various farm operations. In the study area, although most of the farm operations are performed by the growers, their family members and local labour, but in harvesting season, growers largely depend upon hired labour from outside the union territory of Jammu and Kashmir. During field work, it has been observed that horticulture is an important source of livelihood among the rural masses, but there are variations in different districts. It is clearly visible from Table 2.4 that percentage share of income generated from horticulture increases as we move from comparatively hot plain areas to cool high altitude which show that horticulture is playing significant role in the economic development of the horticulture growers of hill zone.

The average income share from horticultural activities to the total income is 48%. Shopian district has maximum income from horticulture (65%), while Srinagar (39%) has the least income share from horticulture. The reasons are that Shopian heavily relies on horticulture and very least area is under other crops, while in Srinagar less area is devoted to horticulture and other sources of income contribute to total income basket. Budgam district being nearest to the city generates huge revenue from vegetable gardens, so its percentage share is high (56%). The district-level variations in

TABLE 2.4
Socio-Economic Impact of Horticulture

District	No. of Household Surveyed	Average Income	Income Share from Horticulture (%)	House				Domestic Assets					Other Assets	
				Kachha	Pacca	Single	Double Storey	Refrigerator	Inverter	Washing Machine	Geyser	Air Conditioner	Two Wheeler	Car
Kupwara	174	25,000	51	05[a] (2.87)	169 (97.13)	84 (48.28)	90 (51.72)	102 (58.62)	50 (28.74)	60 (34.48)	100 (57.47)	01 (0.57)	150 (86.21)	54 (31.03)
Budgam	126	30,000	56	03 (2.38)	123 (97.62)	56 (44.44)	70 (55.56)	86 (68.25)	60 (47.62)	62 (49.21)	70 (55.56)	01 (0.79)	98 (77.78)	43 (34.13)
Baramulla	180	32,000	49	05 (2.78)	175 (97.22)	80 (44.44)	100 (55.56)	112 (62.22)	94 (52.22)	80 (44.44)	100 (55.56)	02 (1.11)	150 (83.33)	65 (36.11)
Bandipora	43	26,000	45	03 (6.98)	40 (93.02)	17 (39.53)	26 (60.47)	22 (51.16)	24 (55.81)	20 (46.51)	29 (67.44)	00 (0.00)	32 (74.42)	16 (37.21)
Srinagar	18	35,000	39	0 (0.00)	18 (100)	03 (16.67)	15 (83.33)	16 (88.89)	15 (83.33)	16 (88.89)	17 (94.44)	00 (0.00)	14 (77.78)	12 (66.67)
Ganderbal	66	27,000	41	04 (6.06)	62 (93.94)	30 (45.45)	36 (54.55)	38 (57.58)	26 (39.39)	25 (37.88)	35 (53.03)	00 (0.00)	40 (60.61)	28 (42.42)
Pulwama	102	34,000	44	03 (2.94)	99 (97.06)	40 (39.22)	62 (60.78)	62 (60.78)	45 (44.12)	55 (53.92)	76 (74.51)	01 (0.98)	85 (83.33)	46 (45.10)
Shopian	157	35,000	65	05 (3.18)	152 (96.82)	70 (44.59)	87 (55.41)	54 (34.39)	70 (44.59)	64 (40.76)	95 (60.51)	01 (0.64)	100 (63.69)	65 (41.40)
Anantnag	194	32,000	42	06 (3.09)	188 (96.91)	79 (40.72)	115 (59.28)	118 (60.82)	96 (49.48)	84 (43.30)	116 (59.79)	02 (1.03)	166 (85.57)	96 (49.48)
Kulgam	140	34,000	43	03 (2.14)	137 (97.86)	60 (42.86)	80 (57.14)	64 (45.71)	51 (36.43)	58 (41.43)	80 (57.14)	01 (0.71)	90 (64.29)	60 (42.86)
Total	**1200**	**31,000**	**48**	**37 (3.08)**	**1163 (96.92)**	**519 (43.25)**	**681 (56.75)**	**674 (56.17)**	**531 (44.25)**	**524 (43.67)**	**718 (59.83)**	**9 (0.75)**	**925 (77.08)**	**485 (40.42)**

Source: Primary Survey/Field Survey (2019). Survey performed by Showkat Ahmad Ganie.

[a] The numbers withinparentheses indicate the percentage share.

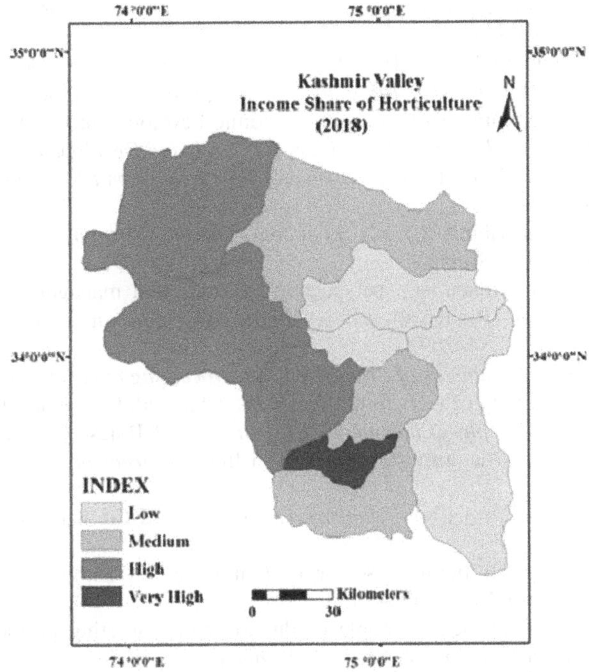

FIGURE 2.2 Income share from horticulture.

income from horticulture sector are presented in Figure 2.2. In brief, we can say that in Kashmir Valley, horticulture has contributed significantly in the economic development of the farmers. But because of the varied agro-climatic conditions, all areas do not experience same degree of profitability.

2.7 CONCLUSION

The input–output analysis is central for analysing the economic viability of any sector of economy. This kind of study proves fruitful for the decision makers to common people. The current study exposed the input–output analysis of horticulture sector of state in general and Kashmir Valley in particular as this sector is considered a backbone of economy. The results indicated the positive impact of this sector on livelihood of the people which is evident from the housing structure/status, household facilities, possession of essential and luxury items, etc. The results of this study can be replicated to other areas of the country especially in Himalayan states having the same topography.

REFERENCES

Ahmad, R., Hussain, B., Ahmad, T. (2021). Fresh and dry fruit production in Himalayan Kashmir, Sub-Himalayan Jammu and Trans-Himalayan Ladakh, India. *Heliyon*, 7(1), e05835.

Akhter, R. (2015). An analysis on production and export of fresh and dry fruits in Jammu and Kashmir. *Journal for Studies in Management and Planning*. 1(11): 747.

Bamzai, P. N. K. (1987). *Socio-Economic History of Kashmir, 1846-1925*. New Delhi: Metropolitan Book Co.

Dahiya, P., Malik, D. P.Singh, I. J.(2001).Trend and growth rate of area, production and yield of fruit crops in Haryana. *The Bihar Journal of Agricultural Marketing*. 9(2)1: 70–176.

Darzi, I. (2016).Horticulture sector towards economic development of Jammu & Kashmir. *International Journal of Multidisciplinary Research and Development*. 3(4): 238–240.

Govt. of Jammu and Kashmir. (2010). *Horticulture in Jammu and Kashmir*. Department of Horticulture, J&K.

Govt. of Jammu and Kashmir. (2014). *Economic Survey, Directorate of Economics and Statistical Planning*. Srinagar.

Hanan, E. (2015). Entrepreneurship perspective for trade and management of horticulture sector in Kashmir Himalayan valley. *International Journal of Social Sciences and Management*. 2(3): 284–289.

Handiganur, S. (1995). *Economics of Production and Processing of Grapes in Bijapur district, Karnataka*. University of Agricultural Sciences, Dharwad, Karnataka (India).

Hussain, M. (2006). *Geography of Jammu and Kashmir*. 5th Ed. Rajesh Publication, New Delhi.

Jadhav, M. T. (2013): Horticulture development in India. *Journal of Current Science*. 1(1): 13–17.

Khullar, D.R. (2014). *India-A Comprehensive Geography*. 3rd Ed. Kalyani Publishers, New Delhi.

Lone and Sen. (2014). Horticulture sector in Jammu and Kashmir economy. *European Academic* Research. II(2): 2405–2432.

Malik, Z.A. (2013). Assessment of apple production and marketing problems in Kashmir Valley. *Journal of Economic & Social Development*. 9: 25–30.

Masoodi, M. A. (2003). *Agriculture in Jammu and Kashmir—APerspective*. Mohisraw Book Series, Srinagar.

Mehta, P. (1966). *A Study in Input-Output Relationship in Apple Industry*.Annual Research Report. Department of Agricultural Economics& Social, Punjab Agricultural University, Ludhiana, pp. 139–141.

Mittal, S. (2007). Can Horticulture be a success Story for India. *Working Paper*No. 197. Published by Indian council for Research of International Economic Relations.

Patil, G. R.1997. Area, production and productivity of vegetables and fruits in India. *Indian Journal of Agricultural Economics*. 52(3): 629–630.

Primary Survey/Field Survey. (2019). Survey performed by Showkat Ahmad Ganie. Conducted by the Researchers /Authors using a Questionnaire.

Rather, N. A., Lone, P. A., Reshi, A.A., Mir, M.M. (2013). An analytical study on production and export of fresh and dry fruits in Jammu and Kashmir. *International Journal of Scientific and Research Publications*. 3(2): 1–7.

Rauf, A., Saraf, S.A., Malik, M.A., Sidique, S.A. (2010). Growth performance of apple in Jammu & Kashmir and Himachal Pradesh—A comparative study. *Agricultural Situation in India*. 121–126.

Raza, M. A., Mohammad, A. (1978). *The Valley of Kashmir: A Geographical Interpretation*. Vol. I: The Land. Vikas New Delhi; p. 59

Showkat, A. G., Parry, J. A., Bhat, M. S.(2017). Spatio-temporal analysis of food imports in Jammu and Kashmir, India. *International Journal of Interdisciplinary Research and Innovations*. 5(4): 54–61.

Thakur, D.S. (1971a). Economics and operational efficiency in marketing of Himachal Pradesh apples in India. *Indian Journal of Agricultural Marketing*. 28(1): 96–104.

Thakur, D. S. (1971b). Marketing of apple in Mahasu District of Himachal Pradesh. *Indian Journal of Agricultural Bets*IXL (7):619.

Wani, M.H., Wani, S.A., Mir, N.A. (1993). Economic analysis of different age orchards in apple. *Agricultural Situation in India*. 48: 657–660.

3 Recent Advances in Rootstock and Varietal Improvement in Horticulture Crops

M. A. Kuchay and Javid Iqbal Mir
Central Institute of Temperate Horticulture, Srinagar, J&K, India

Mohammad Tauseef Ali
Division of Fruit Science, SKUAST- K, J&K, India

Bhawna Kaushal and Shivali Chauhan
Dr YSP University of Horticulture and Forestry Nauni, Solan, HP, India

3.1 INTRODUCTION

Global fruit production grew 59% from 2000 to 2021, reaching 910 million metric tonnes in 2021. Global vegetable production surged by 68% to 1.15 billion metric tonnes in 2021 (FAO 2021). The EU accounts for 31.0% of global cut flower and decorative pot plant sales, followed by China (18.6%) and the United States (12.5%). In the European Union in 2016, the largest sales of cut flowers and ornamental pot plants were recorded in the Netherlands, followed by France and Italy (Darras 2020), even though decorative plants are economically beneficial (Yin et al. 2023). Fruits and vegetables (F&V) are recommended for their high fibre, vitamin, mineral, electrolyte, and phytochemical content, notably antioxidants. The health advantages of fruits and vegetables (F&V) are becoming recognised, yet adults consume less than the recommended amount (Pem and Jeewon 2015). Rootstock is merely an underground plant component from which the above-ground portion of the plant grows. The scion is the cutting or branch of the plant that is used to attach to the rootstock during grafting. Plant genotypes are used to take advantage of natural traits including dwarfism, resistance to biotic and abiotic stresses, and improved fruit quality. Rootstocks have been used to adapt to various soil conditions, which has led to an increase in overall fruit quality. Some rootstock breeding initiatives have introduced clonal rootstocks to address salt stress, suction, infections, graft failure, high vigour, low yield, poor nut quality, bad soil, climate change, drought, etc. Both conventional

DOI: 10.1201/9781003412489-3

breeding and genetic engineering are being utilised to generate both rootstock and fruit cultivars with altered traits to meet consumer demand.

3.2 ROLE OF ROOTSTOCKS IN HORTICULTURE CROPS

In the field of horticulture, rootstocks are a vital component. One of the most important aspects is their capacity to influence characteristics like as canopy structure, nutrient intake, flowering, fruit set, and yield. It can also withstand biotic pressures from soil pathogens as well as abiotic challenges from extreme temperatures, salinity, and a lack of nutrients (Reddy et al. 2003). In addition to a lack of suitable land, temperature stress, drought, flooding, and pollution from salt and organic pollutants are all factors that can negatively affect fruit crop production. Using rootstocks that are able to decrease the impacts of external stress on the scion is one strategy to prevent or alleviate these yield losses (Higgs and Jones 1991). Interstems are small segments of stem that fit between the rootstock and the scion to promote the development of the plant (Hayat et al. 2022). It is a tool for coping with the difficulties of graft incompatibility. Utilising Flying Dragon (FD) as a rootstock greatly reduces the size of the plant when compared to trifoliate orange (*Poncirus trifoliata* L.) (Castle 2010; Moreira et al. 2019; Lavagi-Craddock et al. 2022). Rootstocks of grapes that are resistant to the effects of salt include *Vitis champeni* and *Vitis vinifera*. The rootstock of guavas (*Psidium pumilum*) resists wilting and has a dwarfing effect. The fruits grown from this rootstock have the highest total sugar content, the highest Total Soluble Solids (TSS), and numerous seeds.

The Geneva series apple rootstock improved the scion cultivar. G.41, the Geneva series' most common rootstock, yields more and develops fewer root suckers than M.9, although they are similar in size. It has a high level of resistance to *Phytophthora*, woolly apple aphids, and fire blight. It grows in high-pH soils and resistant to replanting disease. With the use of M9 rootstocks, it is possible for them to take in nitrogen, manganese, and iron; however, they are unable to take in potassium or calcium (Cong et al. 2014). The most efficient P-uptake and transport rootstock is MM106, while MM111-grafted plants yielded 44% more fruit than M9-grafted plants (Amiri et al. 2014). Semi-vigorous *Pyrus calleryana* seedlings are able to thrive in many soil types. It can tolerate heat and flames, but not cold. *Pyrus betulaefolia* seedlings can survive winter in many soil types. The trees are somewhat resistant to fire blight and tend to be slightly larger than the "Bartlett seedling" variety. *Pyrus ussuriensis*, which is resistant to cold, exhibits resistance to fire blight and pear *psylla*. Peach, plum, apricot, and almond seedlings are grown on their own, while cherry plants are cultivated on "Paja" (*Prunus cerasoides*) seedlings. Beehmi, also known as *P. mira*, is a well-known almond rootstock (Singh and Gupta 1971). Peach seedlings are susceptible to nematodes; however, "Nemaguard" and "Okinawa" were resistant. Some peach nurseries employ "Behmi" (*P. mira*) rootstock. In Kashmir, Himachal Pradesh, and Kumaon Hills of UP, "Paja" (*P. cerasoides*) is the most prevalent rootstock used in sweet cherry; on the other hand, old plantings are always grafted onto Mazzard or Mahaleb. Russian sour cherry seedlings can be used for *Prunus besseyi* for hardiness (Ananda 1993). While growing vegetables, certain rootstocks can be used to change the marketable top part by grafting from spontaneous genetic variation.

With the suitable rootstock, grafting can change the shape of the scion and protect it from biotic (air- and soil-borne diseases) and abiotic (extreme temperature, drought, waterlogging, and soil acidity or alkalinity) stresses. Select the scion and rootstock at the 2-true leaf stage. The scion and rootstock must have the same stem diameter. Watermelon grafts well with squash and interspecific hybrids due to its high compatibility, robust root system, *Fusarium* resistance, and extreme temperature.

Genus *Rosa* is a popular ornamental plant in the cut flower cultivation, food, pharmaceutical, cosmetic, and perfume industries. Growing roses from their own roots led to slow growth and a short lifespan, so gardeners started using wild roses as rootstocks to speed up development and extend the life of their blooms. It is well known that rootstock and superior variety roses are often physiologically incompatible. *Rosa canina* L. and *Rosa multiflora* Thumb. are commonly used as rootstocks for all rose groups and types, with the exception of the climbing group. However, opinions on the use of this species as a rootstock are varied and frequently at variance with one another. The practice of grafting cut roses onto the rootstocks Natal Briar (*Rosa multiflora*), Major (*Rosa indica*), and Hort No. 1 (*Rosa multiflora*) is relatively prevalent in Korea (Monder et al. 2023). Rootstock changes the physiology of scion growth like photosynthesis, stomatal, and non-stomatal leaves to avoid abiotic stress. It is vital to take into consideration the crops, cultivars, soil conditions, growing season, and cultivation practices in order to choose the rootstock that is the most appropriate for use (Kwon et al. 2022). An outline of the characteristics of a suitable rootstock is presented in Table 3.1.

TABLE 3.1
The Main Characteristics of Commercial Rootstocks of Horticulture Crops

Rootstocks	Main Characteristics
Apple	
A. Seedling rootstock	
Crab apple	
B. Clonal rootstocks	
M9	Dwarf, suitable for high density planting (HDP)
M27 (M13 × M9), P22	Ultra-dwarf suitable for HDP
MM.104 and MM. 106	Resistant to apple woolly aphids
MM.111 and MM.104	Resistant to drought
MM.111	Vigorous
G.41 (Malling 27 × Robusta 5)	New dwarf rootstock (HDP), highly resistant to fire blight and phytophthora
G.16 (Ottawa 3 × M. floribunda)	New fully dwarf rootstock, resistant to fire blight. (alternative to M9)
G. 935	Winter hardy, highly resistant to fire blight and phytophthora (alternative to M26)
Northern spy	Resistant to apple woolly aphids

(*Continued*)

TABLE 3.1 (*Continued*)
The Main Characteristics of Commercial Rootstocks of Horticulture Crops

Rootstocks	Main Characteristics
Pear	
Seedling rootstock	
Kainth (*P. pashia*)	Commercially used in India
B. Clonal rootstocks	
EMLA Quince C, Quince C	Ultra-dwarf, suitable for HDP
OH × F333 (OHF)	Semi-dwarf
D-4 (*P. calleryana*)	Most vigorous rootstock
Old Home, Farmingdale	Resistant to fire blight
OHF series (Old Home × Farmingdale)	Resistant to fire blight
Plum	
A. Seedling rootstock	
Wild apricot	
B. Clonal rootstocks	
Myrobalan B	Resistant to cold. Collar rot and nematodes
Mariana	Resistant to nematode
St. Julien (*Prunus insititia*)	Most versatile rootstock for plums
Pixy (Selection from St. Julien)	Ultra-dwarf rootstock
St. Julien	Vigorous
Cherry	
Gisela	Cold hardy
Mazzard (*P. avium*)	Dwarf, commercial rootstock for sweet cherry
Colt	Semi-dwarf resistance to *Phytophthora* root rot, bacterial canker
F 12/1	Tolerant to bacterial canker, recommended for H.P
Mahaleb (*P. mehaleb*)	Most drought-tolerant, suited to deep, well-drained loams
Peach	
Nemaguard, Nemared, Flordaguard, Guardian	Resistant to nematode (RKN)
GF 677 (Natural peach-almond hybrid), Cadaman, Barrier 1	Tolerant to calcareous soils
Sharpe	Tolerant to Armillaria fungus
Julior, Penta, Tetra	Tolerant to water-logged soil
Rubira, Tetra	Semi-dwarfing
Titan Hybrid	Highly vigorous, root-knot nematode resistance, tolerant of calcareous soil
Mango	
Kurukkan	Salt resistant
Muvandan, Olour	Vigorous
Vellaikolamban	Dwarfing rootstock for Alphonso and Dasheri
13/1 (Israel)	Tolerant to calcareous soils
Bappakai	Best for Neelum variety

(*Continued*)

TABLE 3.1 (*Continued*)
The Main Characteristics of Commercial Rootstocks of Horticulture Crops

Rootstocks	Main Characteristics
Citrus	
Sour orange (*Citrus aurantum*)	Tolerant to xyloporosis, exocortis, and gummosis, standard rootstock for phytophthora root rot
Rough lemon (*C. jambhiri*)	Highest rank citrus rootstock in India Tolerant to tristeza xyloporosis, exocortis, and gummosis, suitable for deep sandy soils
Trifoliate orange	Cold hardy, tolerant to phytophthora, tristeza, RKN, and dwarfing
Flying Dragon (FD) trifoliate orange (*C. trifoliata* var. monstrosá)	Potential dwarfing rootstock (HDP) Resistant to tristeza virus Tolerant to tristeza xyloporosis, exocortis, and gummosis
Troyer citrange	Most promising rootstock for Nagpur mandarin and Kinnow mandarin in India. Suitable dwarfing rootstock for kinnow in India
Karna Katta	Suitable semi-vigorous rootstock for kinnow in India
Sohsarkar	Suitable vigorous rootstock for kinnow in India
Rangpur lime	Promising rootstock for mandarin and sweet orange (central and south India) Tolerant to tristeza, salt, soil toxicity resistant to drought, high vigorous
Rough lemon, lemon, and Cleopatra mandarin	Suitable rootstocks for sweet orange in India. Resistant to tristeza, xyloporosis, exocortis phytophthora, and greening
C. macrophylla	Resistant to tristeza virus
C. unshiu	Tolerant freezing condition
Citron, Kumquat	Highly resistant to citrus cankar
Cleopatra mandarin	Suitable for sweet orange (north-east India)
Volkamer lemon (*C. volkameriana*)	Suitable for navel orange, Valencia orange, and grape fruit
Grapes	
Vitis champini, *Vitis berliendieri*, *V. riparia*	Tolerant to soil salinity
Salt creek cv. of *V. champini*), 1619,1616, Teleki 5-A	Resistant to nematodes
Dogridge cv. *V. champini*	Popular nematode resistant rootstock Suitable for drought and salinity prone areas
Riparia Gloire cv. *V. riparia*, St. George cv. *V. rupestris*	Phylloxera-resistant rootstocks
Vegetable Crops	
Cucumber	
Fig leaf Gourd	Good Fusarium resistance and low-temperature resistance
Cucurbita moschata	Improves stress tolerance

(*Continued*)

TABLE 3.1 (*Continued*)
The Main Characteristics of Commercial Rootstocks of Horticulture Crops

Rootstocks	Main Characteristics
Interspecific hybrid squash (*Cucurbita maxima* x *C. moschata*)	Fusarium Wilt and low-temperature resistance
TRC 11550 (Intraspecific Hybrid)	Nematode resistance
Watermelon	
Bottle Gourd (*Lagenaria siceraria*)	Fusarium wilt and chilling tolerance
Shintoza	Fusarium wilt and chilling tolerance
Pumpkins (*Cucurbita pepo*)	Vigorous root system, resistant to fusarium, and low temperature
African horned cucumber (*Cucumis metuliferus*)	Fusarium resistance and good nematode tolerance
Interspecific hybrid squash (*Cucurbita maxima* x *C. moschat*)	Resistant to fusarium wilt and vigorous root system, high-temperature tolerance

Source: Davis, A.R., Perkins-Veazie, P., Sakata, Y., Lopez-Galarza, S., Maroto, J.V., Lee, S.G., et al. Cucurbit grafting. *Plant Science* 27: 50–74; Kazemi, S., Zakerin, A., Abdossi, V., and Moradi, P. 2021. Fruit yield and quality of the grafted tomatoes under different drought stress conditions. *Asian Journal of Agriculture and Biology* 1: 1–14.

3.3 ROOTSTOCK BREEDING

The utilisation of rootstocks has an effect on the precocity, juvenility, tree size, productivity, and longevity of fruit grown orchards (Domingues et al. 2021). The rootstocks are the structures that are important for providing support to plants and efficiently absorbing nutrients from the soil. Rootstocks have an effect on a variety of characteristics, including the growth of the scion, the yield and quality of the fruit, and the resistance to both biotic and abiotic stress. The selection of rootstocks for a certain region based on environmental parameters, such as temperature, irrigation, soil pH, and fertility, as well as biological considerations, which include viruses, bacteria, nematodes, apple aphids, and other organisms that have the potential to cause damage to the plant (Warschefsky et al. 2016). The rootstock that is capable of inhibiting the growth of the grafted variety is referred to as a dwarfing rootstock. This is in contrast to the growth of the grafted variety on its own roots. In the latter part of the 20th century and the beginning of the 21st century, rootstock breeders achieved remarkable achievements. The increased activity of rootstock breeders has supplied fruit growers with numerous new rootstocks (Cousins 2005). Pathogens that are transmitted through the soil, such as nematodes, fungi, and bacteria, are key components that are involved with replant disease. According to Gainza et al. (2015), there are many other abiotic stress variables that are implicated. Rootstocks from the Geneva® family are able to adapt to the apple replant disease (ARD) conditions in southern Brazil, producing apple varieties that are fruitful. *Malus* species classified as “Gala Select” and “Fuji Suprema” were grafted onto rootstocks designated as “G.202,” “G.210,” “G.213,” and “G.814.” The G.210 semi-dwarfing rootstock can be used to immediately convert “Gala Select” and “Fuji Suprema” apple orchards under non-fallow conditions without affecting vigour or fruit quality. Two rootstocks widely cultivated in China, "M.9-T337" and "M.26," were compared to the premier rootstock selection "12-2" for their potential apple replant disease (ARD) resistance. The most prominent rootstock, known as "12-2," has been found to be a valuable genetic resource for developing apple replant disease (ARD)-resistant apple rootstocks in China. "SuperSour," a hybrid rootstock of citrus (*Citrus sinensis* L.) exhibits superiority in fruit quality and yield as compared to other rootstocks. The use of cuttings and *in vitro* technologies for rootstock propagation eliminates the need for nucellar seeds and their juvenility phase, while simultaneously increasing testing replication and doing away with the 6- to 15-year wait for new hybrids to bear fruit. Consequently, numerous "SuperSour" hybrid rootstocks showed considerable improvements over conventional rootstocks in terms of fruit quality, yield efficiency, canopy health, and overall fruit production.

In China, watermelon, cucumber, pumpkin, melon, luffa, wax gourd, bottle gourd, bitter gourd, and melon cv. Akekekouqi (*Cucumis melo*) were grafted onto eight different species of Cucurbitaceae. The graft compatibility of these eight species was determined. The technique of starch-iodine staining was utilised in order to make a prediction regarding graft compatibility. Melon is graft compatible with cucumber and pumpkin; however, luffa, wax gourd, bottle gourd, bitter gourd, and watermelon are not. Melon is graft-incompatible with watermelon for the most part. As an additional benefit, the starch-iodine staining technique allows for earlier evaluation of graft compatibility, which is beneficial to breeding programmes (Roberto et al. 2022). *In vitro* culture studies are a more economical and precise way of elucidating the physiological and molecular basis of salt tolerance. The reaction of the sweet cherry rootstock Gisela 5 (*Prunus cerasus* × *Prunus canescens*) to increasing concentrations of NaCl (0–150 mM) was examined *in vitro* (Erturk et al. 2007). Screening for salt tolerance in *Pistachio vera* genotypes is possible using *in vitro* selection techniques. When subjected to extreme salinity stress (120 mM NaCl), the genotype Akbari 2 performed the best in terms of growth, biomass, proline content, and K+/Na+ percentage (Raoufi et al. 2021).

Rootstocks are now used in many economically important fruit crops. Almond (*Prunus amygdalus* (L.) Batsch, syn. *P. dulcis* (Mill.)) orchards, and cultivars are typically compatible with peach (*P. persica* (L.) Batsch) rootstocks and interspecific hybrids, according to their graft compatibility. There have been a number of new dwarfing rootstocks created in the past 10 years that impart low to medium vigour, allowing for the establishment of more intense and sustainable cropping systems (Montesinos et al. 2021). Citrus rootstocks *Poncirus trifoliata* and *Citrus macrophylla* were used to quantify the expression of cold response genes CAMTA1, CAMTA3, CAMTA5, CBF1, ICE1, and COR413 IM1. The most effective method for choosing rootstocks was to look for tolerant genotypes of *P. trifoliate*, which showed increased expression of the genes CAMTA3, CAMTA5, CBF1, and COR413 IM1 (Primo-Capella et al. 2021). Species of *Corylus colurna* are more resistant to cold and drought than cultivars of *C. avellana*. One advantage of *C. colurna* as a rootstock is that it does not produce suckers. Two clonal selections ("Dundee" and "Newberg") from open pollination of *C. colurna*, seeds from chosen *C. avellana* trees, and strong, suckers-free cultivars of *C. colurna* are the three hazelnut rootstocks that have been studied. Dundee and Newberg, the first two non-suckering clonal rootstocks, were produced from *C. colurna* seeds that were open-pollinated (Rovira 2021). It is critical to choose rootstocks that can withstand drought. As a plant's hydraulic conductivity decreases in response to diminished water availability, it becomes more resistant to water flow and experiences less evaporation loss as it adapts to water deficit conditions. Developing new interspecific hybrids with tolerances to a variety of abiotic stressors is a difficult task for rootstock breeding programmes. The "Garnem" almond-peach hybrid is a cross between two *Prunus* rootstocks, Myrobalan 'P.2175' (*P. cerasifera* Ehrh.) and *'P. amygdalus'* Batsch × *'P. persica'* Batsch. Comparison of 'Garnem' with Myrobalan 'P.2175' for drought tolerance shows that 'Garnem' has a lot of promise as a breeding stock for future drought-tolerant varieties. The use of the transcription factor Myb25-like as a biomarker of drought sensitivity could be beneficial to the breeding process for *Prunus* rootstock (Bielsa et al. 2021).

3.3.1 Rootstock Breeding through Genetic Transformation

In recent years, transgrafting has established itself as a potentially innovative method that has the ability to mitigate some of the concerns that are associated with transgenic technology. For the purpose of transgrafting, one of the graft components can be either a genetically engineered rootstock that has the ability to give resistance to stress conditions or a rootstock that has been genetically modified (Haroldsen et al. 2012). The overexpression of MdWRKY9 in apple rootstock M26 caused the rootstock to become even more dwarf. This was accomplished by inhibiting the synthesis of endogenous brassinosteroid. Additionally, the overexpression of the MdNHX1 gene resulted in a significant increase in the transgenic rootstock's tolerance to salt. Furthermore, the overexpression of MsDREB resulted in an improvement in drought tolerance by lowering the concentrations of naturally occurring cytokinin. The overexpression of the *rolB* gene significantly enhanced the rooting of apple rootstocks M26 and Jork 9. A cost-effective genetic transformation method is required for the efficient application of genetic engineering in targeted rootstock breeding. Agrobacterium-mediated transformation is simple and cheap, but it has only been established in a few cultivars like M26, which inhibits transgenic apple rootstock breeding (Wang et al. 2019). Stella Mini Tomatoes were planted in GM Micro-Tom tomato plants that had been genetically modified with beta-glucoronidase (GUS). The transcriptome, proteome, metabolome, and general food ingredient profiles of fruit samples were examined using next-generation sequencing (NGS)-illumina, shotgun LC-MS/MS, LC-ESI-MS, and GC-EI-MS, respectively. Stella scion fruits grown on GM and non-GM Micro-Tom rootstocks, and general food element analysis revealed no statistically significant differences (Kodama et al. 2021). A transgenic plum rootstock that is resistant to plum pox virus (PPV) has recently been suggested as a way to protect cultivars or species that are resistant to transformation from contracting the virus. Overexpressing the cytosolic superoxide dismutase (SOD) from spinach and/or ascorbate peroxidase (APX) from pea are used to create transgenic European plum lines that are tolerant to salt. Extreme water stress tolerance by one transgenic line had high APX activity (Diaz-Vivancos et al. 2013; Faize et al. 2013). Genetic engineering depends on transformation efficiency; thus, researchers have tried many methods to boost it. The efficiency of the *Malus* transformation appears to be dependent on the genotype, the potential for regeneration of the explant, and the age of the explant. The Japan Morioka 2 (JM2) apple rootstock was introduced in order to achieve a transformation that is both accurate and effective. In experiments with relatively low levels of transformation efficiency, the results ranged from 0.25% to 3.8% (Wada et al., 2020).

3.4 CROP IMPROVEMENT IN HORTICULTURE CROPS

The term "fruit breeding" refers to the process of enhancing the genetics of fruit crops by a variety of methods, including selection, hybridization, mutation, induction, and molecular techniques, in order to meet the needs of humankind in terms of food, fibre, and animal feed (Janick 2011). However, certain limitations, including a lengthy juvenile phase, massive plant size, environmental concerns, and flower and fruit drop, make it difficult to breed for quality enhancement in perennial fruit crops (Karanjalker and Begane 2016). Varietal advances have been at the cutting edges of agricultural improvement for a very long time, primarily for the purpose of increasing yields and increasing resistance to diseases and pests. Develop crops with increased resource use efficiency, nutritional content, and biotic/abiotic tolerance. Through the process of selective breeding for desirable features, new cultivars can be produced that have the ability to produce more. They are more resistant to the conditions of the local environment, require less water, and have a greater nutrient content. In the vegetable improvement programme, more focus is being placed on the production of hybrids to take advantage of heterosis and combine multiple diseases and stress resistance. Additionally, the quality of the produce is being improved in order to make it more acceptable. Combining conventional and innovative methods of breeding could be an effective step towards finding a solution to this issue. Biotechnology improves precision and reliability and may shorten the breeding cycle in long-juvenility crops. Marker-assisted selection (MAS), genomics, target genes, transgenics, and cisgenics work effectively for challenging crops.

3.4.1 Progress in Genome Editing of Fruit Crops

Genome editing (GE) includes insertions, deletions, and base substitutions at target loci in plants and microorganisms. ZFNs, TALENs, and the clustered regularly interspaced short palindromic repeats (CRISPR)/Cas system have been employed for genome editing (GE) in plants for 20 years. To introduce new traits into apple cultivars, crosses between established apple cultivars and uncultivated apple types with desired characteristics, like disease resistance, have been made. Disease resistance is one such characteristic. Even the best hybrid choices have a good bit of the undesirable trait along with the desirable trait. Since apple trees usually reach maturity between the ages of 5 and 10, the process of eradicating undesirable features by backcrossing might take a long time. Apple cultivars can now be improved in novel ways through genetic transformation, which allows for the controlled introduction of desirable genes into current varieties without threatening the spread of undesirable features. Genes that prevent fungal black spot are a likely selection for transfer to elite cultivars because of their resistance to the disease. A particular gene in a plant can be made inactive by the application of transformation through the introduction of several copies of the gene in a sense or antisense direction (Yao et al. 2000). Techniques for adapting the CRISPR-Cas9 technology to apple and grapevine plants. In order to effectively cause particular modifications without the need for DNA, we achieve this by delivering CRISPR-Cas9 ribonucleoproteins (RNPs) directly to apple and grapevine protoplasts in conjunction with plasmid-mediated genome editing. Sequential method is for creating and transferring CRISPR-Cas9 components to apple and grapevine protoplasts, checking targeted mutagenesis, and regenerating plants after plasmid delivery. Direct and plasmid-mediated delivery of CRISPR-Cas9 RNPs can correctly and efficiently modify apple and grapevine characteristics and those of other crop species. Direct CRISPR-Cas9 RNP delivery takes 2–3 weeks, while plasmids take >3 months to grow plants (Osakabe et al. 2018).

Banana streak virus is a significant problem for banana production. Under conditions of stress, the dsDNA virus causes infectious particles to be produced by integrating itself into the genome of the host. Under the influence of CRISPR/Cas9, mutations were introduced into the banana streak virus (BSV) sequences that were incorporated into the host genome of the plantain cultivar Gonja manjaya. CRISPR/Cas9 was recently utilised in order to generate mutations in the MaGA20ox2 gene, which is responsible for banana dwarfism. The generation of semi-dwarf or dwarf banana cultivars, both of which are in high demand, is a potential use of this technology in the future. As a result of the application of the CRISPR/Cas9 technology, a wide range of crops have been transformed in recent years. These crops are listed in Table 3.2.

Many horticultural crops have been modified with CRISPR-Cas9. The approach offers great potential for fruit crop manipulation. Citrus, apple, banana, *Vaccinium corymbosum* (blueberry), *Theobroma cacao* (cacao), *Castanea sativa* (sweet chestnut), coffee, grape, pear, *Actinidia deliciosa* (kiwi), and *Punica granatum* have all shown stable CRISPR-Cas9 genome edits in a number of proof-of-concept studies. By editing the SQUAMOSA PROMOTER BINDING PROTEIN-LIKE 6 (MdSPL6) gene, tissue culture shoot development improved. This may work for challenging species or cultivars. During regeneration, albino cells result from phytoene desaturase (PDS) gene inactivation. Tissue culturing does not show phenotypic alterations from most relevant gene modifications (Jacobson et al. 2023) which successfully modified the apple CRISPR gene utilising gene-specific sgRNAs. Editing multiple genes simultaneously is beneficial. Two apple homologs of the floral development gene AGAMOUS are targeted by CRISPR-Cas9. Male and female fertility, floral determinacy, and homeotic stamen-to-petal conversion were reduced by MADS15 and MADS221 RNAi in the "Galaxy" cultivar. This reduces fertility and invasiveness while keeping blooms. The "Galaxy" apple variety used in RNAi is uncommon in that it does not require pollination to produce fruit. Domestic apple cultivars "Royal Gala" and "M.26" were used by CRISPR-Cas9, which more precisely targeted genes and should only produce fruit after pollination and seed establishment. For example, the invasive *Pyrus calleryana* (Bradford pear), which was formerly widely planted as an ornamental plant and used to develop fire blight–resistant varieties, could have its invasive potential diminished in some areas. Red-fleshed domestic apples are either

TABLE 3.2
List of Fruit Crop CRISPR/Cas9-Modified Genes and Traits

Species	Target Gene	Target Trait	Delivery Method
Apple and Pear	PDS TFLI	Albino phenotypes Early flowering	Vacuum infiltration in a suspension of *Agrobacterium tumefaciens*
Banana	MaGA20ox2	Semi-dwarf size	*Agrobacterium*-mediated suspension cell transformation
Banana	*eBSV*	Control of virus pathogens	*Agrobacterium*-mediated suspension cell transformation
Citrus (Grapefruit)	CsLOB1	Canker disease	*Agrobacterium*-mediated epicotyl transformation
Citrus (Sweet Orange)	CsLOB1	Canker disease resistance	*Agrobacterium*-mediated epicotyl transformation
Grape and Apple	IdnDH	Biosynthesis of tartaric acid	*Agrobacterium*-mediated suspension cell transformation
Strawberry	APETALA3 (AP3)	Flowering control	*Agrobacterium*-mediated leaf disc
Strawberry	Auxin response factor 8 (FvARF8) and auxin biosynthesis gene (FveTAA1)	Auxin biosynthesis	*Agrobacterium*-mediated transformation

bred or achieved by overexpressing the apple transcription factor *MYB10* for anthocyanin production. Scientists have made purple tomatoes with more anthocyanin by adding genes from a wild species or making a transgenic variety with DNA from Antirrhinum (snapdragon), a plant that can't be bred with other plants. A marketable genetically engineered fruit coloration trait is the pink-fruited *Ananas comosus* (pineapple) that Del Monte Fresh Produce has released.

Most horticulture crops have difficulty with shoot regeneration via plant hormones, especially auxin and cytokinin responses. Because species genotypes respond differently to plant hormones for shoot regeneration *in vitro*, plant hormones, especially auxin and cytokinin responses, hinder shoot regeneration in most horticulture crops. Genotypes of a species can respond differently to plant hormones for shoot regeneration *in vitro*. Increased shoot regeneration can be achieved through the overexpression of native and heterologous baby boom (BBM) genes. Extracellularly overexpressing *Brassica napus* baby boom (*Bn*BBM) from *Brassica napus* L. in Arabidopsis thaliana and *Brassica napus* L. showed that it promoted cell growth and improved the plants' ability to heal itself. Transgenic lines of *Arabidopsis thaliana* that overexpressed *Rc*BBM1 and RcBBM2 from *Rosa canina* L. were better at growing new shoots in tissue culture.

Investigated on *in vitro* phenotype difference between MBM and GUS lines of apple, respectively. Figure 3.1 shows that the multi- target binding motifs (MBM)

FIGURE 3.1 The phenotypic effects of MdBBM1 overexpression in apple transgenic lines studied in vitro.

lines exhibited two distinct shoot architectures were type-1 and type-2. Type-2 MBM shoots were denser and had more branches (Figure 3.1c). Type-1 MBM shoots, on the other hand, had the same number of branches and plant architecture as GUS shoots when they were grown in shoot maintenance medium (Figure 3.1a,b). There were 29 MBM transgenic lines. Twenty-one had type-1 MBM shoots and eight had type-2 MBM shoots (MBM2–3, MBM7–9, MBM14–15, and MBM19) (Figure 3.1d). Compared to the GUS and type-1 transgenic lines, which generated around five branches per shoot, the type-2 transgenic lines produced a substantial number of branches (10–15 branches per shoot) (Figure 3.1d). The MBM transgenic lines also impacted root morphology and development. More roots were formed by the MBM lines, particularly the type-2 lines, after 3 weeks of *in vitro* culture on rooting medium compared to the GUS lines (Fig ure3.1 e-g) (Chen et al. 2022).

3.4.2 Rapid Cycle Breeding

Fruit crop breeding takes 15–25 years to get new cultivars to the market. This would greatly simplify and shorten difficult programmes like the introgression of features from genetically distant wild relatives, which often require several generations of crossing. Overexpression of flowering genes or silencing of repressor genes that control flowering genes during the juvenile phase can shorten the apple juvenile phase (Schropfer et al. 2022). One parent (green) with good fruit quality and a transgenic T-DNA insertion for early blooming is crossed with another parent (red) that has a trait that is wanted, like disease resistance. Unfortunately, only *Malus* wild species, which produce poor fruit quality, have many resistances. The red colour represents

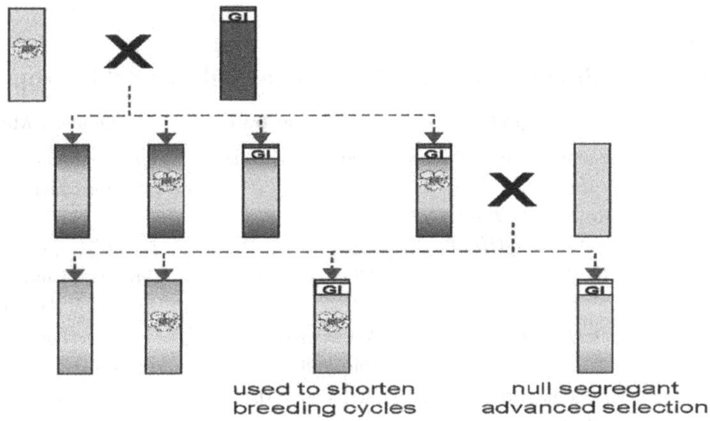

FIGURE 3.2 Schematic representation of the idea behind the rapid cycle breeding program.

the high percentage of harmful alleles in these genotypes. The frequency of unfavourable alleles is decreased through repeated backcrossing with green apple cultivars of excellent quality. The flower is a representation of the transgene that allows for early flowering. An example of a desirable attribute is disease resistance, and the acronym "GI" stands for "gene of interest," which gives this trait. A quarter of the progeny will have both the early flowering and resistance features if they are inherited monogeneously. To further reduce breeding cycles, these genotypes can be utilised in subsequent crosses. A further 25% of the progeny will be deficient in the target gene. Such non-transgenic null segregants can be released from the breeding programme as advanced selections, as shown in Figure 3.2.

3.4.3 Progress in Genome Editing of Vegetable Crops

A wide variety of abiotic and biotic stressors can reduce the production of vegetable crops; thus, resistant and tolerant cultivars are required. Several vegetables could benefit from improvement in terms of flavour, nutrient content, plant structure, and shelf life. The genomes of crops with potential commercial value have been modified using CRISPR/Cas9 in Table 3.3.

For both fresh and processed tomato cultivars, different features are prioritised during tomato breeding. It is difficult to transfer complex features, including abiotic stress tolerance, via traditional breeding procedures. Phenotypic stress tolerance is a complex feature that results from multiple physiological, biochemical, and molecular processes. Its application has led to enhanced fruit quality and resistance to diseases. The use of recombinant DNA technology, sometimes called genetic engineering, has allowed researchers to avoid these issues by transferring genetic features from one species to another. Early gene transfer research used *Agrobacterium tumefaciens* to introduce foreign genes into plants. Due to its repeatability, *Agrobacterium* is an

TABLE 3.3
CRISPR/Cas9-Modified Genes and Traits in Several Vegetable Crops

Species	Target Gene	Target Trait	Delivery Method
Brassica campestris	Pectin-methylesterase genes *Bra003491*, *Bra007665*, and *Bra014410*	Methylation of pectin	*Agrobacterium*-mediated transformation
Cabbage	*BoPDS*	Albino phenotypes	*Agrobacterium*-mediated hypocotyl transformation
Carrot	*DcMYB113-like*	Anthocyanin biosynthesis	*Agrobacterium*-mediated transformation
Potato	Acetolactate synthase1 (StALS1)	Herbicide resistance	*Agrobacterium*-mediated leaf transformation
Potato	*SBE1*, *SBE2*	Starch quality	Polyethylene glycol (PEG)-mediated protoplast
Tomato	Aux/IAA9 (*SlIAA9*)	Parthenocarpic fruits	*Agrobacterium*-mediated leaf
Tomato	Ripening inhibitor (RIN)	MADS-box transcription factor regulating fruit ripening	*Agrobacterium*-mediated transformation
Tomato	Self-pruning 5G (SlSP5G)	Day-length-sensitive flowering	*Agrobacterium*-mediated transformation
Tomato	NPR1	Drought tolerance	*Agrobacterium*-mediated cotyledon segment transformation
Tomato	Enzyme pectate lyase (PL) polygalacturonase 2a (PG2a), and β-galactanase (TBG4)	Pectin Degradation control	*Agrobacterium*-mediated transformation
Water melon	PDS	Albino phenotype	*Agrobacterium*-mediated callus

excellent plant transformation agent for solanaceous species. The biolistic technique is an easy and promising transformation method (Krishna et al. 2019).

Eggplant fruit setting is sensitive to suboptimal temperatures, humidity, and drought. These circumstances impact aubergine yield. Fruit setting without fertilisation (parthenocarpy) may boost yield in unpredictable environments. Exogenous plant growth regulators, or genes, produce parthenocarpy. Due to chemical and labour costs, production hormone treatments are expensive. The polygenic parthenocarpic trait causes low-frequency fruit sets and smaller fruit sizes in several cultivars. Transgenic genetic engineering is used to generate parthenocarpic eggplant. Either direct synthesis or transcriptional control in upstream pathways is used to develop parthenocarpy in aubergine. The tryptophan monooxygenase-coding iaaM gene is isolated from *Pseudomonas syringae* pv. Savastanoi generates indole acetamide, which is chemically or enzymatically transformed into indole-3-acetic acid.

The *Antirrhinum majus* DefH9 gene yielded a placenta and ovule-specific promoter. Transgenic aubergine possesses the DefH9-promoted iaaM gene to increase auxin production in the ovule and placenta. Emasculated flowers produced seedless fruits. Early fruit output was sixfold higher in DefH9-iaaM transgenic hybrids. The CRISPER-CAS system was used to remove three particular polyphenol oxidase (PPOs) genes from the genome: SmelPPO4, SmelPPO5, and SmelPPO6. PPOs are responsible for the oxidation of polyphenols, which colour the flesh of aubergines after they have been chopped. Less browning is beneficial for both the industrial processing and the fresh consumption markets. This demonstration would surely pave the way for more targets in eggplant (Alam and Salimullah et al. 2021).

Flavr Savr™ tomatoes, introduced by Calgene in 1994, were the first transgenic crops to be produced for commercial purposes. There is an antisense polygalacturonase (PG) gene in this particular tomato. Plants generate ethylene under the regulation of three enzymes: S-adenosylmethionine (SAM) synthase, 1-aminocyclo propane-1-carboxylate (ACC) synthase, and ACC oxidase. Genomic manipulation of ethylene production genes, as well as genes encoding SAM or ACC metabolisers, has made it possible for the modulation of fruit ripening. Still, it's possible to manipulate tomato fruit ripening by crossing heterozygous genotypes with anti-ripening genes (rin and nor). These genes have been incorporated into many fresh and processed tomatoes. New Leaf®, New Leaf Y®, and New Leaf Plus® were the names of different transgenic potato cultivars that were marketed by Nature Mark®, which is a part of Monsanto®. It is possible that the New Leaf®, New Leaf Y®, and New Leaf Plus® potatoes have been so successful in the market because of the mild winters that occur in the Pacific Northwest, a large number of aphids and the viruses that are associated with them, and the difficulties in eradicating the Colorado potato beetle. Increased yields with reduced pesticide use were an advantage to commercial producers' seed manufacturers (Thomas et al. 1997; Thornton, 2003).

3.4.4 Genome Editing in Ornamental Plants

Problems may arise for the horticultural industry if flower yields are low. Cut flowers, such as roses, tulips, carnations, and chrysanthemums, are highly prized for their aesthetic value in the floral market. Although flowers do not provide any nutritional benefit to people, they have been produced for a long time regardless. Due to the high demand for attractive new plant cultivars with novel morphological features, colours, and scents, the commercial floriculture industry has a far shorter life cycle compared to other crops (Shibata 2008). New ornamental plant varieties have been created through a number of breeding techniques. Traditional breeding methods, including hybridisation and mutation breeding, have produced plants with a wide range of colours, altered shapes, improved structures, and resistance to diseases. These usual methods are hard to use because ornamental plant species have complicated genetic inheritance and polyploidy, which is caused by a high level of heterozygosity. Consequently, new methods are required to enhance the harvest of decorative plants (Azadi et al. 2016; Sharma and Messar 2017). The technology for sequencing genomes has advanced greatly in the last several decades. Particularly, noteworthy is the publication of genomic sequence information on several important ornamental plant

TABLE 3.4
Targeted Genes and Traits Changed by CRISPR/Cas9 in Ornamental Crops

Species	Target Gene	Target Trait	Delivery Method
Chrysanthemum morifolium	Yellowish-green fluorescent (CpYGFP)	Fluorescence protein disruption	*Agrobacterium*-mediated leaf sections transformation
Phalaenopsis orchid	*MADS*	Floral initiation and development	*Agrobacterium*-mediated protocorm transformation
Petunia	*Nitrate reductase* (*PhNR*)	Deficiency in nitrate assimilation	PEG-mediated protoplast transfection

species since 2005. These include *Chrysanthemum nankingense*, *Dendrobium officinale*, *Dianthus caryophyllus*, *Helianthus annuus*, *Hibiscus syriacus*, *Ipomoea nil*, *Petunia hybrid*, *Halaenopsis*, *Primula veris*, *Rosa chinensis*, *Rosa multiflora*, and *Rosa roxburghii*. Plant breeding and whole-genome sequencing have both shown tremendous improvements in the last several years. CRISPR, which stands for clustered regularly interspaced short palindromic repeat, is a better way to change the genome than meganucleases (MNs), zinc fingers (ZFNs), and transcription activator-like effector nucleases (TALENs). Base editors and prime editors are two new CRISPR systems that have improved gene editing abilities, lowered activity that doesn't target DNA, and raised DNA selectivity (Ahn et al. 2020). In the life cycle of ornamental plants, the initiation of bloom production and development is an important phase that requires careful attention. There are significant financial benefits to be gained from enhancing the properties of flowers through the application of more recent techniques, such as CRISPR/Cas9. Table 3.4 lists several ornamental cultivars modified with CRISPR/Cas9.

3.5 CONCLUSION

Rootstocks have a substantial influence on the tree's ability to grow early, produce fruit, and maintain a desired size. Rootstock selection and application can be made more informed if one has a greater awareness of the interactions between rootstock and scion, as well as all the factors that pertain to soil and climate. Horticultural rootstocks that are resistant to both biotic and abiotic stress should be developed as soon as possible. This is of the utmost priority. Using rootstocks in several horticultural crops is crucial to the successful cultivation of fruit trees. Canopy structure, fertiliser consumption, flowering, fruit set, and yield are all factors that should be considered carefully. Because of their short growing seasons and restricted availability, fruit crops require a hybrid of conventional and modern breeding techniques to achieve the desired characteristics. Biotechnology has the ability to reduce the breeding cycle for crops with a prolonged juvenile phase by increasing accuracy and uniformity during the breeding process.

REFERENCES

Ahn, C.H., Ramya, M., An, H.R., Park, P.M., Kim Y.J., Lee, S.Y., and Jang, S. 2020. Progress and challenges in the improvement of ornamental plants by genome editing. *Plants* 9: 687.

Alam, I., and Salimullah, M. 2021. Genetic engineering of eggplant (*Solanum melongena* L.): progress, controversy and potential. *Horticulturae* 7: 78.

Amiri, M.E., Fallahi, E., and Safi-Songhorabad, M. 2014. Influence of rootstock on mineral uptake and scion growth of 'golden delicious' and 'Royal Gala'apples. *Journal of Plant Nutrition* 37: 16–29.

Ananda, S.S. 1993. Rootstocks of Sone Fruits. Advances in Horticulture: Fruit Crops. Malhotra Publishing House, New Delhi, India.

Azadi, P., Bagheri, H., Nalousi, A.M., Nazari, F., and Chandler, S.F. 2016. Current status and biotechnological advances in genetic engineering of ornamental plants. *Biotechnology Advance* 34, 1073–1090.

Bielsa, B., Sanz, M.Á., and Rubio-Cabetas, M.J. 2021. 'Garnem' and Myrobalan 'P.2175': two different drought responses and their implications in drought tolerance. *Horticulturae* 7: 299.

Castle, W.S. 2010. A career perspective on citrus rootstocks, their development, and commercialization. *Hort Science* 45: 11–15.

Chen, J., Tomes, S., Gleave, A.P., Hall, W., Luo, Z., Xu, J., and Yao, J.L. 2022. Significant improvement of apple (*Malus x domestica* Borkh.) transgenic plant production by pre transformation with a Baby boom transcription factor. *Horticulture Research* 9: uhab014.

Cong, C., Chao, L., Cui-ying, L., Xiao-yu, K., Yang-jun, Z., and Feng-wang, M. 2014. Differences in the efficiency of potassium (K) uptake and use in five apple rootstock genotypes. *Journa of Integrative Agriculture* 13: 1934–1942.

Cousins, P. 2005. Rootstocks breeding: an analysis of intractability. *HortScience* 40, 1945–1946.

Darras, A.I. 2020. Implementation of Sustainable Practices to Ornamental Plant Cultivation Worldwide: a critical review. *Agronomy* 10(10): 1570.

Davis, A.R., Perkins-Veazie, P., Sakata, Y., Lopez-Galarza, S., Maroto, J.V., Lee, S.G., et al. Cucurbit grafting. *Plant Science* 27: 50–74.

Diaz-Vivancos, P., Faize, M., Barba-Espin, G., Faize, L., Petri, C., Hernández, J.A., and Burgos, L. 2013. Ectopic expression of cytosolic superoxide dismutase and ascorbate peroxidase leads to salt stress tolerance in transgenic plums. *Plant Biotechnology Journal* 11: 976–985.

Domingues, A.R., Marcolini, C.D.M., Gonçalves, C.H., Resende, J.T.V., Roberto, S.R., and Carlos, E.F. 2021. Rootstocks genotypes impact on tree development and industrial properties of 'Valencia'sweet orange juice. *Horticulturae* 7: 141. doi: 10.3390/horticulturae7060141.

Erturk, U., Sivritepe, N., Yerlikaya, C., Bor, M., Ozdemir, F., and Turkan, I. 2007. Responses of the cherry rootstock to salinity in vitro. *Biologia Plantarum* 51: 597–600.

Faize, M., Faize, L., Petri, C., Barba-Espin, G., Diaz-Vivancos, P., Clemente-Moreno, M.J., Koussa, T., Rifai, L.A., Burgos, L., and Hernandez, J.A. 2013. Cu/Zn superoxide dismutase and ascorbate peroxidase enhance in vitro shoot multiplication in transgenic plum. *Journal of Plant Physiology* 170: 625–632, doi: 10.1016/j.jplph.2012.12.016.

FAO, 2021. World Food and Agriculture – Statistical Yearbook 2021. Rome, Italy. doi: 10.4060/cb4477en

Gainza, F., Opazo, I., Guajardo, V., Meza, P., Ortiz, M., Pinochet, J., and Muñoz, C. 2015. Rootstock breeding in Prunus species: ongoing efforts and new challenges. *Chilean Journal of Agricultural Research* 75: 6–16.

Haroldsen, V.M., Szczerba, M.W., Aktas, H., Lopez-Baltazar, J., Odias, M.J., Chi-Ham, C.L., Labavitch, J.M., Bennet, A.B., and Powell, A.L. 2012. Mobility of transgenic nucleic acids and proteins within grafted rootstocks for agricultural improvement. *Frontiers* in *Plant Science* 39: 1–12.

Hayat, F., Li, J., Liu, W., Li, C., Song, W., Iqbal, S., Khan, U., Umer Javed, H., Ahsan Altaf, M., and Tu, P. 2022. Influence of citrus rootstocks on scion growth, hormone levels, and metabolites profile of 'Shatangju' Mandarin (Citrus reticulata Blanco). *Horticulturae* 8: 608.

Higgs, K.H., and Jones, H.G. 1991. Water relations and cropping of apple cultivars on a dwarfing rootstock in response to imposed drought. *Journal of Horticultural Science* 66: 367–379.

Jacobson, S., Bondarchuk, N., Nguyen, T.A., Canada, A., McCord, L., Artlip, T.S., Welser, P., and Klocko, A.L. 2023. Apple CRISPR-Cas9-a recipe for successful targeting of AGAMOUS-like genes in domestic apple. *Plants* 12: 3693. doi: 10.3390/plants12213693.

Janick, J. 2011. History of fruit breeding. Fruit, Vegetable Cereal Science and *Biotechnology* 5: 1–7.

Karanjalker, G., and Begane, N. 2016. Breeding perennial fruit crops for quality improvement. *Erwerbs-Obstbau* 58: 119–126.

Kazemi, S., Zakerin, A., Abdossi, V., and Moradi, P. 2021. Fruit yield and quality of the grafted tomatoes under different drought stress conditions. *Asian Journal of Agriculture and Biology* 1: 1–14. doi: 10.1007/s13205-019-1665-0.

Kodama, H., Miyahara, T., Oguchi, T., Tsujimoto, T., Ozeki, Y., Ogawa, T., Yamaguchi, Y., and Ohta, D. 2021. Effect of transgenic rootstock grafting on the omics profiles in tomato. *Food Safety* 9(2), 32–47. doi: 10.14252/foodsafetyfscj.D-20-00032

Krishna R., Karkute, S.G., Ansari, W.A., Jaiswa, D.K., Verma, J.P., and Singh, M. 2019. Transgenic tomatoes for abiotic stress tolerance: status and way ahead. *Biotech* 9: 143.

won, O.H., Choi, H.G., Kim, S.J., Lee, Y.R., Jung, H.H., and Park, K.Y. 2022. Changes in yield, quality, and morphology of three grafted cut roses grown in a greenhouse year-round. *Horticulturae* 8: 655. doi: 10.3390/ horticulturae8070655

Lavagi-Craddock, I., Dang, T., Comstock, S., Osman, F., Bodaghi, S., and Vidalakis, G. 2022. Transcriptome analysis of citrus dwarfing viroid induced dwarfing phenotype of sweet orange on trifoliate orange rootstock. Microorganisms 10: 1144.

Monder, M.J., Bąbelewski, P., and Sołtan, S. 2023. Diversity in anatomical features of rose rootstock root necks: *Rosa canina* 'Inermis', 'Pfänder', 'Schmid's Ideal', *Rosa laxa* Retz. and *Rosa multiflora* Thunb. *Scientia Horticulturae* 316: 112004. doi: 10.1016/j.scienta.2023.112004.

Montesinos, A., Thorp, G., Grimplet, J., and Rubio-Cabetas, M.J. 2021. Phenotyping almond orchards for architectural traits influenced by rootstock choice. *Horticulturae* 7: 159. doi: 10.3390/ horticulturae7070159.

Moreira, A.S., Stuchi, E.S., Silva, P.R., Bassanezi, R.B., Girardi, E.A., and Laranjeira, F.F. 2019. Could tree density play a role in managing Citrus Huanglongbing epidemics? *Tropical Plant Pathology* 44: 268–274.

Osakabe, Y., Liang, Z., Ren, C., Nishitani, C., Osakabe, K., Wada, M., et al. 2018. CRISPR-Cas9-mediated genome editing in apple and grapevine. *Nature Protocols* 13: 2844–2863. doi: 10.1038/s41596-018-0067-9.

Pem, D., and Jeewon, R. 2015. Fruit and vegetable intake: benefits and progress of nutrition education interventions-narrative review article. *Iran Journal of Public Health* 44(10): 1309–1321.

Primo-Capella, A., Martínez-Cuenca, M.-R., and Forner-Giner, M.A. 2021. Gene expression under short-term low temperatures: Preliminary screening method to obtain tolerant citrus rootstocks. *Horticulturae* 7: 447. https:// doi.org/10.3390/horticulturae7110447

Raoufi, A., Salehia, H., Rahemi, M., Shekafandeha, A., and Khalili, S. 2021. In vitro screening: The best method for salt tolerance selection among pistachio rootstocks. *Journal of the Saudi Society of Agricultural Sciences* 20: 146–154.

Reddy, Y.T.N., Reju, M., Kurian, P.R., Ramachander, G.S., and Kohli, R.R. 2003. Long term effects of rootstocks on growth, fruit yielding patterns of Alphonso mango (M. Indica). *Scientia Horticulturae* 97: 95–108.

Roberto, S.R., Novello, V., and Fazio, G. 2022. New rootstocks for fruit crops: breeding programs, current use, future potential, challenges and alternative strategies. *Frontiers in Plant Science* 13: 878863.

Rovira, M. 2021. Advances in hazelnut (*Corylus avellana* L.) rootstocks worldwide. *Horticulturae* 7: 267. doi: 10.3390/horticulturae7090267

Schropfer, S., Lempe, J., Emeriewen, O.F., and Flachowsky, H. 2022 Recent developments and strategies for the application of agrobacteriummediated transformation of apple (Malus × domestica Borkh). *Frontiers in Plant Science* 13: 928292. doi: 10.3389/fpls.2022.928292.

Sharma, R., and Messar, Y. 2017. Transgenics in ornamental crops: creating novelties in economically important cut flowers. *Current Science* 113: 43–52.

Shibata, M. 2008. Importance of genetic transformation in ornamental plant breeding. *Plant Biotechnology* 25: 3–8.

Singh, R.N., and Guptha, P.N. 1971. Rootstock problem in stone fruits and potentialities of wild species found in India. *Punjab Horticulture Journal* 11: 157–175.

Thomas, P.E., Kaniewski, W.K., and Lawson, E.C. (1997) Reduced field spread of potato virus in potatoes transformed with the potato leafroll virus coat protein gene. *Plant Disease* 81: 1447–1453. doi: 10.1094/PDIS.1997.81.12.1447

Thornton, M. 2003. The rise and fall of new leaf potatoes. *NABC Report* 15: 235–243.

Wada, M., Nishitani, C., and Komori, S. 2020. Stable and efficient transformation of apple. *Plant Biotechnology* 37: 163–170. doi: 10.5511/plantbiotechnology.20.0602a.

Wang, Y., Li, W., Xua, X., Qiua, C., Wua, T., Wei, Q., Mac, F., and Hana, Z. 2019. Progress of apple rootstock breeding and its use. *Horticultural Plant Journal* 5: 183–191.

Warschefsky, E.J., Klein, L.L., Frank, M.H., Chitwood, D.H., Londo, J.P., von Wettberg, E.J., and Miller, A.J. 2016. Rootstocks: diversity, domestication, and impacts on shoot phenotypes trends. *Plant Science* 21: 418–437.

Yao, J.L., Cohen, D., Atkinson, R., and Morris, B. (2000). Transgenic Apple (Malus x domestica). In: Bajaj, Y. P. S. (eds) *Transgenic Trees. Biotechnology in Agriculture and Forestry* 44. Berlin, Heidelberg: Springer.

Yin, X., Feng, L., and Gong, Y. 2023. Mitigating ecotoxicity risks of pesticides on ornamental plants based on life cycle assessment. *Toxics* 11, 360. doi: 10.3390/toxics11040360.

4 Biotechnological Intervention in Horticulture

Uzma Noor Shah, Nayana Borah, Sandeep Kumar, C., and Jayashree, V.H.
Department of Life Sciences, School of Sciences, Jain (Deemed -to-be-University), Bangalore, India

4.1 INTRODUCTION

Horticulture is the science and art of cultivating plants for food, medicine, beauty, or cultural significance. Horticulture plants include a wide range of species such as vegetables, fruits, ornamentals, and herbs, among others. These plants have been selectively bred and propagated for thousands of years to optimize their yield, taste, colour, texture, and other desirable traits.

For example, they include, amongst others, Rosaceae (apple [Malus 9 domestica], rose [Rosa spp.], etc.), Rutaceae (*Citrus reticulata* Blanco, etc.), Orchidaceae (*Cymbidium sinense*, moth orchid [*Phalaenopsis* spp.], etc.), Asteraceae (*Lactuca sativa*, *Bellis perennis*, etc.), Lilium (onion [*Allium cepa*], cornflower [*Centaurea cyanus*], etc.), Solanaceae (tomato [*Solanum lycopersicum*], pepper [*Capsicum annuum*], etc.), Cucurbitaceae (watermelon [*Citrullus lanatus*], melon [*Cucumis melo*], pumpkin [*Cucurbita* spp.], etc.), Liliaceae (*tulipa gesneriana*, etc.), Cruciferae (cabbage [*Brassica oleracea*], violet [*Viola odorata*], etc.), and Camellia (tea tree, etc.) (Jiang et al. 2002b).

Horticultural crops are diverse and can serve a variety of purposes, such as food, beverages, ornamental products, and medicine. Fruits and vegetables are particularly beneficial for human health, as they contain carbohydrates, minerals, and specialized metabolites that promote health. Flowers and tea are also valuable horticultural crops because they can improve our daily lives with their pleasant aroma and soothing effects. The market value of horticultural crops is high, with fruits and vegetables alone worth over $825 billion in 2018, and tea accounting for $16.75 billion. Surprisingly, horticultural crops have a higher production value per unit area compared to most staple crops (Zhang et al. 2019).

 DOI: 10.1201/9781003412489-4

4.2 CHALLENGES ASSOCIATED WITH PRODUCTION OF HORTICULTURAL CROPS

Advancements in agricultural technology have led to significant improvements in the yield and quality of horticultural crops in the 21st century. However, these gains are still inadequate to meet the needs of a rapidly growing population. The world population, estimated at around 7 billion in 2020, is projected to reach 9 billion by 2050. Moreover, challenges such as global warming, desertification, and environmental pollution are making it increasingly difficult to find enough cultivable land to meet the demand for food. Ensuring food supply and security is already a significant challenge. Sustainability has become a crucial issue for agriculture as it encounters mounting constraints. Compared to staple crops like rice, wheat, maize, and soybean, horticultural crops face unique difficulties and problems in production (Grierson et al. 2011; Jamnadass et al. 2020; Tian et al. 2020).

Breeding cycles for most horticultural crops are typically longer, and improving quality has become increasingly important. Traditional methods such as mutation and crossing are often used in the breeding of perennial fruit trees, which have extended juvenile periods and require cutting or grafting for genetic stability or to obtain desirable traits. This process is time-consuming and labour-intensive. For instance, in the breeding of tea varieties, lines are initially selected via hybridization or natural populations and then propagated through cutting. After 3 years of planting, lines with superior traits are identified, a process that can take over a decade. Quality traits, such as aroma, taste, and colour, which are crucial determinants of nutritional and economic value, are prioritized breeding targets for horticultural crops. While yield has traditionally been the primary focus of breeding, flavour characteristics have been overlooked (Gao et al. 2019; Zhu et al. 2019).

Secondly, horticultural crops pose more technical challenges in cultivation and management. Horticultural crops are a diverse group of species, most of which are perennials that are cultivated continuously for several years, even decades. As a result, they are more susceptible to pests and diseases. The growth and harvest of horticultural crops often require the use of greenhouses or other specialized facilities, which makes their cultivation more labour-intensive and complex. Therefore, it can be concluded that horticultural crops generally require more intensive management and labour input than other crops (Tay et al. 2021; Yu et al. 2019).

Lastly, horticultural crops face significant postharvest loss, with many fruits, vegetables, and flowers requiring fresh storage to maintain their quality (Alenyorege et al. 2019; Ichimura and Goto 2002). Some horticultural crops also require complex postharvest processing before consumption. For example, tea leaves need to undergo different processing steps to become green, black, or oolong tea, involving a comprehensive metabolism of substances in fresh tea leaves. Similarly, bitter almonds must undergo processing to remove toxic cyanogenic glucosides before consumption. The processing, storage, transportation, and sale of horticultural crops often lead to significant losses in both flavour and quality (Cortes et al. 2019).

4.3 TRADITIONAL BREEDING TECHNIQUES

New cultivars are created by top crossing, mass selection, recurrent selection, and synthetic variety production. After evaluating the source population, mass selection selects appropriate plants or seed from those parent plants. Traditional breeding is the selective breeding-based modification of both plants and animals. Traditional plant breeding techniques may use elements of biotechnology like tissue culture and mutational breeding. Using recombinant DNA technology, a transgene is a gene that has been transferred from one organism to another. Wheat, barley, rice, potatoes, soybeans, and onions are a few examples of plants that have been created by mutation breeding. The advantage of conventional plant breeding is that it produces more genetic resources that may be used to improve crops by introducing the desired features. Conventional farming relies on the use of chemicals that are virtually definitely hazardous and were probably made artificially in a lab to treat the soil and plants. The plant is protected by these items against disease and pests.

4.3.1 Bridge Cross

When a direct crossing between two species, A and B, is not feasible, the crossing barrier may be overcome by an intermediate crossing with a third species, C, which is compatible with both species. To create an interspecific hybrid, species A and C are first crossed, and then, species B is later bred with the resulting hybrid. So, the genomes of species A and B or portions of them can be joined indirectly.

4.3.2 Utilization of Sub- or Supra-Optimal Stigma Age, or Sub-Optimal Circumstances, for Pollination

To overcome compatibility limitations posed by the stigma and/or the style, mature pollen is transferred to an immature or old stigma to prevent the existence of active components that impede pollen tube growth. As an alternative, pollination is done after the flowering season has ended or with plants that have developed in less-than-ideal environments.

4.3.3 Pollination Using Application of Chemicals

Plant growth regulators (hormones), such as the conventional auxins, cytokinins, and gibberellins as well as more recently discovered ones, like the stress signalling-related salicylic acid, are applied to plants or flowers to promote successful pollination. They can hasten pollen tube development, increase pistil viability, and hasten fruit and seed maturation. Incompatibility has also been broken by the application of additional substances.

4.3.4 Treatment of Pollen and/or Pollen Mixes during Pollination

In rare cases, incompatible pollen can traverse barriers successfully by being heated beforehand. As an alternative, pollination can be accomplished by combining compatible and incompatible pollen. By being able to germinate and enter the style, the incompatible pollen is able to "clear the way" for the compatible pollen. While suitable pollen has a higher likelihood of fertilizing the ovules, it can be pre-treated (either chemically, by irradiation, or mechanically) to lessen its ability to do so. This will improve the efficiency of the difficult process of selecting the desired progeny from the undesirable pollen among the progeny of the "assisting" compatible pollen. The suitable pollen that is "assisting" in such pre-treatment is referred to as mentor pollen. Sequential pollination is another option, in which compatible pollen is spread out a day or two before pollen that is incompatible.

4.3.5 In vitro Pollination

Stigma of in vitro-placed pistils receives pollen after the leafy flower parts and anthers have been separated from the flower bud, and the pistil has been placed on a tissue growth medium. Ovules that have been fertilized can then be separated and cultured further in vitro to prevent incompatibility with the surrounding maternal tissues. The ovary wall, style, and/or stigma may also be removed, and pollen may be applied to (Placental Pollination) the placenta to entirely bypass compatibility obstacles present in stigma. After the ovarian wall has been removed or after the ovules have been placed in in vitro culture, pollen can come into touch with the ovules directly.

4.3.6 Translocation Breeding

Chromosome translocation occurs when nonhomologous chromosomes exchange certain components. After radiation treatment, chromosomal segments from a trisomic addition line with one extra homologous chromosome from a different species were recovered, and the plants had the added chromosome segments incorporated into their genomes. It is possible to cause the translocation. The introduction of "gametocidal" (causing chromosomal breakage) chromosomes from another species can also cause translocation.

4.3.7 Manipulation of Chromosome Pairing in Meiosis

Reduced chromosomal pairing and a lack of recombination are critical factors in the creation of interspecific hybrids. Chromosomes that generally exhibit little pairing (i.e., between homologous chromosomes generated from separate parental species) can couple as a result of specific mutations. Additionally, by applying a number of chemical substances, physical stressors like a temperature shock, or UV light, recombination frequencies can be artificially increased. In order to promote the interchange of segments between homologous chromosomes, it is also possible to repress genes that favour homologous pairing during breeding with complicated between chromosomal complements generated from various parental species, or allopolyploids. However, because of the ensuing reproductive instability, it is preferable to avoid the creation of multivalent during meiosis, especially in recently generated autopolyploid. This may be accomplished through selection for increased progeny fertility and/or chromosomal alterations that cause the polyploid to behave in a "diploid" (or "allopolyploid") manner (diploidization), resulting in the formation of primarily bivalents during meiosis. As opposed to the regular complement of A chromosomes, supernumerary chromosomes can occasionally favour chromosomes that are homologous recombining, rather than between homologous ones that are closely related.

4.3.8 Mutagenesis

Ethyl methyl sulfonate causes mutations in plant components subjected to ionizing radiation (such as X-rays or gamma rays) or chemical mutagens, such as sodium azide or alkylating compounds, such as Ethyl methanesulphonate (EMS). Possible results include chromosome breakage, translocation, doubling, elimination, and single base-pair changes. In addition to chemical mutagens, radiation is also frequently responsible for these events (substitution, insertion, and deletion). Genetically altered plants can be selected through germination, sexual reproduction, and selection for entirely mutated plants, resulting in generations M1 and M2. Seeds can be treated, resulting in the development of chimaeras, which requires selection for completely mutated plants. Pollen can be altered and selection applied to the offspring of plants pollinated by the altered pollen to prevent chimaeras. Despite the fact that pollen has a shorter shelf life, handling it can be more difficult. Clonally generated species can be used to treat vegetative sections, preferably buds where meristems capable of regenerating new plants can be targeted. The plants predominantly produced through mutagenesis could exhibit undesirable phenotypic features next to the desired one; this phenomenon is known as "mutagenic drag," and it is comparable to "linkage drag" with wide crossings. Thus, additional propagation or backcrossing is frequently required, along with selection, to get rid of unwanted features.

4.3.9 Cell Fusion

Somatic hybridization, another name for in vitro cell fusion procedures, can boost the effectiveness of creating hybrids, which can also be created through sexual crossover but only with great difficulty. Crossing obstacles is a common cause of these challenges. Somatic hybridization can also be used to effectively integrate particular nuclear and cytoplasmic (plastid and mitochondrial) genomes in place of time-consuming backcrossing with additional suitable crosses. In order to promote cell union, protoplasts are frequently produced from leaf cells or cell suspension cultures by manually or enzymatically removing the cell walls. After that, during the somatic hybridization process, plant protoplasts are fused chemically (for instance, with polyethylene glycol) or electrically. After being chosen for a successful fusion of the two genomes on artificial media, fusion products are encouraged to reconstruct cell walls and are allowed to divide to produce calli. Using standard tissue culture methods, organogenesis or embryogenesis is used to regenerate plants from calli. By using flow cytometry and cell sorting, desirable hybrids can be chosen (FCM).

4.3.10 Interspecific Grafting

A branch is grafted onto a rootstock from a different plant, in this example another species, to improve growth characteristics or infection resistance. The term "scion" is now used to describe the shoot.

4.3.11 Production of Hybrid Varieties

Two extremely inbred lineages are crossed to create hybrid cultivars. Emasculation of the mother line was initially carried out to ensure that the intended father line pollinates the mother line. Moreover, male sterility may be induced by chemicals or thermal inactivation (such as or more ethanol particular gametocides), but this is not a common practise in seed production. If genetic male sterility is available in the crop species, it is more effective to apply it. The mitochondria or the cytoplasm, or both, can be the source of the sterility. The mother line is kept alive in cases of nuclear male sterility by mating with an isogenic line that harbours a dominant gene for male fertility at the male sterility locus in a heterozygous state. The progeny half that inherited the recessive male sterility allele is chosen to serve as the maternal line in the production of hybrids. The preferred approach for producing hybrids is cytoplasmic male sterility (CMS), which has been made available in crop species because it is not particularly practical. Using isogenic lines with male-fertile cytoplasm as pollen donors, mother lines are multiplied using CMS. In this manner, the pollen from a nearly identical genotype overcomes the male sterility of the maternal line, preserving the inbred line, and the CMS is preserved because cytoplasmic genomes are often not transmitted by the pollen. Male fertility is re-established in the hybrids by using paternal lines with nuclear-based genes that restore fertility (Kumar et al. 2000).

4.3.12 In vitro Tissue Culture

In vitro micropropagation, meristem culture, somatic embryogenesis, and somatic organogenesis are methods used to reproduce plants. They can therefore be employed in preceding sections' discussions of haploidization and cell fusion, or they can be used to produce heterozygous offspring of crosses for more research and selection without the segregation associated with sexual reproduction. By DNA methylation, mutation, and polyploidization, in vitro tissue culture can cause soma clonal variation and can be utilized to create variation similar to mutagenesis. It can be improved by applying mutagenic chemicals; given that chimaera formation is less likely when cell suspensions are used, this is especially beneficial. Hence, selection during tissue culture for disease resistance would be a breeding-specific use (Tripathi et al. 2008). On the other hand, in vitro selection generally did not have much success and is no longer a technique that is applied very regularly. The problems with reproductive instability and the frequently poor correlations between, for instance, resistance in vitro and in conventional culture are related to this. Particularly, changes in DNA methylation patterns resulting from epigenetic processes may cause somaclonal variation, which is not always stable across several generations (Phillips et al. 1994). For use in breeding and propagation, virus-free material can be obtained by tissue culture. Tissue culture is particularly useful for ornamentals like trees and flower bulbs because it helps hasten the multiplication process required for the commercialization of new cultivars.

4.3.13 Apomixis

In essence, apomixis is the production of seeds without fertilization. Breeders would find this trait appealing since it makes it possible to effectively spread elite hybrids through seed. Consequently, the more laborious method of repeatedly mating inbred lines to produce seeds of hybrid kinds might be avoided. At the same time, breeding companies would lose the benefit of hybrid cultivars' built-in protection. Mutagenesis induction of sexual reproduction or heat treatments are required in order to initiate sexual reproduction in the apomictic *Poa pratensis* and permit the execution of such crossover programmes. A few typically dominant loci that may be coupled or form a large linkage block frequently appear to control apomixis. However, it appears that there are enough problems to make introducing the feature into a sexual crop species difficult (Ozias-Akins and Van Dijk 2007; Whitton et al. 2008). The hybrid between apomictic Tripsacum dactyloides and maize, which included a significant crop species, has drawn the most interest in this region. Hybrid apomictic lines could be produced from this cross, and the apomixis trait could be mapped; however, introgression into maize has not yet been effective (Leblanc et al. 2004). In the model plant Arabidopsis, certain apomixis-related genes have even recently been discovered (d'Erfurth et al. 2009). If these genes are helpful in introducing apomixis into crops, more research will be necessary to determine that.

4.4 MOLECULAR BREEDING TECHNIQUES

Plant breeding has a lot to gain from the creation of molecular markers, which are the result of molecular genetics and genomics research. In the developing discipline of "molecular breeding," known as "marker-assisted selection" (MAS), DNA markers are used in plant breeding (Collard and Mackill 2008). It can be used in plant breeding to both transmit specific genomic regions of interest and hasten the recovery of the recurrent parent genome (background selection; Ibitoye and Akin-Idowu 2011). A molecular marker gives traditional breeding a logical foundation while enhancing its precision and genotype-based selection. Thanks to the identification and application of molecular markers for desirable traits, the production of new commercial varieties now takes less time and necessitates less intricate phenotypic study. DNA markers that are closely related serve as molecular tools for MAS in plant breeding. Gene tagging, sometimes referred to as gene association, is a technique that can be applied (Collard and Mackill 2008). According to Ibitoye and Akin-Idowu (2011), DNA markers are especially useful for mapping the genetic composition of plants, identifying genetic variation in parental germplasm, creating genetic linkage maps, tagging the genes controlling important traits, and choosing quantitative traits that are difficult to select from phenotypic assessment alone. Furthermore, understanding of how molecular markers enable genetic and genomic control of horticultural features may also help in the creation of more successful breeding strategies and map-based gene isolation methods. Additionally, markers can provide clones of certain genes that can be used to genetically modify horticultural crop species (Ibitoye and Akin-Idowu 2011). Numerous horticultural crops have been produced, and MAS is being used to identify markers that are tightly linked to the main genes responsible for the expression of key traits (such as disease/pest resistance, fruit/nut quality, and self-incompatibility) (Zhu and Barritt 2008). The most frequently used markers in main cereals are simple sequence repeats (SSRs) or microsatellites (Gupta and Varshney 2000). Expressed sequence tags have been used in various horticultural crops to create SSRs and single nucleotide polymorphism (SNP) markers (Ekué et al. 2009; Frary et al. 2005). SNPs have been successfully used in a range of species, including apples and grapes, to build genetic linkage maps. Linkage maps have been developed over the past 10 years for numerous horticultural crops using a variety of mapping populations and molecular markers (Sonah et al. 2011). Numerous quantitative trait loci (QTL) have been discovered in fruit and vegetable crops for a number of significant traits, such as fruit shape, fruit weight, and fruit sugar content in tomatoes; resistance to Rol *Globodera rostochiensis*; scab resistance gene in apples (Jung et al. 2005); nematode resistance in sugar beets; and the virus resistance el (Lu et al. 2006). Therefore, more key horticultural traits can be studied using this strategy in the future.

4.5 TRANSGENIC BREEDING

4.5.1 Creation and Assessment of Transgenic Events

Genetic modification is a time-consuming, expensive, and technically challenging process for adding alleles. The response of a species to tissue culture (callus induction and regeneration) serves as the foundation for most genetic transformation protocols. Only a small subset of genotypes in the majority of organisms are capable of transformation (Saikishore et al. 2005). However, in planta transformation techniques are currently only applicable to a few model species, such as Arabidopsis and Medicago, and have not yet reached the same level of success as their tissue culture-derived transgenic counterparts. To introduce a foreign gene during genetic transformation into a little plant tissue capable of regenerating into a full plant, two techniques are used: agrobacterium and particle bombardment. A transgene is made up of nucleotide sequences such as a promoter, the coding sequence of the gene, an enhancer, and a terminator. The transgene's selective spatial and temporal expression can be used to determine the promoter. The presence of enhancer sequences increases transgenic expression levels. The required gene is randomly introduced under carefully monitored nutrient and culture conditions, a small number of target tissue cells are transformed, and these altered cells are then selectively cultured to eventually regenerate entire plants. These transgenic plants (T0) have effectively incorporated the target gene(s) using techniques including polymerase chain reaction (PCR) and Southern blotting. In order to assess the levels of trait expression, ELISA (enzyme-linked immunosorbent assay) or comparable techniques are utilized. Western blotting is used to confirm the expression of the target characteristic. The incorporation of the transgene at a particular locus or set of loci is known as a transgenic event. The same transgenic integration pattern is present in clone plants. All the clones are derived from one converted cell. As an example, if we produce 50 transgenic plants containing the desired gene, the transgene may only integrate at 10 loci, indicating that the desired gene has integrated into 10 different patterns in the plant genome, leading to 10 events. The efficacy of transformation procedures is assessed based on transgenic events generated without taking into account clones. Clones are important even though they do not improve the effectiveness of the technique for planning future transgenic breeding operations. So, it's critical to assess clone performance before drawing any conclusions from a hopeful event (Figure 4.1).

There are examples of single, double, and multiple copy transgene insertions. A number of molecular tools are also used to confirm the integration and expression of the transgene, including PCR (for preliminary transgene integration verification), Southern blotting (for transgene integration and copy number), Northern blotting (for transgene transcription to mRNA), Western blotting (for transgene translation into protein), and ELISA (for quantitative transgene expression in target).

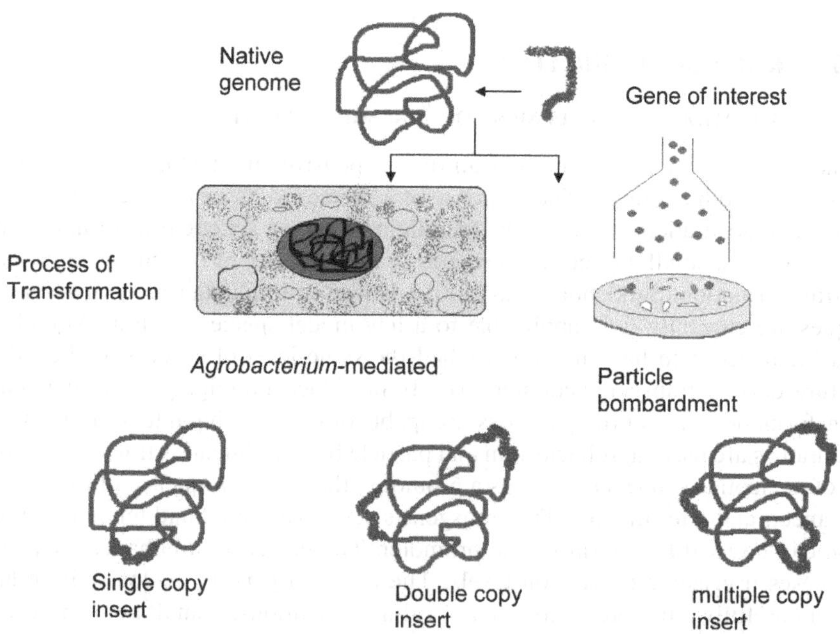

FIGURE 4.1 Process of transgene integration (Saikishore et al. 2005).

4.5.2 Transgenic Breeding Strategies

Via backcrossing, event evaluation in genetic backgrounds that are typical of future commercial goods, and the development of candidate products, the transgenic breeding programme's main goal is to produce better versions of inbred.

4.6 BIOFARMING

Large-scale agricultural production of proteins and macromolecules including antigens, antibodies, and enzymes that are essential for medical, pharmaceutical, and industrial purposes is referred to as "biofarming" (Dalal et al. 2006). Potatoes and tomatoes are the most often used host systems for biofarming in horticulture. Plant distribution of vaccinations has drawn a lot of attention since it has several benefits over vaccine administration via injection. Vaccines that are edible are those manufactured from plant tissues that may be eaten, including grains, tubers, or fruits. Several attempts were made to express antigens related to various diseases in edible plant parts in order to investigate the expression of the hepatitis E virus (HEV) open reading frame 2 partial gene in transformants and the immunoactivity of expressed products, as well as to determine the viability of creating a new type of plant-derived HEV oral vaccine. One such approach involved overexpressing the HEV open reading frame 2 partial gene in tomato plants (Ma et al. 2003). Similar to hepatitis B surface antigen (HBsAg) in potato and cholera toxin B subunit with an endoplasmic reticulum retention signal in tomato, the loop-forming B-cell epitope (H386–400) of the measles virus hemagglutinin protein, rabies virus epitopes fused with tobacco mosaic virus in spinach, the VP1 protein gene, and the coat protein of enterovirus 71 (EV71) were all successfully expressed in tomato. Potato and tomato have also undergone genetic modification to produce proteins and antibodies that have medicinal potential. The creation of human alpha-interferon protein in transgenic potato plants, the expression of a recombinant human acetylcholinesterase isoform in transgenic tomato plants, and the manufacture of antibodies in transgenic tobacco, soybean, and potato plants are a few examples (Yadav et al. 2013). By introducing a foreign protein gene or by changing their metabolic pathways to create an acceptable end product, by product, or a distinctive biomolecule, plants can be exploited to produce industrial enzymes, chemicals, and raw materials. Endoglucanase (E1) from *Acidothermus cellulolyticus* was efficiently converted in transgenic potatoes, accumulating up to 2.6% of the total leaf soluble protein. Transgenic potatoes have also expressed human milk proteins such lactoferrin and casein. The Nephila clavipes dragline protein gene has also been overexpressed in potato and tobacco to produce significant amounts of spider dragline silk. In a different study, adenylate kinase overexpression in potatoes increases the adenylate content, and transgenic lettuce and tomato plants have been found to express monellin. Additionally, potatoes have been successfully genetically modified by introducing either the microbial fructosyltransferase genes from Bacillus subtilis (sacB) or Streptococcus mutans (ftf), or the plant genes 1-SST (sucrose: sucrose 1-fructosyltransferase) and 1-FFT (fructan:fructan 1-fructosyltransferase) from Sharma and Khurana (2016).

4.7 NEED FOR TRANSGENIC HORTICULTURAL CROPS: PRESENT SCENARIO

The ongoing global population growth trend is cause for concern since it will limit the supply of horticultural products in decades to come. Currently at 7.7 billion, the globe's population is anticipated to increase to 8.5 billion by 2030 and 9.7 billion by 2050 (United Nations 2019). Demand for horticulture crops has increased as a result of population growth and healthier diets. But other abiotic and biotic factors, like as climate change, have a significant impact on horticulture productivity and could get worse in the future (Gruda et al. 2019). In an effort to maximize yields and profits, producers often misuse agrochemicals, which has a severe impact on crop quality and food safety and even endangers the environment and people's health through the food chain (Qin et al. 2021). In order to accomplish sustainable horticultural production, it is necessary to investigate alternative eco-friendly protective agents due to the rising hazards linked with chemical controls. The majority of consumers like organically cultivated fruits and vegetables free of pesticides. In order to preserve the excellent crop quality and safety, prices for the primary vegetables and fruits are rising. There is an urgent need to increase production of horticulture crops with high nutritional qualities, greater production, and increased insect and disease resistance to satisfy the needs of the global population, which is quickly expanding and has a high demand for affordable, fresh, and nourishing food. Genetic constraints are mostly to blame for how little progress has been made in horticulture crop quality improvement through conventional plant breeding. The main drawback of conventional plant breeding for fruits and vegetables is the long breeding period needed to carry out numerous backcrosses in order to introduce the desired feature. Because the desired qualities are frequently available in wild cultivars, numerous backcrosses are required to introduce such resistant genes by conventional breeding, which ultimately lowers the quality of fruits and vegetables.

With high accuracy, transgenic technology provides an accurate instrument for introducing desired features in target plants. Genes, the foundation of all plant breeding, regulate traits. Currently, using transgenic breeding technology, which combines recombinant DNA and genome editing techniques, researchers may precisely alter the gene (addition/deletion/modification) encoding a particular characteristic to develop novel phenotypes.

Transgenic plants, genetically modified (GM) plants, or biotech plants are all terms used to describe plants created using this technology. The main method used in transgenic breeding is the introduction of desirable foreign genes into the genome of the intended host plant. The key benefits of this method are its accuracy, short development times, and ability to enhance a characteristic without altering the genotype's other genetic makeup, preserving the supremacy of the top germplasm. Due to their extreme heterozygosity, many horticultural plants benefit the most from this. According to current societal trends, people want more variety of healthful vegetables and fruits in their daily meals that are also more nutrient-dense, more flavourful, and fresher. Consumers expect those fresh horticultural items to be more easily accessible and less expensive. By introducing advantageous genes into current elite

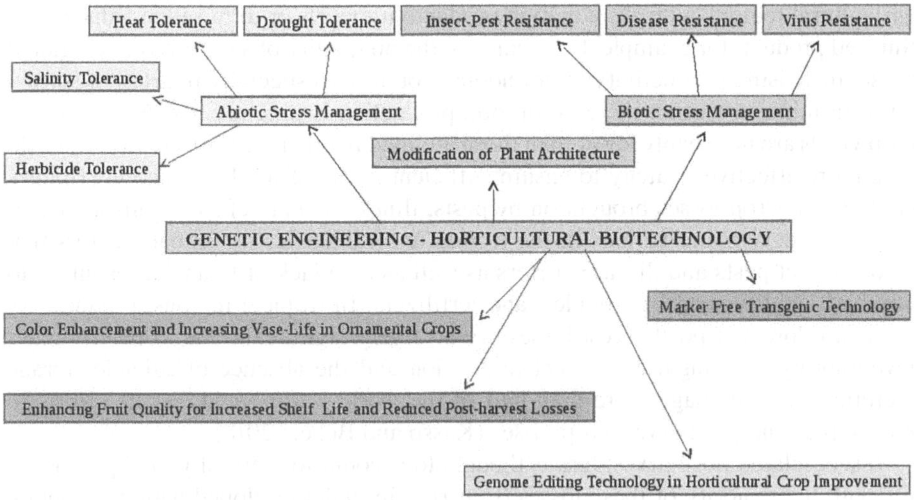

FIGURE 4.2 Summary of numerous transgenic techniques for horticulture crop improvement (Parmar et al. 2017).

cultivars that were frequently previously inaccessible, transgenic horticulture crops provide solutions to all of these problems. Despite the introduction of a small number of transgenic horticulture crops into the market, potential risks make the widespread commercial success of this technology quite concerning. The US government approved the Flavr Savr tomato for commercialization in 1994, and it was the first transgenic horticultural crop to receive approval (Bruening and Lyons 2000). The genes of cell wall of Flavr Savr tomatoes were altered, delaying ripening and improving flavour and shelf life. The transgenic papaya with increased virus resistance, which was approved by the USA in 1996, was another significant success story, after the original success of the Flavr Savr tomato (Azad et al. 2014). Currently, a sizable number of transgenic horticulture crops are being sold in a variety of nations, and many more are in the final stages of commercialization. Figure 4.1 provides a summary of numerous transgenic techniques for horticulture crop improvement (Figure 4.2).

4.7.1 Transgenic Horticultural Crops for Crop Productivity

Transgenic plant cultivars have previously proven to be effective at increasing yield and reducing costs. A 5%–10% increase in yield, as well as up to 40% and $60–$120 (US dollars) per acre in herbicide and insecticide savings, were all attributed to the cultivation of virus-, insect-, and herbicide-resistant plants between 1996 and 1997 (James 1999). Nevertheless, despite how great they are, these productivity gains are likely to have a limited impact on the world's food supply because the items currently on the market are only appropriate for sizable mechanized farms that engage in intensive agriculture. In reality, the majority of transgenic crops developed to date, particularly by the private sector, are intended to either lower production costs in agricultural regions with already high productivity levels or to raise the value of the finished product (for example, by enhancing the oil quality of seed crops). In a global sense, increasing productivity in developing countries, especially in areas of subsistence farming, where an increase in food production is urgently needed and where crop yields are noticeably lower than those attained in other parts of the world, would be a more effective strategy to ensure sufficient levels of food production (Hodges et al. 2011). Crop losses brought on by pests, illnesses, and deficient soils are exacerbated in developing nations in the tropics and subtropics by climate factors that favour insect pests and disease vectors as well as by a lack of financial resources to buy high-quality seeds, pesticides, and fertilizers. In tropical regions, productivity levels are low, but postharvest losses are also very significant due to the climate's favourability for fungal and insect infestation and the absence of suitable storage facilities. Pests damage more than half of the world's crop yield despite efforts to reduce pre- and postharvest crop losses (Kasso and Bekele 2018).

Insect-related postharvest loss is thought to account for 15% of global productivity, with the majority of these losses occurring in underdeveloped nations (Gautam et al. 2022). It's feasible that plant biotechnology will help solve many of these issues. Plant biotechnology has the significant benefit of frequently producing solutions for crop enhancement that may be used with a wide range of crops. A variety of crops may benefit from techniques including delayed ripening, insect resistance, and GM viral resistance. Although only explored on a commercial scale for the tomato, genetically engineered delayed ripening has enormous promise for usage in tropical fruit crops, which suffer significant losses in developing nations due to their early ripening and lack of suitable storage conditions and effective transportation methods to get them to the final consumer.

4.7.2 Transgenic Horticultural Crops for Improved Nutrition

Micronutrient malnutrition (deficiencies in iron, iodine, vitamin A, and zinc) affects the majority of people in underdeveloped nations. It is one of the most significant risk factors for sickness, and in particular for the death of young children and pregnant women. It is now possible to create transgenic crops with improved nutritional value utilizing transgenic technologies. The following techniques can be used to transgenically modify the nutritional value of plants: (1) enhancing the quality, composition, and concentrations of nutrients like protein, starch, and fatty acids in various crops; and (2) raising the concentrations of antioxidants (like carotenoids and flavonoids) (Gerszberg et al. 2015). Golden rice, meant to treat vitamin A deficiency, is a prime example of a transgenic plant with higher nutritional value.

Crops can now be GM thanks to recombinant DNA technology, mostly to increase agricultural yield and quality. Numerous studies have documented the transmission of genes through this pathway between various species, including plants, insects, and microbes. Contrary to normal breeding methods, interspecies genetic barriers are not a limitation of transgenic technology. Therefore, the genes of interest might be shared across plant species with various levels of ploidy or genetic complexity. Dietary plants, such as fruits and vegetables, are a vital source of nutrients for maintaining human health.

Only a small number of GM fruit and vegetable crops, such as common bean, eggplant, squash, apple, papaya, and pineapple, are currently in production, despite the fact that several transgenic research types on numerous plant species have been conducted. Due to the perennial nature of these crops as well as the challenges associated with genetic transformation and in vitro regeneration methods, there is less transgenic research on fruit crops than on vegetables and field crops.

In the context of producing wholesome, nutritious food, we'll talk about some of the promising uses of transgenic fruits and vegetables. Carotenoids, such as lycopene and beta-carotene, which are significant health-promoting metabolites, are abundant in tomatoes. Lycopene and beta-carotene contents were being improved by scientists of tomatoes using transgenic technology. Romer et al. (2000) created a transgenic tomato by overexpressing the bacterial phytoene desaturase gene, which increased the quantity of beta-carotene by about 45%. Transgenic cauliflower with increased beta-carotene accumulation was created by Lu et al. in 2006. Additionally, apple and kiwifruit have been GM to have the grape stilbene synthase gene, which produces the phytoalexin resveratrol (Szankowski et al. 2003). The enzyme sorbitol-6-phosphate dehydrogenase (S6PDH) converts glucose-6-phosphate into sorbitol, and Teo et al. (2006) created transgenic apple plants that expressed an antisense copy of the S6PDH gene. The cultivar "Ailsa Craig" of tomatoes has higher levels of the carotenoids phytoene, lycopene, beta-carotene, and lutein, according to Fraser et al. (2002). In the tomato cultivar Ailsa Craig, phytoene synthase (crtB) from the bacteria *Erwinia uredovora* has been overexpressed. Similar to this, Wahlroos et al. (2004) used transgenic *Brassica rapa* to yield oilseed with a higher proportion of histidine. Calcium (Ca) is a nutrient that is equally as important as folate. In order to boost calcium levels and improve bioavailability, transgenic carrots were created (Park et al. 2004). An extremely high zinc concentration was produced by the transgenic lettuce cultivar "Salinas 88" (Mou 2009). Consuming polyphenols, which are significant nutritional components found primarily in fruits and vegetables, is directly associated with a lower risk of developing chronic diseases. In order to increase the synthesis of flavonoids, a transgenic tomato was created through heterologous expression of flavonoid pathway genes (Schijlen et al. 2007). The biotech sector now aims to expand its market with second-generation genetically modified (GM) crops that have been nutritionally upgraded and are said to reduce malnutrition and promote health. Many supporters of biotechnology even go so far as to assert that common diet-related disorders would be prevented by foods that will be available to customers.

4.7.3 Transgenic Horticultural Crops for Biotic Stress

Ten percent to sixteen percent of the world's yield is lost due to pathogens (Nalley et al. 2016). Utilizing transgenic technology can reduce this loss. Only members of the same species, or eventually closely related species, may be crossed in traditional agriculture. Traditional breeders cannot develop biotic stress tolerance or introgress this trait into new varieties if it does not already exist in this naturally occurring gene pool. It is therefore vital to look for other sources of genes in microbial organisms or in completely unrelated plant species. Additionally, the use of chemical sprays may have a negative impact on the environment, including beneficial creatures, and human health, as well as promote the emergence of chemical-resistant weeds and insects (Rout and Peter 2018). The previous three decades have made plant genetic engineering conceivable. More than 175.2 million ha of land are being used to grow transgenic plants that are resistant to pesticides, insects, and diseases, compared to just 1.7 million ha in 1996 (Singh et al. 2021). The following are the main advancements in biotic stress management.

4.7.3.1 Insect Resistance

Insects can harm plants in the field as well as when they are being stored as plant products. Currently, many crop plants lack enough insect-pest resistance. The usage of chemical control techniques has shown to be unsafe for consumers and unsustainable for the ecosystem.

Any genetic advancement that could lower the price of applying chemicals to control pests would be beneficial from the grower's perspective. The Bt (Cry) gene, which was obtained from the soil bacterium *Bacillus thuringiensis*, has effectively been used to control a variety of lepidopteran insects in numerous crops. Using the Bt gene, insect resistance was first noted in tomatoes in 1987. According to Fischhoff et al. (1987), transgenic Bt. tomato plants exhibited resistance to *Heliothis virescens* and *Spodoptera litura*. Brinjal is one of the most widely consumed vegetables in Asia, especially in particular in the Indian subcontinent. However, a lepidopteran bug called *Leucinodes orbonalis* has severely harmed it. By co-cultivating cotyledons with A. tumefaciens, Kumar et al. (1998) successfully inserted a synthetic cry1Ab gene for an insecticidal crystal protein (ICP) into the brinjal (*Solanum melongena* cv. Pusa PurpleLong) plant. By using a double-antibody sandwich ELISA technique, gene expression was assessed. Insect mortality in fruit bioassays showed considerable differences amongst the transgenic lines. A synthetic chimeric gene called SbtCry1Ac that codes for the protein btCry1Ac was used by Zhang et al. (2020a) to modify the *Actinidia chinensis* kiwifruit plant. At 10 days following infection, an average Oraesia excavate inhibition rate of 75.2% was observed when the transgenic plants were tested for insect resistance using insect bioassays. In a different study, transgenic chrysanthemum plants that overexpress the transcription factor *Chrysanthemum morifolium* WRKY48 (CmWRKY48) were observed to suppress the population expansion of aphids (Li et al. 2015). Papolu et al. (2016) developed transgenic brinjal plants that produced the modified rice cystatin (OC-IDD86) gene under the control of the TUB-1 promoter in order to produce resistance against the root-knot nematode (RKN). In transgenic plantains (Musa sp.) cv. "*Gonja manjaya*" plants, Roderick et al. (2012) produced a synthetic peptide that interferes with nematode chemoreception and a maize cystatin gene that inhibits the digesting cysteine proteinases.

4.7.3.2 Disease Resistance

The spread of horticultural crop diseases brought on by various fungus, bacteria, and viruses is one of the main factors restricting crop yield. Owing to the absence of resistance gene/s in the gene pools of a given crop's, the uses of traditional breeding seem to be limited. Crops can now be genetically engineered to withstand disease, which is advantageous in terms of both effectiveness and cost.

The CP-mediated strategy to engineer virus resistance has been applied to fruit crops to generate resistance to diseases including PPV, citrus tristeza virus, and grape fan leaf virus. Many tropical nations cultivate papaya, but the disease PRSV, which significantly reduces productivity, threatens this industry. The virus's CP gene has been introduced to papaya using biotechnological methods to provide PRSV resistance. By introducing an AC4 gene-RNAi construct into transgenic tomato plants, Praveen et al. (2010) were able to reduce the activity of the tomato leaf curl virus. Commercial watermelon cultivars were modified by Yu et al. (2011) using an untranslatable chimeric construct expressing PRSV WCP and truncated zucchini yellow mosaic virus CP genes. By knocking off the production of some endogenous genes, RNAi technology is being utilized rather successfully to treat a variety of bacterial and viral diseases in plants. Clarke et al. (2008) developed transgenic poinsettia plants resistant to the poinsettia mosaic virus by silencing genes with hairpin RNA (hpRNA). Girhepuje and Shinde (2011) developed transgenic tomato plants that produced a wheat endochitinase gene, and disease screening demonstrated the transgenic plants exhibited improved resistance to *Fusarium oxysporum*. Mango (*Mangifera indica*) cv. Ataulfo embryos were genetically modified by Rivera-Domnguez et al. (2011) using the defensin J1 gene.

4.7.3.3 Herbicide Resistance

In higher plants, 5-enolpyruvylshikimate-3-phosphate synthase (EPSP) is significantly inhibited by the herbicide glyphosate. Shah et al. (1986) produced a cDNA clone producing EPSP synthase from a cDNA library of a glyphosate-tolerant *Petunia hybrida* cell line (MP4-G) that overproduces the enzyme. The twenty-fold gene amplification in this cell line has been shown to result in an overproduction of EPSP synthase messenger RNA. A chimeric EPSP synthase gene was made and introduced into petunia cells utilizing the 35S promoter of the cauliflower mosaic virus in order to produce large levels of EPSP synthase. Both regenerated transgenic plants and transformed petunia cells were resistant to glyphosate. Sripaoraya et al. (2006) created transgenic pineapple that had the bar gene for Biala Phos resistance and tested them for Basta tolerance. Plants were discovered to be tolerant to 1600 mL/rai of Basta X (stock concentration 15% w/v glufosinate ammonium), which is double the amount advised for the field application of the herbicide. More efficient weed management in pineapple farms should be made possible by transgenic plants that are tolerant to glufosinate ammonium without causing crop damage.

4.7.4 Transgenic Horticultural Crops for Abiotic Stress

Horticultural crops have serious challenges from abiotic stresses such as drought, temperature extremes, flooding, pH, and salinity. Global warming's increase in temperature causes less frequent but more intense precipitation, the melting of snowpacks, and the worsening of drought conditions. These conditions cause stress-related problems such flooding, increased evapotranspiration, fertile soil loss, nutrient leaching, lower oxygen levels in the soil, decrease in production, and delayed planting. Finding opportunities and developing solutions to these problems would be made easier with an understanding of the stress physiology, molecular biology, genomics, and assessment of the stress tolerance in horticulture crops.

The majority of abiotic stresses are interrelated and influence the genes involved in these pathways, increasing the tissues' antioxidant capacity or causing them to accumulate suitable solutes (Chaudhry and Sidhu 2022). Introducing genes encoding enzymes that catalyse the conversion of a naturally occurring substrate into a product with osmo-protective characteristics and transferring genes involved in signalling and regulatory pathways are the methods that consistently yield successful results (Divya et al. 2019). To boost their ability to withstand abiotic stress, many horticulture crops have undergone genetic modification. Shekhawat et al. (2011) discovered that bananas have different abiotic stress tolerance due to overexpression of the MusaWRKY71 gene from *Musa* spp. cv. Karibale Monthan (ABB group), which produces a WRKY transcription factor protein. Similar to this, grapevine was able to increase its resistance to cold stress through overexpression of (DREB)1b, a cold-inducible transcription factor from Arabidopsis thaliana (Jin et al. 2009). Ren et al. (2021) demonstrated an improved tolerance to cold and drought stress in transgenic apple by overexpressing the cold-inducible MdCIB1 gene from rice, which encodes a bHLH transcription factor. According to Prabhavathi et al. (2002), the bacterial mannitol-1-phosphodehydrogenase (mtlD) gene, which is involved in the manufacture of mannitol, experienced a transformation in eggplants. Bhattacharya et al. (2004) were able to produce salt tolerance in cabbage by inserting the bacterial glycine betaine biosynthesis (BetA) gene, which is involved in the manufacture of glycine betaine.

It is necessary to create cultivars with enhanced drought tolerance utilizing biotechnology methods since apples have a longer juvenile phase. Six GH3 genes were silenced using the RNAi method in the apple. The MdGH3 RNAi plants outperformed wild-type ones under long-lasting drought stress and exhibited more robust root systems, a higher root-to-shoot ratio, improved hydraulic conductivity, enhanced photosynthetic capacity, and more efficient use of water (Jiang et al. 2022a).

4.7.5 Transgenic Horticultural Crops for Shelf Life and Qualitative Traits

The Flavr Savr tomato is one example of how gene technology is being used to extend the shelf life of plants and plant components after they have been harvested. It was discovered that fruit storage and shelf life are mostly impacted by excessive softness. Other objectives include increasing the amount of protein or vitamins. Output features are designed to give consumers personal benefits and to improve the processing quality for businesses that take on the additional processing of the items. Using transgenic plants modified in the expression of cell wall-modifying enzymes, it has been determined the role of certain activities in the softening of fruit during ripening. Fruit ripening has been manipulated by altering the activity of cell wall enzymes such polygalacturonases, which are involved in tissue softening and disintegration. In order to delay fruit ripening, it has also been possible to prevent the manufacture of ethylene, the hormone that causes fruit to mature. The first transgenic plant to be successfully commercialized was a long-lasting tomato called Flavr Savr, which was created by Calgene Inc. in the USA in 1994. The Flavr Savr tomatoes have increased shelf life, better flavour, and total soluble solids. However, because to its vulnerability to disease and low productivity, 3 years later, this Flavr Savr cultivar was taken off the market. Later, other tomato cultivars with longer shelf lives were created by inhibiting two ethylene precursors and oxidase or 1-aminocyclopropane-1-carboxylate (ACC) synthase with antisense RNA. By controlling cytokinin production, it has been possible to delay leaf senescence in tobacco and petunia. The genes responsible for strawberry flavour, aroma, and colour have been found by scientists working at Horticultural Research International in the UK. As a result, utilizing transgenic techniques, it could now create super strawberries with a sweeter flavour. In *Solanum lycopersicum* (tomato), the yeast spermidine synthase (ySpdSyn) gene was expressed by Nambeesan et al. (2010) using both constitutive (CaMV35S) and fruit-ripening-specific (E8) promoters. In comparison with fruits of the wild type, the transgenic ySpdSyn fruits displayed reduced shrivelling, a longer shelf life, and a later onset of decay signs. A CaMV35S-ySpdSyn genotype had a delayed crop maturity as evaluated by the proportion of ripening fruits on the vine due to fruits collecting higher quantities of the antioxidant lycopene. Remarkably, whole-plant senescence was also delayed in the transgenic plants in comparison with wild-type plants. By using an RNAi technique to silence the mitochondrial APX gene's expression, Krishna et al. (2019) created transgenic tomato plants and discovered that the transgenic tomato fruits contained more vitamin C.

4.8 GENE SILENCING

Gene silencing is cell's internal control mechanism that prevents the expression of a specific gene by either cleaving or repressing the mRNA molecules that are encoded by the gene. Horticultural plants have developed various pathways for gene silencing that consists of small RNAs (20–30 nucleotides), which inhibit the expression of genes. RNA interference, transcriptional gene silencing, and virus-induced gene silencing are commonly used techniques for gene silencing in horticultural crops. When double-stranded RNA is present in a cell, it triggers the production of siRNAs, shRNAs, and miRNAs, which leads to the silencing of the mRNA and, consequently, the gene itself (Hamilton and Baulcombe et al. 1999).

RNA interference (RNAi) is a method for suppressing specific genes in a sequence-specific manner. It depends on the use of double-stranded RNAs (dsRNAs) that can inhibit the expression of the targeted gene at the transcriptional or post-transcriptional level. RNAi has been observed in diverse organisms, from worms to mammals, and has become a widely used tool in the study of gene function. It requires only a few molecules of dsRNA per cell to silence expression. The effect is inherited in the subsequent generations. RNAi has opened up new avenues of research into the molecular pathways that trigger gene silencing and has provided a powerful tool for understanding biological function (Fire et al. 1998).

In plants, silencing occurred at the post-transcriptional level, with the accumulation of small RNAs that were complementary to the silenced mRNA. This phenomenon was termed post-transcriptional gene silencing (PTGS) and is now recognized as a precursor to RNAi. The discovery of PTGS in plants led to the realization that RNAi was a conserved biological mechanism present in diverse organisms. The study of PTGS also helped to reveal some of the key players in RNAi, such as Dicer, an RNase III enzyme responsible for processing dsRNA into siRNAs. The study of RNAi and PTGS has since led to a better understanding of the regulation of gene expression and has provided a powerful tool for reverse genetics and functional genomics (Vaucheret 2001).

The term "co-suppression" was coined to explain the process by which exogenous nucleic acid inhibited the gene expression of the cell's corresponding genes, and this inhibition involved two different mechanisms, which caused the suppression of both the endogenous and transgene's corresponding genes loci. In one set of plants, gene inhibition was associated with heavy methylation of the gene's promoter and other subsequent regions. In another class of plant, PTGS was systemic phenomenon that spread from the transgenic plant to non-transgenic plant, after the grafting process. The observation that potato virus X that uses dsRNA in its genome replication process triggered a similar response to PTGS provided evidence that PTGS is similar to RNAi in the worms (Palauqui et al. 1997).

These silencing mechanisms in the plants are stable and inherited; it involves the cross-talk between processes that affect the genomic copy of the gene and processes that affect the corresponding mRNA. The phenomenon of transgene co-suppression is not limited to plants, as it has also been observed in fungi, flies, worms, and rodent fibroblasts (Palauqui et al. 1997).

The silencing can occur at the transcriptional or post-transcriptional level, and that the presence of an inverted repeat structure in the transgene caused stem-loop conformation in the RNA that trigger RNAi pathway. Genetic analysis has identified that conserved biomolecules are mediating RNAi and transgene co-suppression, suggesting a link between the two phenomena. However, more research is needed to fully understand the particulars of these controlling processes and the role of the RNAi pathway in diverse types of silencing events (Palauqui et al. 1997).

4.9 GENOME EDITING TECHNOLOGIES

The recent developments in recombinant DNA technology allowed manipulations of DNA sequences in sequence-specific manner. In 1987, gene targeting technology was invented by animal scientists, which relied on homologous recombination (HR) allowed researchers to create sequence-specific changes in the endogenous gene but at a reduced frequency than normal. To circumvent this problem, researchers used DNA double-stranded breaks (DSBs) that commonly result in HR. Besides HR, in somatic cells, error-prone nonhomologous end-joining (NHEJ) pathway is also used that cause mutations. Over the years, researchers discovered various types of engineered endonucleases, such as meganucleases (MNs), zinc finger nucleases (ZFNs), transcription activator-like effector nucleases (TALENs), clustered regularly interspaced short palindromic repeats (CRISPR)/CRISPR-associated 9 (Cas9), and CRISPR from Prevotella and Francisella 1 (CRISPR/Cpf1), to create site-specific DSBs, allowing easy manipulations of genome of various organisms.

Significantly advances were made in horticultural crops after the discovery of CRISPR/Cas9 technology (Hsu et al. 2014).

Genome editing technologies rely on the ability to protein to bind DNA in a sequence-specific manner such as the interaction between a promoter and its corresponding transcription factor. The nucleases mentioned earlier can be used for genome editing by targeting specific sequences and generating double-stranded breaks (DSBs) through protein–DNA interactions (for meganucleases, zinc finger nucleases (ZFNs), and transcription activator-like effector nucleases (TALENs)) or RNA-DNA base-pairing (for CRISPR/Cas9 and CRISPR/Cpf1) (Gaj et al. 2013).

Meganucleases or homing nucleases were the first nucleases used to manipulate the genome. These nucleases were initially discovered in microorganisms or organelles and are capable of binding to DNA sequence elements that range from 12 to 40 base pairs, allowing them to cleave both DNA strands in a site-specific manner. However, the limited number of recognizable sites for these nucleases, especially for the I-CreI protein, has restricted their ability to edit desired target sites. To solve this problem, scientists have used combinatorial assembly to generate MN variants, which can target the desired DNA sequence. However, modified MNs often have overlapping recognition and catalytic domains that can compromise their catalytic activity. Consequently, plant scientists have not widely used MNs for genome editing (Gaj et al. 2013).

ZFNs and TALENs are two classes of genome editing nucleases that are manipulated to create DNA sequence-specific changes in DNA. ZFNs, or zinc finger nucleases, are designed by fusing the endonuclease FokI's DNA cleavage domain with zinc finger domain that recognizes specific DNA sequences. TALENs, or transcription activator-like effector, employ DNA-binding domains that are produced by the bacterial plant pathogen *Xanthomonas*, to achieve sequence specificity. The TALENs central binding domain consists of repeat sequences with each repeat binding to one nucleotide through variable di-residues at positions 12 and 13. Overall, TALENs are considered to be more precise and accessible than ZFNs (Gaj et al. 2013).

CRISPR/Cas9 and CRISPR/Cpf1 are two genome editing systems that use RNA-DNA binding to accomplish sequence specificity. In the CRISPR/Cas9 system, the sgRNA binds to a target DNA sequence and is recognized by the Cas9 protein, which cuts the DNA strands at a specific location. The Cas9 system recognizes a protospacer adjacent motif (PAM) that averts the use of sgRNAs in T-rich regions. On the other hand, Cpf1 is an endonuclease that recognizes T-rich PAM sequences and produces cohesive ends upon cutting. Both systems have been used for site-specific mutations and gene regulation, and can be adapted for precise base editing (Zetsche et al. 2015).

4.10 CURRENT STATUS OF HORTICULTURAL CROP IMPROVEMENT

In recent years, the use of genome editing in horticultural crops has been rapidly increasing. Several studies were in the model systems and staple crops, such as tomatoes and potatoes, which were successful. The edited genes are pleiotropic affecting development, metabolism, and stress responses. However, other crops, including citrus, apple, grape, and oilseed crops, have also been successfully edited for needed traits, like improved oil quality and resistance to plant pathogens. In the coming decades, functional characterization of genes in diverse crops will help identify valuable targets for potential horticultural improvement, such as increased productivity, marketing quality, and nutritional value (Zhang et al. 2020a).

4.11 GENOME EDITING IN HORTICULTURE AND ITS APPLICATIONS

Genome editing technology is now an important arm in horticultural crop improvement. It allows for targeted modifications of genes, either by introducing desired traits or removing undesirable ones. Compared to other traditional breeding methods, genome editing offers greater precision, efficiency, and versatility. Recent developments in the gene editing tools and techniques has enabled the researchers to target a wide range of genetic modifications, such as small insertions/deletions, large fragment deletions, gene replacements, and precise base substitutions (Hussain et al. 2021).

One advantage of genome editing over traditional breeding methods is the ability to introduce desired traits without compromising other properties. This can be achieved by editing specific genes responsible for the target trait while leaving other genes intact. In addition, genome editing can introduce genetic variations into elite or heirloom lines without the need for prolonged breeding efforts. Wild relatives of cultivated plants can also be utilized for genome editing since they may contain unique genetic variations for important traits. By editing the genomes of these wild species, researchers can transfer desirable traits to commercial cultivars without introducing unwanted genes (Chen 2019).

Moreover, genome editing technology can be employed to produce epialleles that are not caused by changes in DNA sequence but by the modifications to the DNA or chromatin structure. Epialleles are major origin of new variation required for crop improvement, as they can affect gene expression and traits. Researchers have shown that combining genome editing with epigenetic regulation causes significant changes in gene expression and phenotype (Gao et al. 2021).

Overall, genome editing technology opens new avenues for crop improvement, allowing for targeted genetic modifications that can introduce desirable traits, remove undesirable ones, and create new sources of genetic variation.

4.12 CHALLENGES AND FUTURE PERSPECTIVES FOR THE IMPROVEMENT OF HORTICULTURAL CROPS THROUGH GENOME EDITING

Furthermore, the complex genetics and breeding systems of many horticultural crops also pose challenges for genome editing. Some crops, such as fruit trees, have long generation times, making it difficult to rapidly introduce and test genetic changes.

Others, such as vegetable crops, may have polyploid genomes or exhibit self-incompatibility, further complicating genetic manipulation. Additionally, regulatory frameworks for genome-edited horticultural crops may vary between countries and regions, with some countries having stricter regulations than others (Barone and De Palma 2020).

Despite these challenges, genome editing holds great promise for the improvement of horticultural crops, and ongoing research is working to overcome these obstacles. The development of efficient transformation and regeneration protocols, as well as improved methods for identifying target genes, will be key in horticultural crops improvement through gene editing technology (Barone and De Palma 2020).

Another important consideration is the potential for off-target effects, which can lead to accidental variations in the genome. To minimize off-target effects, researchers can use bioinformatics tools to predict potential off-target sites and select specific editing reagents that have low off-target effects. Additionally, the use of base editors and prime editors, which have a higher specificity than Cas9 nucleases, can further reduce off-target effects (Purnhagen et al. 2018).

Finally, it is important to consider the regulatory framework for genome-edited horticultural crops. In many countries, genome-edited crops are regulated as genetically modified organisms (GMOs), which can be subject to strict regulations and lengthy approval processes. However, some countries, such as Japan and Australia, have recently exempted certain types of genome-edited crops from their GMO regulations (Purnhagen et al. 2018).

The use of in vitro transcripts or ribonucleoproteins (RNPs) can result in transgene-free edited plants. In vitro transcripts are produced by in vitro transcription of the Cas9 and sgRNA templates, which are then delivered into plant cells for genome editing. RNPs, on the other hand, are pre-assembled complexes of Cas9 protein and sgRNA that are injected into the cell. Both methods have been successfully used for genome editing in various plant species, including horticultural crops such as tomato, cucumber, and lettuce. However, the use of in vitro transcripts or RNPs can be challenging in some crops due to low transformation efficiency and/or low editing efficiency. Therefore, the development of new delivery methods and the optimization of existing methods are essential to overcome these challenges and enable efficient and transgene-free genome editing in horticultural crops (Zhang 2016).

4.13 CONCLUSION

Horticultural crops are vital in human life, providing us with food, beverages, and ornamental items. We have seen significant programme-based development and triumphs in the horticulture research sector as a result of astounding technology improvements and high-value joint efforts, which is promising. In terms of precision and efficiency, mutagenesis via genome editing outperforms spontaneous and induced mutations. Although this technology is increasingly being used in many crops, its widespread use in horticultural crop breeding will necessitate the resolution of three challenges. First, through communication among consumers, breeders, and biologists, clear breeding traits of the horticultural crop in question should be identified. Second and third, appropriate mechanisms for administering editing reagents and afterwards regenerating mutants must be established. Given the enormous potential of genome editing and the significance of horticulture crops, we anticipate that these obstacles will be solved in the near future.

REFERENCES

Alenyorege, E. A., Ma, H., Ayim, I., Lu, F., & Zhou, C. (2019) Efficacy of sweep ultrasound on natural microbiota reduction and quality preservation of Chinese cabbage during storage. *Ultrasonics Sonochemistry*, *59*, 1–10

Azad, M., Kalam, A., Amin, L., & Sidik, N. M. (2014). Gene technology for papaya ringspot virus disease management. *The Scientific World Journal*, *2014*, 768038.

Barone, A., & De Palma, M. (2020). Genome editing of horticultural crops: Advances and perspectives. *Frontiers in Plant Science*, *11*, 590

Bhattacharya, R. C., Maheswari, M., Dineshkumar, V., Kirti, P. B., Bhat, S. R., & Chopra, V. L. (2004). Transformation of Brassica oleracea var. capitata with bacterial betA gene enhances tolerance to salt stress. *Scientia Horticulturae*, *100*(1–4), 215–227.

Bruening, G., & Lyons, J. (2000). The case of the FLAVR SAVR tomato. *California Agriculture*, *54*(4), 6–7.

Chaudhry, S., & Sidhu, G. P. S. (2022). Climate change regulated abiotic stress mechanisms in plants: A comprehensive review. *Plant Cell Reports*, *41*(1), 1–31.

Chen, K., Wang, Y., Zhang, R., Zhang, H., Gao, C., & CRISPR/Cas system: A Powerful Tool for Genome Engineering. (2019). *Plant Diversity*, *41*(5), 1–13.

Clarke, J. L., Spetz, C., Haugslien, S., Xing, S., Dees, M. W., Moe, R., & Blystad, D. R. (2008). Agrobacterium tumefaciens-mediated transformation of poinsettia, *Euphorbia pulcherrima*, with virus-derived hairpin RNA constructs confers resistance to Poinsettia mosaic virus. *Plant Cell Reports*, *27*, 1027–1038.

Collard, B. C., & Mackill, D. J. (2008). Marker-assisted selection: An approach for precision plant breeding in the twenty-first century. *Philosophical Transactions of the Royal Society B: Biological Sciences*, *363*(1491), 557–572.

Cortes, V., Talens, P., Barat, J. M. & Lerma-Garc ıa, M. J. (2019) Discrimination of intact almonds according to their bitterness and prediction of amygdalin concentration by Fourier transform infrared spectroscopy. *Postharvest Biology and Technology*, *148*, 236–241

d'Erfurth, I., Jolivet, S., Froger, N., Catrice, O., Novatchkova, M., & Mercier, R. (2009). Turning meiosis into mitosis. *PLoS Biology*, *7*(6), e1000124.

Dalal, M., Dani, R. G., & Kumar, P. A. (2006). Current trends in the genetic engineering of vegetable crops. *Scientia Horticulturae*, *107*(3), 215–225.

Divya, K., Bhatnagar-Mathur, P., Sharma, K. K., & Reddy, P. S. (2019). Heat shock proteins (Hsps) mediated signalling pathways during abiotic stress conditions. In *Plant Signaling Molecules* (pp. 499–516). Editors: Iqbal et.al., Woodhead Publishing, Elsevier, Cambridge.

Ekué, M. R., Gailing, O., & Finkeldey, R. (2009). Transferability of simple sequence repeat (SSR) markers developed in *Litchi chinensis* to *Blighia sapida* (Sapindaceae). *Plant Molecular Biology Reporter*, *27*, 570–574.

Fire, A., Xu, S., Montgomery, M. K., Kostas, S. A., Driver, S. E., & Mello, C. C. (1998). Potent and specific genetic interference by double-stranded RNA in *Caenorhabditis elegans*. *Nature*, *391*(6669), 806–811.

Fischhoff, D. A., Bowdish, K. S., Perlak, F. J., et al. (1987). Insect tolerant transgenic tomato plants. *Biotechnology*, *5*(8), 807–813.

Frary, A., Xu, Y., Liu, J., Mitchell, S., Tedeschi, E., & Tanksley, S. (2005). Development of a set of PCR-based anchor markers encompassing the tomato genome and evaluation of their usefulness for genetics and breeding experiments. *Theoretical and Applied Genetics*, *111*, 291–312.

Fraser, P. D., Romer, S., Shipton, C. A., et al. (2002). Evaluation of transgenic tomato plants expressing an additional phytoene synthase in a fruit-specific manner. *Proceedings of the National Academy of Sciences*, *99*(2), 1092–1097.

Gaj, T., Gersbach, C. A., & Barbas, C. F. (2013). ZFN, TALEN, and CRISPR/Cas-based methods for genome engineering. *Trends in Biotechnology*, *31*(7), 397–405.

Gao, L., Gonda, I., Sun, H., et al. (2019) The tomato pan-genome uncovers new genes and a rare allele regulating fruit flavor. *Nature Genetics*, *51*, 1044–1051.

Gao, X., Chen, J., Dai, X., Zhang, D., & Zhao, Y. (2021). An overview of epigenome editing in plants. *Plant Communications*, *2*(3), 100169.

Gautam, R. B., Goyal, S. K., & Kumar, L. (2022). Post-harvest techniques for quality seed production: An approach. *The Pharma Innovation Journal*, *11*, 3069–3078.

Gerszberg, A., Hnatuszko-Konka, K., Kowalczyk, T., & Kononowicz, A. K. (2015). Tomato (*Solanum lycopersicum* L.) in the service of biotechnology. *Plant Cell, Tissue and Organ Culture (PCTOC)*, *120*, 881–902.

Girhepuje, P. V., & Shinde, G. B. (2011). Transgenic tomato plants expressing a wheat endochitinase gene demonstrate enhanced resistance to Fusarium oxysporum f. sp. *lycopersici*. Plant Cell, Tissue and Organ Culture (PCTOC), *105*, 243–251.

Grierson, C. S., Barnes, S. R., Chase, M. W., et al. (2011) One hundred important questions facing plant science research. *The New Phytologist*, *192*, 6–12

Gruda, N., Bisbis, M., & Tanny, J. (2019). Influence of climate change on protected cultivation: Impacts and sustainable adaptation strategies-A review. *Journal of Cleaner Production*, *225*, 481–495.

Gupta, P. K., & Varshney, R. K. (2000). The development and use of microsatellite markers for genetic analysis and plant breeding with emphasis on bread wheat. *Euphytica*, *113*(3), 163–185.

Hamilton, A. J., & Baulcombe, D. C. (1999). A species of small antisense RNA in posttranscriptional gene silencing in plants. *Science*, *286*(5441), 950–952.

Hodges, R. J., Buzby, J. C., & Bennett, B. (2011). Postharvest losses and waste in developed and less developed countries: Opportunities to improve resource use. *The Journal of Agricultural Science*, *149(S1)*, 37–45.

Hsu, P. D., Lander, E. S., & Zhang, F. (2014). Development and applications of CRISPR-Cas9 for genome engineering. *Cell*, *157*(6), 1262–1278

Hussain, W., Baig, M. A., & Ahmad, S. (2021). Genome editing: A powerful tool for crop improvement. *Biotechnology and Applied Biochemistry*, *68*(1), 52–63.

Ibitoye, D. O., & Akin-Idowu P. E. (2011). Marker-assisted-selection (MAS): A fast track to increase genetic gain in horticultural crop breeding. *African Journal of Biotechnology* *10*(55), 11333–11339.

Ichimura, K., & Goto, R. (2002) Extension of vase life of cut narcissus tazetta var. chinensis flowers by combined treatment with sts and gibberellin a3. *Journal of the Japanese Society for Horticultural Science*, *71*, 226–230.

James, C. (1999). Global review of commercialized transgenic crops: 1998, *ISAAA Brief* No. 8 www.isaaa.org.

Jamnadass, R., Mumm, R. H., Hale, I., et al. (2020) Enhancing African orphan crops with genomics. *Nature Genetics*, *52*, 356–360.

Jiang, L., Shen, W., Liu, C., et al. (2022a). Engineering drought-tolerant apple by knocking down six GH3 genes and potential application of transgenic apple as a rootstock. *Horticulture Research*, 11, 3069–3078

Jiang, X., Zhang, W., Fernie, A. R., & Wen, W. (2022b). Combining novel technologies with interdisciplinary basic research to enhance horticultural crops. *The Plant Journal for Cell and Molecular Biology*, *109*(1), 35–46.

Jin, W., Dong, J., Hu, Y., Lin, Z., Xu, X., & Han, Z. (2009). Improved cold-resistant performance in transgenic grape (*Vitis vinifera* L.) overexpressing cold-inducible transcription factors AtDREB1b. *HortScience*, *44*(1), 35–39.

Jung, C. S., Griffiths, H. M., De Jong, D. M., Cheng, S., Bodis, M., & De Jong, W. S. (2005). The potato P locus codes for flavonoid 3′, 5′-hydroxylase. *Theoretical and Applied Genetics*, *110*, 269–275.

Kasso, M., & Bekele, A. (2018). Post-harvest loss and quality deterioration of horticultural crops in Dire Dawa Region, Ethiopia. *Journal of the Saudi Society of Agricultural Sciences*, *17*(1), 88–96.

Krishna, R., Karkute, S. G., Ansari, W. A., Jaiswal, D. K., Verma, J. P., & Singh, M. (2019). Transgenic tomatoes for abiotic stress tolerance: Status and way ahead. *3 Biotech*, *9*, 1–14.

Kumar, P. A., Mandaokar, A., Sreenivasu, K., et al. (1998). Insect-resistant transgenic brinjal plants. *Molecular Breeding*, *4*, 33–37.

Kumar, S., Banerjee, M. K., & Kalloo, G. (2000). Male sterility: Mechanisms and current status on identification, characterization and utilization in vegetables. *Vegetable Science*, *27*(1), 1–24.

Leblanc, O., Grimanelli, D., Hernandez-Rodriguez, M., Galindo, P. A., Soriano-Martinez, A. M., & Perotti, E. (2004). Seed development and inheritance studies in apomictic maize-Tripsacum hybrids reveal barriers for the transfer of apomixis into sexual crops. *International Journal of Developmental Biology*, *53*(4), 585–596.

Li, P., Song, A., Gao, C., et al. (2015). Chrysanthemum WRKY gene CmWRKY17 negatively regulates salt stress tolerance in transgenic chrysanthemum and Arabidopsis plants. *Plant Cell Reports*, *34*, 1365–1378.

Lu, S., Van Eck, J., Zhou, X., et al. (2006a). The cauliflower or gene encodes a DnaJ cysteine-rich domain-containing protein that mediates high levels of β-carotene accumulation. *The Plant Cell*, *18*(12), 3594–3605.

Ma, Y., Lin, S. Q., Gao, Y., Li, M., Luo, W. X., Zhang, J., & Xia, N. S. (2003). Expression of ORF2 partial gene of hepatitis E virus in tomatoes and immunoactivity of expression products. *World journal of gastroenterology*, *9*(10), 2211.

Mou, B. (2009). Nutrient content of lettuce and its improvement. *Current Nutrition & Food Science*, *5*(4), 242–248.

Nalley, L., Tsiboe, F., Durand-Morat, A., Shew, A., & Thoma, G. (2016). Economic and environmental impact of rice blast pathogen (*Magnaporthe oryzae*) alleviation in the United States. *PloS One*, *11*(12), e0167295.

Nambeesan, S., Datsenka, T., Ferruzzi, M. G., Malladi, A., Mattoo, A. K., & Handa, A. K. (2010). Overexpression of yeast spermidine synthase impacts ripening, senescence and decay symptoms in tomato. *The Plant Journal*, *63*(5), 836–847.

Palauqui, J. C., Elmayan, T., Pollien, J. M., & Vaucheret, H. (1997). Systemic acquired silencing: Transgene-specific post-transcriptional silencing is transmitted by grafting from silenced stocks to non-silenced scions. *The EMBO Journal*, *16*(14), 4738–4745.

Papolu, P. K., Dutta, T. K., Tyagi, N., Urwin, P. E., Lilley, C. J., & Rao, U. (2016). Expression of a cystatin transgene in eggplant provides resistance to root-knot nematode, Meloidogyne incognita. *Frontiers in Plant Science*, *7*, 1122.

Park, S., Kim, C. K., Pike, L. M., Smith, R. H., & Hirschi, K. D. (2004). Increased calcium in carrots by expression of an Arabidopsis H+/Ca 2+ transporter. *Molecular Breeding*, *14*, 275–282.

Parmar, N., Singh, K. H., Sharma, D., et al. (2017). Genetic engineering strategies for biotic and abiotic stress tolerance and quality enhancement in horticultural crops: A comprehensive review. 3 Biotech, *7*, 1–35.

Phillips, R. L., Kaeppler, S. M., & Olhoft, P. (1994). Genetic instability of plant tissue cultures: Breakdown of normal controls. *Proceedings of the National Academy of Sciences, 91*(12), 5222–5226.

Prabhavathi, V., Yadav, J. S., Kumar, P. A., & Rajam, M. V. (2002). Abiotic stress tolerance in transgenic eggplant (*Solanum melongena* L.) by introduction of bacterial mannitol phosphodehydrogenase gene. *Molecular Breeding*, *9*, 137–147.

Praveen, S., Ramesh, S. V., Mishra, A. K., Koundal, V., & Palukaitis, P. (2010). Silencing potential of viral derived RNAi constructs in tomato leaf curl virus-AC4 gene suppression in tomato. *Transgenic Research, 19*, 45–55.

Purnhagen, K. P., Wesseler, J., & Kleter, G. A. (2018). Regulation of genome-edited crops must consider health and environmental risks. *Nature Biotechnology*, *36*(8), 707–709.

Qin, G., Niu, Z., Yu, J., Li, Z., Ma, J., & Xiang, P. (2021). Soil heavy metal pollution and food safety in China: Effects, sources and removing technology. *Chemosphere, 267*, 129205.

Ren, Y. R., Yang, Y. Y., Zhao, Q., Zhang, T. E., Wang, C. K., Hao, Y. J., & You, C. X. (2021). MdCIB1, an apple bHLH transcription factor, plays a positive regulator in response to drought stress. *Environmental and Experimental Botany*, *188*, 104523.

Rivera Domínguez, M., Astorga Cienfuegos, K. R., Vallejo Cohen, S., Vargas Arispuro, I., & Sánchez Sánchez, E. (2011). Transgenic mango embryos (Mangifera indica) cv.'Ataulfo' with the defensin J1 Gen. *Revista Mexicana de Fitopatología*, *29*(1), 78–80.

Roderick, H., Tripathi, L., Babirye, A., Wang, D., Tripathi, J., Urwin, P. E., & Atkinson, H. J. (2012). Generation of transgenic plantain (Musa spp.) with resistance to plant pathogenic nematodes. *Molecular Plant Pathology*, *13*(8), 842–851.

Römer, S., Fraser, P. D., Kiano, J. W., Shipton, C. A., Misawa, N., Schuch, W., & Bramley, P. M. (2000). Elevation of the provitamin A content of transgenic tomato plants. *Nature Biotechnology*, *18*(6), 666–669.

Rout, G. R., & Peter, K. V. (Eds.). (2018). *Genetic Engineering of Horticultural Crops*. Academic Press.

Saikishore, N., Ramanakumari, B., Balakrishna, D., Rao, S. V., Bhat, B. V., Seetharama, N., & Visarada, K. B. R. S. (2005). Constraints and relevant strategies in genetic transformation of sorghum. Plant Biotechnology and Its Applications in Tissue Culture (pp. 33–40). International: New Delhi.

Schijlen, E. G., de Vos, C. R., Martens, S., et al. (2007). RNA interference silencing of chalcone synthase, the first step in the flavonoid biosynthesis pathway, leads to parthenocarpic tomato fruits. *Plant Physiology*, *144*(3), 1520–1530.

Shah, D. M., Horsch, R. B., Klee, H. J., et al. (1986). Engineering herbicide tolerance in transgenic plants. *Science*, *233*(4762), 478–481.

Sharma, M., & Khurana, S. P. (2016). Cell-free biosystems. Applied Molecular Biotechnology (pp. 466–486). CRC Press Taylor & Francis Group: USA.

Shekhawat, U. K. S., Ganapathi, T. R., & Srinivas, L. (2011). Cloning and characterization of a novel stress-responsive WRKY transcription factor gene (MusaWRKY71) from Musa spp. cv. Karibale Monthan (ABB group) using transformed banana cells. *Molecular Biology Reports*, *38*, 4023–4035.

Singh, A., Rajput, V., Singh, A. K., Sengar, R. S., Singh, R. K., & Minkina, T. (2021). Transformation techniques and their role in crop improvements: A global scenario of GM crops. In *Policy Issues in Gmodified Crops* (pp. 515–542). Academic Press, Elsevier, Cambridge, U.S.A.

Sonah, H., Deshmukh, R. K., Singh, V. P., Gupta, D. K., Singh, N. K., & Sharma, T. R. (2011). Genomic resources in horticultural crops: status, utility and challenges. *Biotechnology advances*, *29*(2), 199–209.

Sripaoraya, S., Keawsompong, S., Insupa, P., Power, J. B., Davey, M. R., & Srinives, P. (2006). Genetically manipulated pineapple: Transgene stability, gene expression and herbicide tolerance under field conditions. *Plant Breeding*, *125*(4), 411–413.

Szankowski, I., Briviba, K., Fleschhut, J., Schönherr, J., Jacobsen, H. J., & Kiesecker, H. (2003). Transformation of apple (Malus domestica Borkh.) with the stilbene synthase gene from grapevine (Vitis vinifera L.) and a PGIP gene from kiwi (Actinidia deliciosa). *Plant Cell Reports*, *22*, 141–149.

Tay, A., Lafont, F. & Balmat, J. F. (2021) Forecasting pest risk level in roses greenhouse: Adaptive neuro-fuzzy inference system vs artificial neural networks. *Information Processing in Agriculture*, *8*(3), 386–397.

Teo, G., Suzuki, Y., Uratsu, S. L., et al. (2006). Silencing leaf sorbitol synthesis alters long-distance partitioning and apple fruit quality. *Proceedings of the National Academy of Sciences*, *103*(49), 18842–18847.

Tian, Z., Wang, J. W., Li, J., & Han, B. (2020) Designing future crops: Challenges and strategies for sustainable agriculture. *The Plant Journal*, *105*, 1165–1178

Tripathi, M. K., Tiwari, S., & Khare, U. K. (2008). In vitro selection for resistance against purple blotch disease of onion (*Allium cepa* L.) caused by Alternaria porri. *Biotechnology*, *7*(1), 80–86.

United Nations. (2019). 2019 revision of world population prospects.

Vaucheret, H. (2001). Post-transcriptional gene silencing in plants. *Journal of Cell Science*, *114*(17), 3083–3091.

Wahlroos, T., Susi, P., Solovyev, A., Dorokhov, Y., Morozov, S., Atabekov, J., & Korpela, T. (2004). Increase of histidine content in *Brassica rapa* subsp. oleifera by over-expression of histidine-rich fusion proteins. *Molecular Breeding*, *14*, 455–462.

Whitton, J., Sears, C. J., Baack, E. J., & Otto, S. P. (2008). The dynamic nature of apomixis in the angiosperms. *International Journal of Plant Sciences*, *169*(1), 169–182.

Yadav, D. K., Yadav, N., & Khurana, S. M. P., 2013. Vaccines: Present status and application. In: Verma, A., Singh, A. (Eds.), *Animal Biotechnology* (pp. 491–508). Elsevier Acadmic Press.

Yu, T. A., Chiang, C. H., Wu, H. W., et al. (2011). Generation of transgenic watermelon resistant to Zucchini yellow mosaic virus and papaya ringspot virus type W. *Plant Cell Reports*, *30*, 359–371.

Yu, Y., Zhang, K., Yang, L., & Zhang, D. (2019) Fruit detection for strawberry harvesting robot in non-structural environment based on Mask-RCNN. *Computers and Electronics in Agriculture*, *163*, 104846

Zetsche, B., Gootenberg, J. S., Abudayyeh, O. O., Slaymaker, I. M., Makarova, K. S., Essletzbichler, P., & Zhang, F. (2015). Cpf1 is a single RNA-guided endonuclease of a class 2 CRISPR-Cas system. *Cell*, *163*(3), 759–771.

Zhang, W., Zhang, W., Sun, L., et al. (2019) The effects and mechanisms of epigallocatechin-3-gallate on reversing multidrug resistance in cancer. *Trends in Food Science & Technology*, *93*, 221–233

Zhang, Y., He, X., Zhao, H., et al. (2020a). Genome-Wide identification of DNA methylases and demethylases in Kiwifruit (Actinidia chinensis). *Frontiers in Plant Science*, *11*, 514993.

Zhang, Y., Liang, Z., Zong, Y., et al. (2016). Efficient and transgene-free genome editing in wheat through transient expression of CRISPR/Cas9 DNA or RNA. *Nature Communications*, *7*, 12617

Zhang, Y., Zhang, F., Li, X., Ballerini, E. S., & Rutten, T. (2020b). Genome editing for horticultural crop improvement. *Horticulture Research*, *7*(1), 1–15.

Zhu, G., Gou, J., Klee, H., & Huang, S. (2019b) Next-gen approaches to flavor-related metabolism. *Annual Review of Plant Biology*, *70*, 187–212.

Zhu, Y., & Barritt, B. H. (2008). Md-ACS1 and Md-ACO1 genotyping of apple (Malus x domestica Borkh.) breeding parents and suitability for marker-assisted selection. *Tree genetics & genomes*, *4*(3), 555–562.

5 Potential of Nanotechnology for Improving Nutrient Efficiency of Temperate Horticulture Crops

Kanika Dulta
School of Applied and Life Sciences, Dehradun, Uttaranchal University, 248007, India

Gözde Koşarsoy Ağçeli
Hacettepe University, Faculty of Science, Department of Biology, 06800 Beytepe Campus, Ankara, Turkey

Priyanka Tangariya
Department of Food Science, Amity University, Mohali, India

Yogesh Kumar Awasthi
School of Applied and Life Sciences, Dehradun, Uttaranchal University, 248007, India

Renu Bala Sharma
Dev Bhoomi Uttarakhand University, India

Tripti Negi
Dev Bhoomi Uttarakhand University, India

Mohd Mazhar
Integral University, Lucknow, India

Keshav Kumar
Dr. Yashwant Singh Parmar University of Horticulture and Forestry, Solan

DOI: 10.1201/9781003412489-5

Deepika Negi
Dr. Yashwant Singh Parmar University of Horticulture and Forestry, Solan

5.1 INTRODUCTION

The foundation of the economics of developing nations is the food and agriculture industries. The food and agricultural industries are under increased strain due to the rapid rise in global population and the effects of climate change. In India, the agriculture sector contributes 18% to the nation's Gross domestic product (GDP) and relies on 50% human resources. India is the world's second-biggest supplier of vegetables and fruits, behind China. The accessibility of a vast variety of fresh fruits and vegetables is attributed to India's diverse climate. Based on information from National Horticulture Database published by the National Horticulture Board, in 2019–2020, India harvested 191.77 million metric tonnes of vegetables and 99.07 million metric tonnes of fruits. 6.66 million hectares were used for fruit agriculture, whereas 10.35 million hectares were used for vegetable farming. According to FAO (2019), India produces the most bananas (26.08%), papaya production (44.05%), and mango production (45.89%). India produces less fruit than other nations, while being the second-largest producer of fruits globally. India's lower fruit production is caused by a multitude of issues, which includes population pressure, unprofitable assets, and declining soil fertility. The demand for fruits worldwide is growing along with the global population. Over the past 50 years, farmers have depended significantly on genetically modified crops, insect pests, disease-resistant cultivars, fertilizers, and pesticides to satisfy this need (Yadav et al. 2013). Fruit output has increased significantly owing to the usage of fertilizers in agriculture generally. But over use of these chemical-based fertilizers has impacted on condition of fruit and soil (Bai et al. 2020). Like all other crops, fruit crops' growth, development, production, and quality are greatly impacted by the accessibility of nutrients. The regulation of nutrients is a crucial component in fruit production. Sixteen key nutrients must be supplied to fruit trees for optimum growth and development. Fruit crops require a lot of nutrients; thus, an effective nutrition management plan is necessary for their development and output (Kumari et al. 2020).

Human health is directly impacted by climate change, for example, by increasing the number of fatalities from floods, heat stress, and storms, and indirectly by altering the water supply, air pollutants, and disease vector ranges (mosquitoes), aquatic viruses, and food-borne disorders (Wolf and Menne 2007). In addition, discharges of greenhouse gases from agriculture, like those from (CO_2), methanol, and (NO), have an influence on climate change. These emissions are mostly caused by the usage of tillage practises, fossil fuels, livestock waste, and intensively fertilized agricultural soils (Gondal et al. 2021). Additionally, intense plant cultivation, pesticides and fertilizer use, and soil fertility replenishment considerably improve food security and the environment while also unleashing harmful impacts (Gondal et al. 2021). Using chemical fertilizers has increased agricultural yields significantly for quite some time. However, they disrupt soil ecology, soil structure, soil fertility, and mineral distribution, all of which are major long-term obstacles. In order to fix the issue, it is essential to create innovative substances that can deliver nutrients to certain locations while contributing to a healthy atmosphere. According to recent research, graphene is an intriguing material that might be used to transport plant nutrients. It can provide nutrition in a steady, in a regulated way for the benefit of the plants, and it finally enhances crop yield while having no impact on the environment (Kabiri et al. 2017).

Given that the global nanotechnology market reached $1 trillion in 2015, nanotechnology appears to be an alternate which could revolutionize this sector of agriculture Harper (2015). Nanotechnology offers a broad range of applications throughout all phases of manufacturing, processing, and storage, packing, and transportation of agricultural goods, as well as additional selling services (Prasad, n.d.). In order to increase output through pest and nutrient control, decrease the quantity of herbicides needed, and reduce the amount of nutrients lost during fertilization, nanomaterials are utilized in agriculture (Joshi et al. 2019). Nanotechnology provides ecologically friendly techniques to boost crop output by utilizing nanopesticides, nano-fertilizers, nano-coatings, nano-encapsulation, seed germination, stress tolerance, etc. (Figure 5.1.). This improves the plant's ability for nutrient absorption and targeted delivery (Khan et al. 2019). Through the efficient use of fertilizers and pesticides and the minimization of pertinent losses (Shang et al. 2019), agriculture's productivity efficiency is increased via nanotechnology (Guleria et al. 2023).

Nano, which means dwarf in Greek, is the root of the phrase "nanotechnology." Richard Phillips Feynman, winner of the Nobel Prize, once said something important that has a direct bearing on the advancement of nanotechnology (Rana et al. 2021). Nanoparticles display fascinatingly different properties from the substantial surface to volume ratio of mass (Satalkar et al. 2016; Di Sia 2017; Singh and Prasad 2017). Advanced nanomaterials in agriculture have shown a substantial promise for the generation of novel crop species with the aid of genetically engineered hybrid crops, and new, highly effective agrochemicals for agricultural plants' nutrition and safety (Islam 2019; Hossain et al. 2020). Researchers have been working to show the

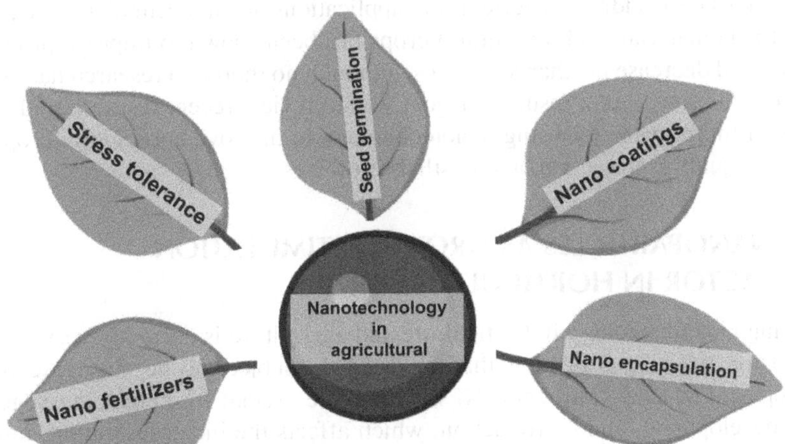

FIGURE 5.1 Role of nanotechnology in agricultural sector.

possibility for agricultural science to benefit from nanobiotechnology in recent years through several studies in various fields. The purpose of nanoparticles has been discussed in numerous scientific publications, whether they are carbon-based (CBNPs) or metal-based (MBNPs), considering their absorption and numerous investigations on the internalization, translocation, perseverance, impact on growth, and broad growth among multiple plant genera, and their commercial value (Misra et al. 2016).

By decreasing agricultural production's important damage and the use of pesticides and fertilizers as effective tools, nanotechnology boosts agricultural productivity. Nanosensors provide significant promise for detecting environmental changes and impairments (Shang et al. 2019). NMs (Nanomaterials) may be sprinkled topically on crops or soaked into the ground. The leaves and branches of fruit plants are typically sprayed with NMs. NMs are commonly utilized in a variety of industries, including contemporary packaging. Utilizing nanotechnology in the manufacturing of fruit may enhance the quality, freshness, shelf life, and nutritional value of these items. In addition to extending mango fruits to be stored for up to standard packaging, nanoscale edible coatings as thin as 5 nm can act as a stopper for the transfer of gases and humidity which prevents converts into colours, flavours, antioxidants, and enzymes.

Chitosan/nano-silica and nano-TiO_2-LDPE (titanium dioxide-low-density polyethylene)-packed fruits have demonstrated higher physiochemical, physiological, and preservation qualities than fruits packaged conventionally. Additionally, the creation, output, and standard of the products made from the strawberry, mango, pomegranate, and date palm as well as reducing the negative outcomes of stress brought on by abiotic factors like salinity in the strawberry and alkalinity in the grape all benefit from the use of nanotechnology. Additionally, fruit plants treated with nanofertilizers were more robust, can withstand abiotic stress, and maintain the quality of their fruit (Zahedi et al. 2020).

There is a lot of information about using nanoparticles which is available to increase crop plant yield, quality, and reduce postharvest losses. Although several publications have addressed agricultural applications of nanotechnology, the application of nanomaterials to horticultural crops has been shown to improve production, quality, and decrease postharvest losses, although no thorough research has yet been done on this topic. As a result, our goal was to review recent research and explore potential future paths for using nanotechnology to improve horticulture crops' productivity, quality, protection, and postharvest losses.

5.2 NANOPARTICLES AS GROWTH STIMULATION FACTOR IN HORTICULTURE CROPS

Utilizing nanotechnology in horticulture and agriculture is currently very common. The factor affecting fruit trees that has been the subject of the most research and development is nano-fertilizers (NFs), which are crucial for enhancing vegetative, floral development, and reproduction, which affects the increases output, improves fruit quality, and eventually lengthens the shelf life by reducing fruit waste. These nanomaterials are typically sprayed on trees at very low concentrations, in multiple sessions over a period of time and are regarded as growth stimulating factors. It has

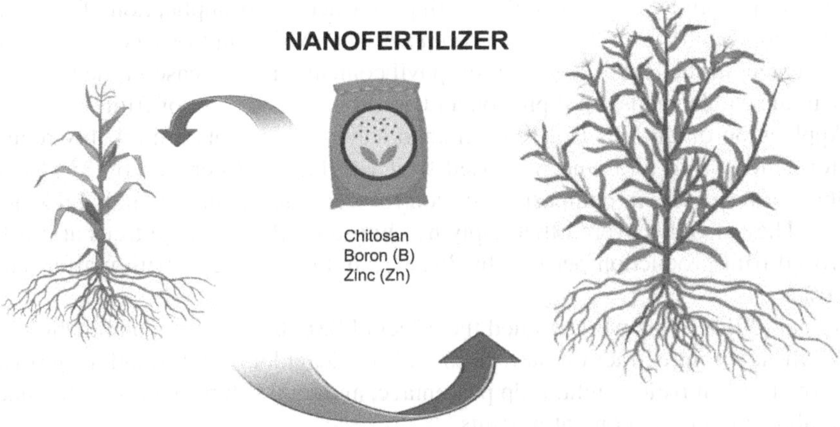

FIGURE 5.2 Macro- and micro-scale NFs such as chitosan, boron (B), and zinc (Zn) have positive effects on plants in agriculture.

been demonstrated that macro- and micro-scale NFs like chitosan, boron (B), and zinc (Zn) (Figure 5.2) and nanocomposite fertilizers like NPKMg (nitrogen, phosphorus, potassium, magnesium), ZnFeMnB (zinc, iron, manganese, boron), and calcite extensively affect the morphological, physical, biochemical, and reproductive characteristics of fruit crops like grape, pomegranate, mango, strawberry, and date (Zahedi et al. 2019).

Contrary to common bulk fertilizers, NFs contain nutrients at nanoscale in high concentrations that can attach to various soil ions effortlessly and liberate various nutrients in a gradual manner by coating them with polymers (Subramanian et al. 2015). Only a small number of reports on the usage of NF technology on fruit trees, despite the fact that it is extremely innovative. Spraying essential macronutrients and micronutrients such as NMs and nanocomposites has a favourable effect on the fruit yield of several fruit crops.

Davarpanah et al. (2016) observed that the foliar spray of Zn and B at low concentrations of 120 and 6.5 mg/L, respectively, was significantly more effective in case of pomegranate. Applied concentration remarkably increased quantity of fruits per branch and fruit yield. Moreover, B-NPs had significantly better impacts on these fruit parameters in comparison with controls and Zn-NPs.

Ranjbar et al. (2020) studied the effect of nanocalcium spray (0%, 1.5%, and 2.0%) on the qualitative parameters of apple (*Malus domestica Borkh*) and observed greatest improvement in the fruit length, diameter, weight, density, and other nutritional parameters, namely starch, fibre content, total antioxidant activity, and titratable acidity using 2% nanocalcium.

Shalan (2020) in his research used potassium sulphate nanopowder (100, 150, and 200 g) on Grapevines (*V. vinifera*) and found that 150 and 200 g of potassium sulphate significantly improved the vegetative growth of plant including leaf surface area, internodal length, and thickness along with increase in the berry weight, length, width, and diameter.

In a similar study, Rahemi et al. (2020) performed foliar application of nano-iron chelate (2000 mg/L) on Quince (*Cydonia oblonga* Mill.) and observed substantial improvement in the leaf area and chlorophyll content with increase in the fruit firmness, total soluble solids, total phenol, and ascorbic acid content of fruits.

Application of Zn-NPs (1 g/L) on mango leaves and shoots remarkably reduced malformation up to 55% and increased fruit weight, yield per tree by 33.74% and 57.36%, respectively, in comparison of control trees, according to Gad and Zagzog (2017). They also discovered that applying chitosan-NPs to mango trees at 5 mL/L improved fruit production per tree by 35.28%, and resistance to fruit malformation by 40%.

Refaai (2014) in his work studied the effect of 0.05% B-NPs in combination with 1.0% wheat sprout extract on date palm (cultivar Zaghloul) and found a significant positive effect on fruit weight, pulp percentage, and total soluble solids with reduced seed, fibre, and tannin content in fruits.

Any product, whether natural or manmade is directly applied to the soil or sprayed on plant tissue in order to supply one or more essential nutrients needed by agricultural plants, is called a fertilizer. Fertilizers may be natural as well as synthetic in origin. However, they are applied to raise the standard of soil and further the crop, but in contrast a large proportion of them are wasted in different ways resulting in loss of crop production as well as causing environmental pollution. A breakthrough in the sector of agriculture for reducing the environmental impact of applied fertilizers came forward in the form of NFs. NFs by fusing plant nutrition with nanomaterials, creating emulsions of nanoscale and covering nutritional molecules with a thin layer of nanomaterials (Babu et al. 2022). Fertilizers made of natural and manmade,boost bioavailability and soil fertility more effectively than standard fertilizers (Kalia and Kaur 2019). Because they may be used in lower amounts than conventional fertilizers, NFs could be a key advancement in the preservation of the environment (Verma et al. 2022), hence reducing leaching, runoff, and gas emissions to the atmosphere (Karthik and Maheswari 2021). Based on the plant nutrient requirements, NFs may be macro- and micronutrient fertilizers.

5.2.1 Macronutrient-Based Nano-fertilizers

These fertilizers provide significant levels of the nutrients needed for traditional farming techniques. They are formed by combining nanomaterials with macronutrients such as N, K, P, Ca, Mg, and S in order to provide giving the body a correct amount of these nutrients crop. The macronutrient-based NFs consist of one or more nutrients that are encapsulated within a particular nanomaterial.

5.2.2 Nitrogen Nano-fertilizers

Nitrogen is the primary and key macronutrient which the majority of agricultural soils lack. Nitrogen fertilizers are manufactures commercially using the Haber–Bosch process (Ekejiuba 2017). Fertilizers high in nitrogen assist to maintain a sufficient supply of nutrient-rich food while also influencing plant growth and development (Amilia Nongbet et al. 2022). Low nitrogen utilization and nitrogen leakage to the environment are two drawbacks of nitrogen fertilizers. This results in high amounts of emissions of greenhouse gases and eutrophication (Mejias et al. 2021). Recently, an advanced N nanofertilizer has been developed using urea-coated hydroxyapatite nanoparticles (rod-shaped structures with an average aspect ratio 10) for targeted delivery and slow release (Raliya et al. 2017). These compound nanoparticles were chosen because to their abundant N and P sources and chemical compatibility.

5.2.3 Phosphorus Nano-fertilizers

Like N, P is essential to all living things, and 90% of P is needed in food production. Although P is frequently present in sufficient concentrations in soil, farmers have a high need for P fertilizers since the plants' P levels are unstable. One of P fertilizers' main downsides is that they quickly transform from the usable form of P into the less accessible ones. They are ineffective at being used and absorbed, because they produce eutrophication (increasing levels of N and P in water) (Nongbet et al. 2022).

Lots of modification were incorporated in traditional methods to increase the efficiency of P NFs. Use of zinc (Z) and zinc oxide (ZnO) nanoparticles was utilized to increase the effectiveness of converting native P into P that is accessible to plants.

5.2.4 Macronutrient-Based Nano-fertilizers

Micronutrients are the substances required by plants in very small or trace amounts. In order to maintain metabolic activities, plants need them. Plant development depends on zinc (Zn), an essential structure or regulating cofactor for a number of enzymes and proteins. Zn is a component of defensive mechanisms against dangerous microorganisms, synthesis of carbs, protein metabolism, and auxin regulation (Sindhu et al. 2019). Additionally, boron (B) contributes to the lignification of plant cell walls and their production, yet it also significantly affects other physiological processes, such as plant development (Basit et al. 2022). Consequently, applying the right proportions of Zn and B is crucial to getting the most productivity out of high-quality horticulture crops.

To ensure healthy growth and development, plants also require trace quantities of iron (Fe); however, Fe availability may result in the breakdown of crucial physiological and metabolic processes, reducing plant production (Sarwar et al. 2010). The capacity to tolerate a variety of environmental restrictions is provided by manganese (Mn), which functions as a cofactor for numerous enzymes. Additionally, Mn plays a crucial role in photosynthesis as well as the production of ATP (Adenosine Triphosphate), chlorophyll, fatty acids, proteins, and secondary metabolites including lignin and flavonoids (Zulfiqar et al. 2019).

5.3 EFFECT OF NANOPARTICLES ON HORTICULTURE SET, GROWTH, AND YIELD

Several recent researches have established that using NFs is superior to in terms of using conventional fertilizers in enhancing the fruit growth, yield, and quality of various fruit crops. Several research types investigated the impact of various nanoparticles on various fruit parameters are summarized in Table 5.1.

TABLE 5.1
Effect of Different Nanoparticles on Various Physical and Biochemical Parameters of Horticulture Crop and Fruits

Nanoparticles	Tested Horticulture Crop and Cultivars	Application	Concentration of Nanoparticles	Effects	References
Nano-zinc oxide (nZnO) and nano-silicon (nSi)	Mango *Mangifera indica L.*, Ewais cultivar	Foliar spray	nZnO: 50, 100, and 150 mg/L; nSi: 150 and 300 mg/L) and combination of nZnO (100 mg/L) and nSi (150 mg/L)	Increase in the total fruit number and weight per tree. Combination of nanoparticles resulted in highest fruit yield	Elsheery et al. (2020)
Se-NPs	Strawberry	Foliar spray	10 and 20 mg/ L	Significantly improved the number of fruits per plant, fruit weight, and yield by 12.54%, 9.55%, and 13.81%, respectively	Zahedi et al. (2019)
Nano (Fe-NPs), sulphate ($FeSO_4$), and chelated Fe	*Citrus sinensis L. Osbeck*	Foliar spray	$FeSO_4$ at 0.1%, $FeSO_4$ at 0.2%, chelated Fe at 0.05%, and chelated Fe at 0.1%	Fe-NPs and chelated Fe (0.1%) increased fruit yield by 32.0% and 25%, respectively	El-Gioushy et al. (2021)
Biosynthesized silver nanoparticles (AgNPs)	Kinnow (*Citrus reticulata* L.)	Exogenous application	AgNPs concentration 10, 20, 30, and 40 ppm	Maximum fruit weight and yield per plant was obtained with 30 ppm concentration in comparison with control and other concentrations	Hussain et al. (2019)
Silver nanoparticles with magnetic field (AgM)	Muskmelon (*Cucumis melo* L.)	Foliar spray	40 g/ha	150% more fruit yield along with larger fruit size when compared with control	Feizi et al. (2013)

(Continued)

TABLE 5.1 (*Continued*)
Effect of Different Nanoparticles on Various Physical and Biochemical Parameters of Horticulture Crop and Fruits

Nanoparticles	Tested Horticulture Crop and Cultivars	Application	Concentration of Nanoparticles	Effects	References
Calcium/yeast nanocomposite	Mango *Mangifera indica* L., Ewais cultivar	Foliar application	1% yeast; 2% yeast; 3% yeast; 1% calcium/yeast nanoparticles; 2% calcium/yeast nanoparticles; 3% calcium/yeast nanoparticles.	Calcium/yeast nanoparticles 3% was the most effective combination and improved fruit set, fruit retention, and no. of fruits per tree by 119%, 150%, and 48%.	Abd El-Razek et al. (2017)
Silver nanoparticles (AgNPs)	Kinnow	Exogenous application	25, 50, 75, and 100 mg/L	75 mg/L AgNPs was most effective treatment and increased the average fruit weight up to 90.78%	Umair Raza et al. (2023)
NPK nanofertilizer	Date palm ("Hillawi")	Foliar spray	Nano super 1, 2 g; NPK 1, 2 g and Nano super 1 g + NPK 1 g	Treatment of nano super 1 g with NPK 1 g was the most effective one which resulted in increased fruit weight and bunch weight by 17.8% and 49.48%, respectively.	Shareef et al. (2020)
Se and Zn nanoparticles	Apple (*Malus domestica* L.)	Foliar spray	Se: 50 ppm; Zn: 250 ppm with edaphic fertilization (SFE) with N, P, and K (150-50-80, 100-60-60) and without edaphic fertilization	SFE + NPZn treatment resulted in highest fruit yield of 58.17 kg and number of fruits 433 fruits per tree	Montaño-Herrera et al. (2022)

(*Continued*)

TABLE 5.1 (*Continued*)
Effect of Different Nanoparticles on Various Physical and Biochemical Parameters of Horticulture Crop and Fruits

Nanoparticles	Tested Horticulture Crop and Cultivars	Application	Concentration of Nanoparticles	Effects	References
Se-NPs	Pomegranate	Foliar spray	10 and 50 nm	Se-NPs of 10 nm and 50 nm increased fruit weight, yield, and diameter by 5/14, 46/50 and 9/26% and 1.4/8.2, 20.6/28.8, and 4.1/20.7%, respectively	Zahedi et al. (2020)
Boron nano-chelated fertilizer	Olive (*Olea europaea* L.)	Spray	105, 158.8, 200, and 300 mg/L	300 mg/L was the most effective concentration which resulted in significantly high fruit set (0.84%–3.13%), weight (26.97–41.02 g), fruit yield (18.50–61.02 kg/tree), oil yield, free fatty acid, and phenol content.	Vishekaii et al. (2019)
Nano-boron and nano-zinc nanoparticles	Olive (*Olea europaea*)	Foliar application	100 and 200 ppm 10 and 20 ppm	20 ppm nano-boron + 200 ppm nano-zinc was the best combination which resulted in increased (31.5%) final fruit set and fruit yield (14.25%)	Genaidy et al. (2020)
Zinc oxide nanoparticles	Sweet cherry (*Prunusavium* L.)	Spray coating	100–1000 mg/L	Fruits coated with 500 mg/L showed less decay on storage along with negligible moisture and sugar, vitamin C content, and increased flavonoid and antioxidant activity	Ouzakar et al. (2023)

(*Continued*)

TABLE 5.1 (*Continued*)
Effect of Different Nanoparticles on Various Physical and Biochemical Parameters of Horticulture Crop and Fruits

Nanoparticles	Tested Horticulture Crop and Cultivars	Application	Concentration of Nanoparticles	Effects	References
Chitosan nanoparticles	Apricot	Edible coating	1% and 1.5%	Fruits with chitosan nanoparticles coating exhibited decreased weight loss, TSS, and lipid peroxidation while retained better carotenoids and ascorbic acid content	Algarni et al. (2022)
Nano-chelate super plus ZFM containing zinc, iron, and manganese	Almonds	Foliar spray	2 g/L	Applied concentration was effective in significantly increasing the fruit set, fresh fruit weight, and yield per branch along with Zn and Mn of leaves	Kamiab and Zamanibahramabadi (2016)
Iron oxide and zinc oxide nanoparticles	Carrot (*Daucus carota* L.)	Foliar spray	ZnO:100ppm and FeO NPs: 50 ppm	Increase in plant height (87.24–91.24 cm), leaf area (65.15–75.73 cm^2), root diameter (2.4–3.53 cm), root length (14.81 to 19.75cm), root yield/treatment (55–72.33 kg), and root yield/hectare (t/ha) (24.25–29.41 tones/ha)	Elizabath et al. (2017)
ZnO and SiO_2 Nanoparticles	Potato	Foliar spray	ZnO: 0, 50, and 100 mg/L SiO_2: 0, 25, and 50 mg/L	100 mg/L of ZnO nanoparticles significantly improved tuber weight (129.53–145.67 g), number of plant tubers (10.67–13.37), and tuber yield (3.45–4.86 kg/m^2)	Seleiman et al. (2023)

(*Continued*)

TABLE 5.1 (*Continued*)
Effect of Different Nanoparticles on Various Physical and Biochemical Parameters of Horticulture Crop and Fruits

Nanoparticles	Tested Horticulture Crop and Cultivars	Application	Concentration of Nanoparticles	Effects	References
Nano-Zn and Fe	Radish (*Raphanussativus* L.)	Foliar spray	0 and 50 mg/L	Foliar spray of nanomaterials increased the shoot and root length by 7%–8%, fresh weight by 27%–41%, and dry matter weight by 50%–60%	Shafiq et al. (2023)
Chitosan nanoparticles loaded with N	Garlic	Foliar spray	0.25 and 0.5 mol/L	0.5 mol/L and 100 kg of N significantly improved plant height from 46.43 to 64.07 cm, plant weight from 97.3 to 197.9 g, and bulbing ratio from 34.7 to 61.4 (mm)	Elsaggan et al. (2018)
Carbon nanoparticles (CNPs) with sulphur	Onion (Allium cepa L.)	Foliar spray	300, 400, and 500 mg/L with 2.5, 3.5, and 4.5 g/L of sulphur	Plants treated with *500* mg/L CNPs + *4.5* g/L *S* resulted in maximum increase in the plant length (28.55–41.33 cm), yield, bulb weight (60.33–110.47 g), and neck diameter (1.5–2.75 cm)	Mahmoud and El-Tanahy (2022)

5.4 NANOTECHNOLOGY IN POSTHARVEST PROCESSES (SHELF LIFE AND IMPROVING ANTIOXIDANT ENZYME ACTIVITY)

The food business may undergo a transformation thanks to nanotechnology by offering numerous applications aimed at enhancing food safety, quality, and nutritional value with shelf life extension. Nano-based food sensors which detect the deformities in food, food packaging, and edible coating are the core areas where nanotechnology successfully applied for the desirable shelf life of food products. One of the more exciting food advancements in recent years is edible coatings, which are thin layers of food ingredients that build directly on fruits, typically by soaking the fruits with a covering substance solution. Edible coatings are capable of carrying ingredients including antimicrobial compounds, anti-browning, nutrients, colour, and tastes and extending shelf life and limiting the growth of pathogens on food surfaces (Figure 5.3). Edible coatings can be used with a number of methods, such as dipping, spraying, or coating, to manage processes like oxidation, humidity, and gas transfer. The barrier characteristics could be improved using nanoparticles and effectiveness of coverings for fruit preservation since these systems have a higher surface area. In packaging system, antimicrobial nanoparticles are used to actively reduce microbiological contaminations, prolonging shelf life and improve storage stability.

5.4.1 Shelf Life Extension of Fruits by Using Nanotechnology

Numerous food applications, including food analysis sensors, biodegradable packaging, eatable packaging for food, smart packaging, and active packaging, significantly apply nanoparticles. In the food industry, nanoparticles are also utilized as anti-caking agents, gelating agents, nanocarriers, nano-additives, nano-capsules, and more. Food safety, preservation, and functionalization are three of the key roles that nanotechnology plays in ensuring food security.

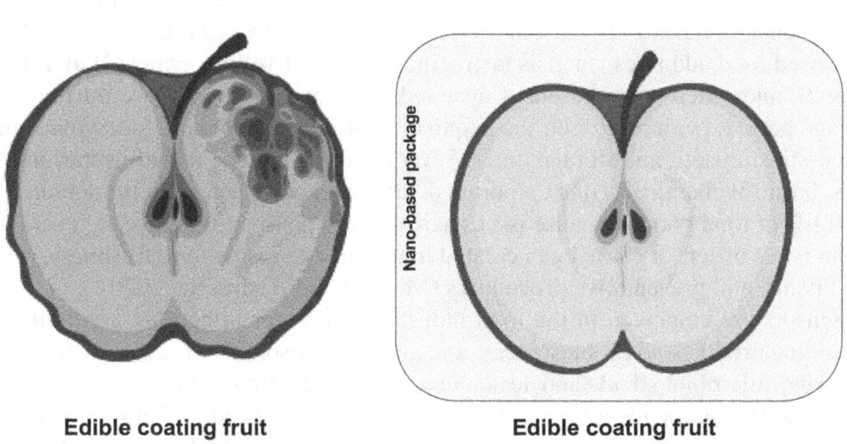

FIGURE 5.3 Nano-based packaging contributes to extending the shelf life of foods thanks to its various features.

Food shelf life is the period of time it takes for its overall quality to deteriorate to the point that it is no longer fit for human consumption. Objective markers, such as the presence of off-flavours and off-odours, changes in textural properties, and changes in appearance, are used to identify these unwanted traits. These modifications may all be influenced by microbial, chemical, or physical processes (Lianou et al. 2016).

Customers nowadays choose fresh products due to their high nutritional value and demand better, highly nutrient-dense fruit that contains phytochemicals, which are elements that promote health.

The healthy physicochemical and sensory properties of these meals must also be preserved, be suitable for consumption and uncontaminated by contaminants and hazardous bacteria in order to support better health conditions (Zambrano-Zaragoza et al. 2021).

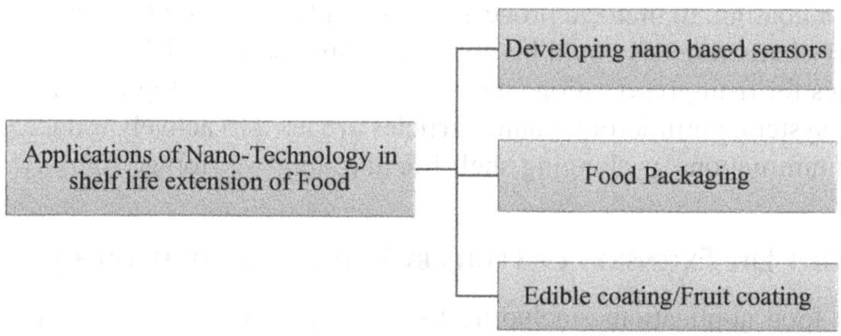

5.4.2 Nano-Based Sensors Used in Shelf Life Analysis of Fruits

The examination of food items' shelf lives is extremely important since it takes into account indicators of food quality (freshness and components that degrade food quality) and safety. However, the nutritional content of food goods during storage is often detected during the shelf life study in comparison with fresh food products, and the presence of hazardous substances (patulin or Staphylococcal Enterotoxin B) and undesired food additives (such as tartrazine and sunset yellow, orange II and rhodamine B, and others) are present. Unwanted food quality might arise during entire storage period, beginning with the acquisition of raw ingredients, following transit, processing impacts, and storage times. Several outside factors, including various poisons, harmful chemicals (like colourants), drug, and pesticide, can also decrease the shelf life of food goods or cause issues across every phase of their cycle. Several different types of sensors have been created in this area to assess food freshness, ruined food items, and preventative procedures (Mustafa and Andreescu 2020).

Sensors are employed in the food industry to look for infections, contaminants, degrading products, toxic substances, and allergies. Also, there are direct (freshness, maturity, microbiological, and leakage) and indirect (time-temperature indication and radio frequency identification) freshness sensors in food. Additionally, biosensors are made up of transducing components (such as electrochemical, optical, and mass-based biosensors) and biorecognition components (such as antibodies, DNA, and enzymes) (Sonkaria, Ahn, and Khare 2012).

5.4.2.1 Types of Nanomaterials Used in Developing Sensors for Shelf Life Assessment of Food Products

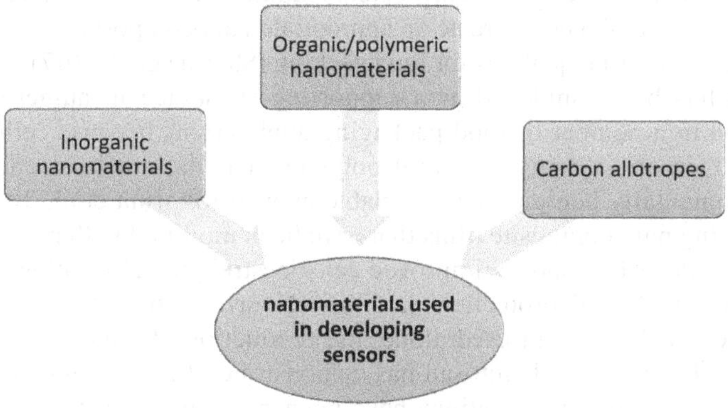

These broad categories of nanomaterials are further divided into subcategories like inorganic nanomaterials (TiO_2, nanoparticles of Au and Ag, ZnO, magnetic nanoparticles, etc.), organic nanomaterials (nanostructured hydrogel, nanostructured films, hyperbranched polymeric nanoparticles, dendrimers, etc.), and carbon-allotrope (carbon nanotube, graphene, reduced graphene oxide, fullerenes, carbon dots, etc.)

5.4.3 Nano-Based Packaging Techniques

Food packing is important crucial role in extending the shelf life of food by providing protection against various factors that can cause spoilage, deterioration, or contamination. The nature and composition of food packaging materials affect food quality. The packing properties of materials with thinner, more robust envelopes, thermal resistance, and other features can be improved by nanomaterials. It is widely used and reported that food packaging uses nanotechnology (Yifen Shang et al. 2019). According to estimates by Duncan (2011), the market for food and beverage packaging made from nanotechnology is expected to rise from 4.13 billion USD in 2008 to 7.3 billion by 2014, representing a yearly expansion rate of 11.65%.

There are numerous active packing applications for different metal and metal oxide nanoparticles. Ag and TiO_2 are extremely useful minerals for food packaging (Li et al. 2017).

Using metal and metal oxide nanoparticles as antimicrobials in active packaging is comparable to using nanocomposites in food packaging.

Due to more efficient optical, electrical, and photosensitive results of TiO_2, it is often utilized as a catalyst, stain, or adsorbent substance (Prakash et al. 2018). Nanoparticles for food packaging provide benefits such as oxygen exchange, enzyme mobilization, antimicrobial potential, and clues for degree exposure to a component associated with deterioration. Zinc oxide nanoparticles and polylactic acid nano-polymer were combined to create a very effective interaction packaging material for food with increased functional properties. When evaluated for food packaging, zinc oxide-treated semolina protein coating significantly reduced oxygen absorption and increased heat resistance. In food packaging, layers of nanoclay, silicates, and moistened alumina silicate have been incorporated. Additionally, polymers and clay with silicate have been combined to create an appropriate nanocomposite contender with exceptional circumscribe qualities for food packing (Sharma et al. 2017).

Cellulose has been employed as a supporting substance in numerous nanomaterials and in a number of food packaging applications because cellulose has a higher surface area. Being a natural polymer, it is extremely strong, affordable, environmentally benign, and recyclable as well (Sharma et al. 2017). Food packaging using nanocomposite films that have been modified with nano-chitosan and nano-cellulose has shown improved tensile strength, elongation to breaking, clearance, and food protection qualities (Noorbakhsh-Soltani et al. 2018). In comparison with chitosan used alone, the production of nano-bio-composted films using polylactides and chitosan has demonstrated their suitability for packaging. Packaging and paper coatings have been made using nanocomposites of cellulose and copper that have antibacterial characteristics (Basu et al. 2018). The layer-by-layer method of coating LDPE films with silver nanoparticles has a considerable antibacterial potential against (gram-positive) *S. aureus* and (gram-negative) *Pseudomonas fluorescens*.

Nanometre-thick films of chitosan were employed to make antibacterial surfaces because of their inherent characteristics particularly focused and used widely. Chitosan nanoparticles are created through the process of ionic gelation, which involves the electrostatic interaction of the positive amino sides of the chitosan with cross-linking polyanions (Sharma et al. 2017). Red grapes' shelf life has been extended by 14 days using a chitosan/gelatin-based nanocomposite with Ag nanoparticles that is executed for food packaging. As effective antibacterial applicants against gram-positive (*Staphylococcus aureus*) and gram-negative (*Pseudomonas aeruginosa*), yeast (*Candida albicans*), and fungi (*Aspergillus niger*) and yeast, chitosan-Ag and chitosan-Au nanocomposites coating are used in packaging (Youssef et al. 2014).

5.4.4 Edible Coatings

The coverings on edible foods are thin or films applied to food items to enhance their appearance, protect them from moisture loss, improve shelf life, and provide additional flavour or texture with additional nutritional and phytochemical support. These coatings are typically made from edible ingredients and can be applied to a range of culinary items, including fruits, vegetables, meats, seafood, confectionery, and bakery items (McHugh and Senesi 2000).

Even after being harvested, fruits and vegetables continue to respiration, completely exhausting the product's oxygen supply. This oxygen is not restored as rapidly as it would be by an eatable covering. In addition, carbon dioxide is produced, which cannot as easily escape through the coating and builds up inside the product. As a result, fruits and vegetables have an almost doubled shelf life and continue to be firm, fresh, and nutrient-dense. The thickness of the coating has an impact on how much the internal atmosphere (O_2 and CO_2) is altered and how much weight loss is minimized. Edible coatings and films have been utilized in the food industry as food product protection since the dawn of time (Madhu et al. 2018).

5.4.4.1 Advantages of Edible Coating

- Slows down respiration rate and generation of ethylene, which delays over ripening.
- Prevents storage issues related to low temperature like chilling injuries in fruit and vegetables.
- Prevents fruit from wilting and maintains it firm to preserve its fresh look.
- Work as an obstruction for environmental gas exchange.
- Incorporates nutrients such vitamins, antioxidants, pigments, and ions that inhibit browning reactions, aroma compounds, and antioxidants.

5.4.5 Nano-Based Edible Coating

Organic nanoparticles are currently being used to create edible barriers that effectively preserve postharvest quality and lengthen the duration of food goods' shelf lives. Nowadays, the most commonly used nano-based materials for edible coatings are chitosan and chitin-based coatings and films, alginate-based films and coatings, nanoliposome, colloidosomes, casein micelles, alginate and chitosan, and nano-cochleate (Miteluț et al. 2021).

5.5 IMPROVING ANTIOXIDANT ENZYME ACTIVITY THROUGH NANOTECHNOLOGY

The human body is protected by antioxidant enzymes from the damaging effects of oxidative stress. Since it causes serious illnesses like diabetes, Alzheimer's disease, cancer, and ageing, oxidative stress is an important health risk factor. It generates reactive oxygen species (ROS), which disrupt the glial cells' structure and function in the brain (Reshmy et al. 2021).

5.5.1 Methods Used for Improving Antioxidant Enzyme Activity

These are the various following methods which are used to improve or preserve the antioxidant enzyme activities.

5.5.1.1 Application of Nanomaterials in Enzyme Immobilization

Previously, the main focus was on immobilizing enzymes using conventional immobilization methods on nanoscale substrates such as physical adsorption, covalent binding, entrapment, and carrier-free immobilization. When an enzyme is immobilized on or into a solid support, its stability, load density, and ease of removal and retrieval for reusing may be improved while preserving catalytic activity. Based on the size of the enzymes at the nanoscale, these techniques were further enhanced, specifically single enzyme nano-encapsulation employing carrier-free immobilization (Madhavan et al. 2017).

5.5.1.1.1 *Physical Adsorption*

This immobilization technique can be employed when the enzyme has to be linked to an inert substance. This procedure comprises adhering a surface-active enzyme material including glass, alginate beads, and others. The immobilizing agent may restrict the enzyme's active site since adsorption is a physical process, significantly lowering the activity of the immobile enzyme. Physical adsorption immobilization is illustrated by laccase adsorption on nanoporous gold particles (Nematian et al. 2020).

5.5.1.1.2 *Covalent Binding*

The enzyme is either cross-linked to the backing material or covalently connected to it. This approach appears to be the most effective, also enabling multipoint immobilized enzyme attachment, and it enhances the effectiveness and stability of immobilized enzymes. Also suggested is that the concave nanopore surface enclosing the enzyme would improve enzyme stability enabling many covalent bonds among the innermost layer of nanopores and enzyme molecules. Nanopores behind the intake that are already blocked by pre-attached enzyme cannot be penetrated by additional enzymes due to a large steric barrier (Misson, Zhang, and Jin 2015).

5.5.1.1.3 *Entrapment*

Entrapment is now a popular, straightforward, and affordable adsorption technique in which an enzyme is bonded to a support by forces such as Vander Waals forces, and it is the most successful technique for immobilization because of ionic interactions and hydrogen bonds. Nanomaterials with different hydrophobic and hydrophilic properties are used to capture enzymes. Because they can capture more enzymes, hydrophobic nanoparticles are preferred, have greater degrees of immobility, and continue to be very active. Normal traps for the enzyme include hydrophobic silane sol gels or insoluble calcium alginate beads (Qamar et al. 2020).

5.5.1.1.4 *Nano-Encapsulation*

Nanosized carriers, including liposomes or polymeric nanoparticles, can contain antioxidant enzymes. The enzymes are enhanced in stability and protected from deterioration by this encapsulation method. Furthermore, it enables controlled release, assuring a continuous and precise delivery of the enzymes to the specified position (Esfanjani and Jafari 2016).

5.5.1.1.5 *Surface Modification*

To improve their interaction with antioxidant enzymes, nanoparticle surfaces might be changed with certain ligands or functional groups. This change can enhance the stability, activity, and immobilization of the enzyme. Additionally, it can promote enzyme recycling, enabling multiple applications of the enzymes (Qiao et al. 2016).

5.6 NANOTECHNOLOGY IN ABIOTIC RESISTANCE OF HORTICULTURE CROPS

The phrase "abiotic stress" refers to how non-living substances adversely affect living things in a specific environment. The abiotic factors affecting agricultural production worldwide include drought, salinity, heavy metals, and extremely hot and low temperatures (Khan et al. 2021). The statistical approximation states that these stressors predominantly affect around 90% of agricultural land areas (Shalaby et al. 2021) causing a 70% decrease in the production of crucial food crops (Chakraborty et al. 2022).

Being the sessile creature in general, the shifting global climate makes plants increasingly susceptible to abiotic stress because they are compelled to endure diverse environmental pressures during their life cycle. Therefore, they alter the molecular, biochemical, and physiological pathways to establish their defence system against environmental challenges at different levels (Husen et al. 2017). Second messengers, including proteins kinases, calcium, reactive oxygen species (ROS), phospholipids, and nitric oxide, relay and enhance the signal when a plant cell recognizes stress (Zhang et al. 2020).

The development of effective and environmentally friendly techniques for the sustainable production of nanoparticles (NPs), which offers great promise for handling issues associated managing abiotic pressures to ensure the sustainability of agriculture, is one area where nanotechnology has found application in recent years (Saxena et al. 2016). Because of their distinct physiochemical characteristics, like the exceptional stability inside cells, incredibly small size (1–100 nm), greater surface area, and reactivity, NPs are becoming increasingly important (Nejatzadeh 2021). NP doses have the potential for both detrimental biological repercussions. Low-level nanomolar concentrations, NPs, function as a critical regulator in controlling plant growth and development. At large dosages, they can cause oxidative damage to biomolecules, which can cause plant cell death (Tariq et al. 2021).

5.6.1 Nanoparticles in Salt-Stress Tolerance

One of the biggest problems in modern agriculture is salinity, which ultimately slows and hinders plant growth and advancement therefore results in plant degeneration (Isayenkov and Maathuis 2019). When the sodium chloride concentration is greater than 200 mM, most plants perish. Salinity affects many horticultural products, including fruits, vegetables, and spices. Salt stress causes ionic strength to be out of balance, which affects a range of biochemical, physiological, and metabolic processes (Mohamed et al. 2022). Salinity affects a wide range of horticultural crops, including fruits, vegetables, and spices, as ionic strength imbalances caused by salt stress have an impact on several biochemical, physiological, and metabolic processes (Sinha and Verma 2021). It was successful in reducing the deleterious outcomes of salt stress in different plants by applying nanoparticles made of Zn, Ag, SiO_2, Cu, Fe, Mn, C, Ti, Ce, and K (Zulfiqar and Ashraf 2021).

5.6.2 Nanoparticles in Drought-Stress Tolerance

According to experts, drought is the environmental stress that has the greatest negative effects on crop productivity. The average temperature will climb by 1.8°C to 4.0°C by 2100, according to the Intergovernmental Panel on Climate Change (IPCC), and drought will afflict large parts of the planet (Ozturk et al. 2021). Agriculture is impacted by drought when there is not enough moisture for plants to grow properly and complete their life cycles. Larger stems and roots, fewer leaves, erratic stomatal oscillations, and changed water and nutrient interactions among poorer agricultural production and ineffective water use are among the primary effects of drought on crop plants (Farooq et al. 2012). It was discovered that applying metal oxide nanoparticles to the leaves, like TiO_2, ZnO, and Fe_3O_4, was successful in boosting the physiological and metabolic activities of the plant enduring a drought (Alabdallah et al. 2021). Additional enhancements were produced to the pomegranate plants' photosynthetic pigments, nutritional status, and physical and chemical characteristics when silicon nanoparticles were administered to drought-enduring plants (particularly those involving fruit shattering), levels of osmolytes, antioxidant enzymes, and abscisic acid, as well as the phenol content (Zahedi et al. 2021).

5.6.3 Nanoparticles in Cold-Stress Tolerance

The adverse effects of plant developing and growing under low- or cold-temperature stress are also a result of global climate change. Chilling and freezing are two common forms of low-temperature stress that affect plants. In relation to the species and degree of endurance, plants can tolerate chilling temperatures between 0°C and 15°C. Other elements that affect chilling temperatures include the ambient temperature and wind speed during exposure. The plant will struggle in frigid temperatures (below 0°C), in contrast to how it reacts to chilling conditions. Low and non-freezing temperatures can harm or kill crop species, which can affect their productivity, survival, and ecological dispersal (Al-Khayri et al. 2023). Low temperatures have an effect on a variety of plant processes, including those related to secondary metabolism, respiration, defence, and the synthesis of proteins and nucleic acids (Aslam et al. 2022). Considering their capacity to increase resistance to cold shock, chitosan nanoparticles and TiO_2 nanoparticles are employed widely in several investigations.

5.6.4 Nanoparticles in Heavy-Metal-Stress Tolerance

One of the harmful elements that lowers agricultural output today is heavy-metal (HM) stress. Global HM pollution is a product of human activities like industry and urbanization (Emamverdian et al. 2015). Increased use of contemporary farming technologies like chemical pesticides and fertilizers has also caused crop plants to experience HM stress. Various heavy metals, including Hg, Pb, Cd, Ni, Co, Cr, and Ag, negatively affect plants (Yadav 2010). Since plants are at the base of trophic systems, there is a great likelihood that these HMs will bioaccumulate along the food chain, which eventually results liver and kidney impairment are persistent health issues that affect both humans and animals. Additionally, HMs directly affect plants through anomalies in their morphology, physiological functioning, and metabolic pathways (Tiwari and Lata 2018). Particularly in cultivated foods and botanical remedies, these have an impact depending on the level of plant-based food production and consumption goods. Using nanoparticles the soil's ability to absorb and change the heavy metals there, limiting their bioaccumulation and mobility. Additionally, NPs cause the development of apoplast barriers reduce the level of heavy metals in the root.

5.6.5 Nanoparticles in Flooding-Stress Tolerance

Due to severe water blockage in soil, the majority of plants are vulnerable to flooding. Extreme rainfall, inadequate soil drainage, or irrigation techniques all contribute to flooding. Thus, one of the major abiotic stresses is floods that influence the availability of food and nations' economies. It affects the vegetation that grows in several ecosystems, including wetlands, salt marshes, riparian zones, floodplains, and tidal zones. Flooding stress will alter soil redox potential and pH, raise carbon dioxide concentration, and mobilize more phytotoxins, which will have an impact on root metabolism, nutrient absorption, and overall plant growth (BAILEY-SERRES and Colmer 2014). It's been suggested that nanoparticles can help plants that are stressed by flooding.

5.6.6 Nanoparticles in Heat-Stress Tolerance

Heat stress can result from high temperatures. Heat stress is frequently thought to be the increase in temperature that lasts longer than necessary and is high enough to permanently impede plant growth (Hu et al. 2020). Heat stress shortens a plant's life cycle and lowers output because it reduces the efficiency of photosynthetic activities (Zhao et al. 2020). Heat stress might become a serious problem, limiting the production of agricultural crops in tropical and subtropical regions.

There are certain well-known NPs used in agriculture to relieve a variety of abiotic stressors on plants, including salt, drought, and metal toxicity. One of the most often utilized NPs in agriculture is silica (Si), which is well established to do a significant impact in plant stress endurance. Exogenous administration of silica to different plants under stress resulted in a rise in the amount of chlorophyll, which boosted photosynthesis, lower transpiration rate, and better product quality (Ahmad and Akhtar 2019). Due to increased nitrogen and potassium intake, Cucumis sativus showed improved growth and production by Alsaeedi et al. (2019). Silicon dioxide (SiO_2) NPs are used under conditions of salt and water scarcity. Utilizing Ag NPs under saline conditions increased seedling length of the shoots and roots, fresh and dried weight, and germination %, all of which helped *Lathyrus sativus* (Grass Pea) survive and grow more successfully (Hojjat 2019). For the purpose of reducing salt stress, cerium oxide (CeO_2) nanoparticles cause shorter root apoplastic barriers, resulting in increased sodium ion (Na+) transit to shoots and decreased Na+ build-up in canola plants (*Brassica napus*) (Rossi et al. 2017). As of now, the crop enhancement programme's widely acknowledged strategy is to use genome editing technology (Bhat et al. 2020; Ansari et al. 2020). As of right now, genome editing technologies are the widely acknowledged method for improving crops. As a result, a team of researchers recently merged this strategy with nanotechnology by also developing the procedure for modifying non-model plant species' genomes with carbon nanotubes and the CRISPR-cas9 gene-editing technique, which brings about a substantially high protein expression level (Demirer et al. 2019). This implies that NP-based genome editing will likely occur shortly may result in a number of improvements to plant traits, including yield, quality, and stress tolerance.

5.7 NANOTECHNOLOGY IN PEST RESISTANCE OF HORTICULTURE CROPS

There are several applications for nanotechnology in agriculture, particularly in the field of pest control. Pest infestation is predicted to cause yearly losses of approximately 26%–30% in barley, sugar beet, wheat, soybean, and cotton; 40% in rice; 39% in potatoes; and 35% in maize (Oerke and Dehne 2004). Pimentel (2009) claimed that pest insects cause around a 13% loss in agricultural production and that roughly 3 billion kg of pesticides is used annually to fight pests. Chemical pesticides and insecticides used to protect crops are exceedingly hazardous to non-target creatures like people, birds, fish, amphibians, and insect pollinators (Gill and Garg 2014).

Numerous nations have shifted from chemical-based agriculture to environmentally sound, socially acceptable, and commercially effective green agriculture. The greatest promise for nano-based pesticides for the management of insect pests comes from the revolution that nanotechnology has brought about in agriculture today.

With its abundance of various biomolecule-containing organisms including plants, algae, fungus, yeast, etc., nature serves as a sizable "bio-laboratory." The use of naturally existing biomolecules allows for the synthesis of nanoparticles in a variety of sizes and forms. These biomolecules are pushing the development of safer, more sanitary, and ecologically friendly procedures for the creation of nanoparticles.

Silica nanoparticles are efficient against bacterial infections (Tripathi et al. 2015) and have been utilized as micronutrients (Song et al. 2016). Some polymeric nanoparticles can also induce resistance to a variety of plant diseases. For instance, chitosan nanoparticles can reduce the *Pyricularia grisea* rice blast fungus (Manikandan and Sathiyabama 2016). Extracted from *Lobelia leschenaultiana*, ZnO nanoparticles provided good dose-dependent mortality of *Lucilia sericata* blowflies (Banumathi et al. 2018). By promoting insecticidal and repellent activities, the combination of nanoparticles with botanical chemicals can aid in pest control (De Oliveira et al. 2019). Copper and silicon nanoparticles can increase the amounts of both enzymatic and nonenzymatic plant defence chemicals in tomato (*Solanum lycopersicum*) plants, enhancing plant tolerance to *Clavibacter michiganensis* and eventually increasing tomato crop output (Cumplido-Nájera et al. 2019). Additionally, CuO and ZnO nanoparticles can significantly alter the rhizosphere's ability to produce important compounds that aid in protecting plants from *Pseudomonas chlororaphis* O6 (Anderson et al. 2017).

The development of nanopesticides aims to improve the efficacy of existing pesticides and reduce the usage of potentially hazardous chemical pesticides (Jhones Luiz de Oliveira et al. 2014; Madbouly et al. 2017; Ureña Saborío et al. 2017) which are associated with changes in pests, breed resistance, and have a detrimental influence on the environment (Campolo et al. 2017; Krutyakov et al. 2017). A powerful vehicle for delivering these nanopesticides to the desired location is nano-encapsulation.

Utilizing artificial nanoparticles as biopesticides is a cutting-edge technique. These nanoparticles are more reactive and can form covalent connections with bio-fungicides. Utilizing this special characteristic of nanomaterials, the plant-pest chain is managed. The aluminosilicate nanotube is among the most effective nanomaterials. Insect hairs pick up the distribution of aluminosilicate nanotubes on plant surfaces. Insects ingest nanotubes laced with pesticides and die as a result. Consequently, nano-bio-fungicides are a potent substitute for environmentally acceptable insect pest control. When it comes to the safe delivery of chemicals like insecticides and fertilizers, certain nanomaterials have smart qualities like delayed liberation (Forgiarini et al. 2001; Kuzma et al. 2006).

Additionally, the creation of environmentally benign pesticides and strain-resistant plants both require nanoparticles. Through the use of nano-biosensors, nanoparticles not only play a significant role in eliminating infections but also in their early identification. Nanotechnology is used in agriculture for a variety of purposes, including crop production and crop protection against insects and other pests. Because of loaded nano-formulated allelochemicals, nanoparticles have been proven to have significant effects on entomology, such as the insecticidal activity of stored grain pests. Silver and TiO_2 nanoparticles are two prevalent examples of nanoparticles with an antibacterial activity. Therefore, early adoption of this nanotechnological achievement will play a significant role in ameliorating the situation of deteriorating food shortage of ever-increasing population.

REFERENCES

Abd El-Razek, E, O A Amin, Amany M El-Nahrawy, and N Abdel-Hamid. 2017. Effect of foliar application of biosimulated nanomaterials (calcium/yeast nanocomposite) on yield and fruit quality of 'ewais' mango trees. *Annual Research & Review in Biology*, 18 (3). 1–11.

Ahmad, Irfan, and Mohd Sayeed Akhtar. 2019. Use of nanoparticles in alleviating salt stress. In Mohd Sayeed Akhtar (ed.), *Salt Stress, Microbes, and Plant Interactions: Causes and Solution: Volume 1*. Springer, 199–215.

Alabdallah, Nadiyah M, Md Mahadi Hasan, Inès Hammami, Azzah Ibrahim Alghamdi, Dikhnah Alshehri, and Hanan Ali Alatawi. 2021. Green synthesized metal oxide nanoparticles mediate growth regulation and physiology of crop plants under drought stress. *Plants* 10 (8). MDPI: 1730.

Algarni, Eman H A, Ibrahim A Elnaggar, Abd El-wahed N Abd El-wahed, Ibrahim M Taha, Huda A Al-Jumayi, Sam M Elhamamsy, Samy F Mahmoud, and Alaa Fahmy. 2022. Effect of chitosan nanoparticles as edible coating on the storability and quality of apricot fruits. *Polymers* 14 (11). MDPI: 2227.

Al-Khayri, Jameel Mohammed, Ramakrishnan Rashmi, Rutwick Surya Ulhas, Wudali N Sudheer, Akshatha Banadka, Praveen Nagella, Mohammed Ibrahim Aldaej, Adel Abdel-Sabour Rezk, Wael Fathi Shehata, and Mustafa Ibrahim Almaghasla. 2023. The role of nanoparticles in response of plants to abiotic stress at physiological, biochemical, and molecular levels. *Plants* 12 (2). MDPI: 292.

Alsaeedi, Abdullah, Hassan El-Ramady, Tarek Alshaal, Mohamed El-Garawany, Nevien Elhawat, and Awadh Al-Otaibi. 2019. Silica nanoparticles boost growth and productivity of cucumber under water deficit and salinity stresses by balancing nutrients uptake. *Plant Physiology and Biochemistry* 139. Elsevier: 1–10.

Anderson, Anne J, Joan E McLean, Astrid R Jacobson, and David W Britt. 2017. CuO and ZnO nanoparticles modify interkingdom cell signaling processes relevant to crop production. *Journal of Agricultural and Food Chemistry* 66 (26). ACS Publications: 6513–24.

Ansari, Waquar A, Sonali U Chandanshive, Vacha Bhatt, Altafhusain B Nadaf, Sanskriti Vats, Jawahar L Katara, Humira Sonah, and Rupesh Deshmukh. 2020. Genome editing in cereals: Approaches, applications and challenges. *International Journal of Molecular Sciences* 21 (11). MDPI: 4040.

Aslam, Mohammad, Beenish Fakher, Mohammad Arif Ashraf, Yan Cheng, Bingrui Wang, and Yuan Qin. 2022. Plant low-temperature stress: Signaling and response. *Agronomy* 12 (3). MDPI: 702.

Babu, Subhash, Raghavendra Singh, Devideen Yadav, Sanjay Singh Rathore, Rishi Raj, Ravikant Avasthe, S K Yadav, et al. 2022. Nanofertilizers for agricultural and environmental sustainability. *Chemosphere* 292: 133451. doi:https://doi.org/10.1016/j.chemosphere.2021.133451.

Bai, Yong-Chao, Ying-Ying Chang, Muzammil Hussain, Bin Lu, Jun-Pei Zhang, Xiao-Bo Song, Xia-Shuo Lei, and Dong Pei. 2020. Soil chemical and microbiological properties are changed by long-term chemical fertilizers that limit ecosystem functioning. *Microorganisms* 8 (5). MDPI: 694.

Bailey-Serres, J., and Timothy D Colmer. 2014. Plant tolerance of flooding stress-recent advances. *Plant, Cell & Environment* 37 (10). Wiley Online Library: 2211–2215.

Banumathi, Balan, Baskaralingam Vaseeharan, Balasubramanian Malaikozhundan, Palaniappan Ramasamy, Marimuthu Govindarajan, Naiyf S Alharbi, Shine Kadaikunnan, Angelo Canale, and Giovanni Benelli. 2018. Green larvicides against blowflies, *Lucilia sericata* (diptera, calliphoridae): Screening of seven plants used in Indian ethno-veterinary medicine and production of green-coated zinc oxide nanoparticles. *Physiological and Molecular Plant Pathology* 101. Elsevier: 214–218.

Basit, Farwa, Sana Asghar, Temoor Ahmed, Usman Ijaz, Muhammad Noman, Jin Hu, Xinqiang Liang, and Yajing Guan. 2022. Facile synthesis of nanomaterials as nanofertilizers: A novel way for sustainable crop production. *Environmental Science and Pollution Research* 29 (34). Springer: 51281–51297.

Basu, Alex, Karen Heitz, Maria Strømme, Ken Welch, and Natalia Ferraz. 2018. Ion-crosslinked wood-derived nanocellulose hydrogels with tunable antibacterial properties: Candidate materials for advanced wound care applications. *Carbohydrate Polymers* 181. Elsevier: 345–350.

Bhat, Mujtaba Aamir, Mudasir Ahmad Bhat, Vijay Kumar, Ishfaq Ahmad Wani, Humayra Bashir, Ali Asghar Shah, Safikur Rahman, and Arif Tasleem Jan. 2020. The era of editing plant genomes using CRISPR/Cas: A critical appraisal. *Journal of Biotechnology* 324. Elsevier: 34–60.

Campolo, Orlando, Asma Cherif, Michele Ricupero, Gaetano Siscaro, Kaouthar Grissa-Lebdi, Agatino Russo, Lorena M Cucci, Patrizia Di Pietro, Cristina Satriano, and Nicolas Desneux. 2017. Citrus peel essential oil nanoformulations to control the tomato borer, tuta absoluta: Chemical properties and biological activity. *Scientific Reports* 7 (1). 13036.

Chakraborty, Swarnavo, Ankur Singh, and Aryadeep Roychoudhury. 2022. Biogenic nanoparticles and generation of abiotic stress-resilient plants: A new approach for sustainable agriculture. *Plant Stress*. Elsevier, 100117.

Cumplido-Nájera, Claudia Francely, Susana González-Morales, Hortensia Ortega-Ortíz, Gregorio Cadenas-Pliego, Adalberto Benavides-Mendoza, and Antonio Juárez-Maldonado. 2019. The application of copper nanoparticles and potassium silicate stimulate the tolerance to Clavibacter michiganensis in tomato plants. *Scientia Horticulturae* 245. Elsevier: 82–89.

Davarpanah, Sohrab, Ali Tehranifar, Gholamhossein Davarynejad, Javier Abadía, and Reza Khorasani. 2016. Effects of foliar applications of zinc and boron nano-fertilizers on pomegranate (*Punica Granatum* Cv. Ardestani) fruit yield and quality. *Scientia Horticulturae* 210. Elsevier: 57–64.

Demirer, Gozde S, Huan Zhang, Natalie S Goh, Eduardo González-Grandío, and Markita P Landry. 2019. Carbon nanotube-mediated DNA delivery without transgene integration in intact plants. *Nature Protocols* 14 (10). Nature Publishing Group UK London: 2954–2971.

Ekejiuba, A. 2017. Evaluation of the exact production quantity of nitrogen fertilizer in real-time from any particular associated gas flare volume in Nigeria. *International Journal of Applied* 7 (3).

El-Gioushy, Sherif Fathy, Rokayya Sami, Amina A M Al-Mushhin, Hanan M Abou El-Ghit, Mohamed S Gawish, Khadiga Ahmed Ismail, and Reda M Y Zewail. 2021. Foliar application of $ZnSO_4$ and $CuSO_4$ affects the growth, productivity, and fruit quality of Washington navel orange trees (*Citrus Sinensis L.*) Osbeck. *Horticulturae* 7 (8). MDPI: 233.

Elizabath, Ambily, Vijay Bahadur, Pragathi Misra, Vipin Mashi Prasad, and T Thomas. 2017. Effect of different concentrations of iron oxide and zinc oxide nanoparticles on growth and yield of carrot (*Daucus Carota* L.). *Journal of Pharmacognosy and Phytochemistry* 6 (4). AkiNik Publications: 1266–1269.

Elsaggan, Mohamed, Mohamed S Khater, and Gebril El Desouky. 2018. Effect of spraying by nitrogen loaded on chitosan nanoparticles on growth and productivity of garlic under sandy soil conditions. *Egyptian Journal of Desert Research* 68 (1). Desert Research Center: 135–155.

Elsheery, Nabil I, Mohamed N Helaly, Hanan M El-Hoseiny, and Shamel M Alam-Eldein. 2020. Zinc oxide and silicone nanoparticles to improve the resistance mechanism and annual productivity of salt-stressed mango trees. *Agronomy* 10 (4). MDPI: 558.

Emamverdian, Abolghassem, Yulong Ding, Farzad Mokhberdoran, and Yinfeng Xie. 2015. Heavy metal stress and some mechanisms of plant defense response. *The Scientific World Journal* 2015. Hindawi: 1–18.

Esfanjani, Afshin Faridi, and Seid Mahdi Jafari. 2016. Biopolymer nano-particles and natural nano-carriers for nano-encapsulation of phenolic compounds. *Colloids and Surfaces B: Biointerfaces* 146. Elsevier: 532–543.

Farooq, Muhammad, M Hussain, Abdul Wahid, and K H M Siddique. 2012. Drought stress in plants: An overview. *Plant Responses to Drought Stress: From Morphological to* Molecular Features. Springer, 1–33.

Feizi, Hassan, Saeed Jahedi Pour, and Kaveh Hassanzadeh Rad. 2013. Biological response of muskmelon (*Cucumis Melo* L.) to magnetic field and silver nanoparticles. *Annual Research & Review in Biology*, 3(4): 794–804.

Forgiarini, A, J Esquena, C Gonzalez, and C Solans. 2001. Formation of nano-emulsions by low-energy emulsification methods at constant temperature. *Langmuir* 17 (7). ACS Publications: 2076–2083.

Gad, Mohamed M, and Osama A Zagzog. 2017. Mixing xanthan gum and chitosan nano particles to form new coating for maintaining storage life and quality of elmamoura guava fruits. *International Journal of Current Microbiology and Applied Sciences* 6 (11): 1582–1591.

Genaidy, Esmail A E, Nazmy Abd-Alhamid, Hassan S A Hassan, Ahmed M Hassan, and Laila F Hagagg. 2020. Effect of foliar application of boron trioxide and zinc oxide nanoparticles on leaves chemical composition, yield and fruit quality of Olea Europaea L. Cv. Picual. *Bulletin of the National Research Centre* 44. Springer: 1–12.

Gill, Harsimran Kaur, and Harsh Garg. 2014. Pesticide: Environmental impacts and management strategies. *Pesticides-Toxic Aspects* 8 (2014). IntechOpen London: 187.

Gondal, Aqarab Husnain, Koko Tampubolon, Muhammad Danish Toor, and Mujahid Ali. 2021. Pragmatic and fragile effects of wastewater on a soil-plant-air continuum and its remediation measures: A perspective. *Reviews in Agricultural Science* 9. The United Graduate Schools of Agricultural Sciences, Japan: 249–259.

Guleria, Geetika, Shweta Thakur, Mamta Shandilya, Sushma Sharma, Sapna Thakur, and Susheel Kalia. 2023. Nanotechnology for sustainable agro-food systems: The need and role of nanoparticles in protecting plants and improving crop productivity. *Plant Physiology and Biochemistry*. Elsevier Masson s.r.l. doi:10.1016/j.plaphy.2022.12.004.

Harper, T. 2015. *The Year of the Trillion Dollar Nanotechnology Market*. AZoNetwork UK Ltd.: Manchester, UK.

Hojjat, S Saeid. 2019. Effect of interaction between ag nanoparticles and salinity on germination stages of *Lathyrus sativus* L. *Journal of Environmental Soil Science*. 2 (2): 186–91.

Hossain, Akbar, Rout George Kerry, Muhammad Farooq, Nawfel Abdullah, and M. Tofazzal Islam. 2020. Application of nanotechnology for sustainable crop production systems. *Nanotechnology in the Life Sciences*, 135–159. Springer Science and Business Media B.V. doi:10.1007/978-3-030-31938-0_7.

Hu, Shanshan, Yanfei Ding, and Cheng Zhu. 2020. Sensitivity and responses of chloroplasts to heat stress in plants. *Frontiers in Plant Science* 11. Frontiers Media SA: 375.

Husen, A, M Iqbal, and I M Aref. 2017. Plant Growth and foliar characteristics of faba bean (*Vicia Faba* L.) as affected by indole-acetic acid under water-sufficient and water-deficient conditions. *Journal of Environmental Biology* 38 (2). Triveni Enterprises: 179.

Hussain, Mubashir, Naveed Iqbal Raja, Zia-Ur-Rehman Mashwani, Muhammad Iqbal, Muhammad Ejaz, and Sumaira Aslam. 2019. Green synthesis and evaluation of silver nanoparticles for antimicrobial and biochemical profiling in kinnow (*Citrus Reticulata* L.) to enhance fruit quality and productivity under biotic stress. *IET Nanobiotechnology* 13 (3). Wiley Online Library: 250–256.

Isayenkov, Stanislav V, and Frans J M Maathuis. 2019. Plant salinity stress: Many unanswered questions remain. *Frontiers in Plant Science* 10. Frontiers Media SA: 80.

Islam, Md. Tofazzal. 2019. Applications of nanomaterials for future food security: Challenges and prospects. *Malaysian Journal of Halal Research* 2 (1). Walter de Gruyter GmbH: 6–9. doi:10.2478/mjhr-2019-0002.

Joshi, Himangini, Correspondence Himangini Joshi, Piyush Choudhary, and SL Mundra. 2019. Future prospects of nanotechnology in agriculture. *~957~ International Journal of Chemical Studies* 7 (2): 957–963.

Kabiri, Shervin, Fien Degryse, Diana N H Tran, Rodrigo C da Silva, Mike J McLaughlin, and Dusan Losic. 2017. Graphene oxide: A new carrier for slow release of plant micronutrients. *ACS Applied Materials & Interfaces* 9 (49). ACS Publications: 43325–42335.

Kalia, Anu, and Harsimran Kaur. 2019. Nano-biofertilizers: Harnessing dual benefits of nano-nutrient and bio-fertilizers for enhanced nutrient use efficiency and sustainable productivity. *Nanoscience for Sustainable Agriculture*. Springer, 51–73.

Kamiab, F, and E Zamanibahramabadi. 2016. The effect of foliar application of nano-chelate super plus ZFM on fruit set and some quantitative and qualitative traits of almond commercial cultivars. *Journal of Nuts* 7 (01). Islamic Azad University, Damghan Branch: 9–20.

Karthik, A, and M Uma Maheswari. 2021. Smart fertilizer strategy for better crop production. *Agricultural Reviews* 42 (1). Agricultural Research Communication Centre: 12–21.

Khan, M Iqbal R, Sudhakar R Palakolanu, Priyanka Chopra, Ashish B Rajurkar, Ravi Gupta, Noushina Iqbal, and Chirag Maheshwari. 2021. Improving drought tolerance in rice: Ensuring food security through multi-dimensional approaches. *Physiologia Plantarum* 172 (2). Wiley Online Library: 645–668.

Khan, Zulqurnain, Sultan Habibullah Khan, Muhammad Zubair Ghouri, Hina Shahzadi, Sarmad Frogh Arshad, Ummara Waheed, Summra Firdous, and Aftab Ahmad. 2019. "Nanotechnology: An Elixir to Life Sciences." Preprints.

Krutyakov, Yurii A, Alexey A Kudrinsky, Alexander A Gusev, Olga V Zakharova, Alexey I Klimov, Alexey D Yapryntsev, Pavel M Zherebin, Olga A Shapoval, and Georgii V Lisichkin. 2017. Synthesis of positively charged hybrid PHMB-stabilized silver nanoparticles: The search for a new type of active substances used in plant protection products. *Materials Research Express* 4 (7). IOP Publishing: 075018.

Kumari, Rani, Manoj Kundu, Anupam Das, Rajiv Rakshit, Sanjay Sahay, Samik Sengupta, and Md Feza Ahmad. 2020. Long-term integrated nutrient management improves carbon stock and fruit yield in a subtropical mango (*Mangifera Indica* L.) orchard. *Journal of Soil Science and Plant Nutrition* 20. Springer: 725–737.

Kuzma, J, P VerHage, and P C Trusts. 2006. Nanotechnology in agriculture and food production: Anticipated applications. Project on Emerging Nanotechnologies. *Journal of Nanotechnology*. 2: 48–53.

Lianou, A, E Z Panagou, and G-JE Nychas. 2016. Microbiological spoilage of foods and beverages. In *The Stability and Shelf Life of Food*, 3–42. Elsevier, Woodhead Publishing.

Madbouly, Adel K, Mohamed S Abdel-Aziz, and Mosaad A Abdel-Wahhab. 2017. Biosynthesis of nanosilver using chaetomium globosum and its application to control fusarium wilt of tomato in the greenhouse. *IET Nanobiotechnology* 11 (6). Wiley Online Library: 702–728.

Madhavan, Aravind, Raveendran Sindhu, Parameswaran Binod, Rajeev K Sukumaran, and Ashok Pandey. 2017. Strategies for design of improved biocatalysts for industrial applications. *Bioresource Technology* 245. Elsevier: 1304–1313.

Madhu, B, S Patel, P V K Jagannadha Rao, and P Sreedevi. 2018. Use of edible coatings to increase the shelf life of jaggery: A review. *International Journal of Current Microbiology and Applied Sciences* 7 (6): 2466–2479.

Mahmoud, S H, and A M M El-Tanahy. 2022. Effect of carbon nanoparticles in biochar and sulphur as a foliar spray on onion plants: A new orientation. *Gesunde Pflanzen*. Springer, 1–8.

Manikandan, Appu, and Muthukrishnan Sathiyabama. 2016. Preparation of chitosan nanoparticles and its effect on detached rice leaves infected with *Pyricularia grisea*. *International Journal of Biological Macromolecules* 84. Elsevier: 58–61.

McHugh, T H, and E Senesi. 2000. Apple wraps: A novel method to improve the quality and extend the shelf life of fresh-cut apples. *Journal of Food Science* 65 (3). Wiley Online Library: 480–485.

Mejias, J H, F Salazar, L Pérez Amaro, S Hube, M Rodriguez, and M Alfaro. 2021. Nanofertilizers: A cutting-edge approach to increase nitrogen use efficiency in grasslands. *Frontiers in Environmental Science* 9. Frontiers: 52.

Misra, Pragati, Pradeep Kumar Shukla, Krishnendu Pramanik, Sanghdeep Gautam, and Chittaranjan Kole. 2016. Nanotechnology for crop improvement. In Chittaranjan Kole (ed.), *Plant Nanotechnology: Principles and Practices*, 219–256. Springer International Publishing. doi:10.1007/978-3-319-42154-4_9.

Misson, Mailin, Hu Zhang, and Bo Jin. 2015. Nanobiocatalyst advancements and bioprocessing applications. *Journal of the Royal Society Interface* 12 (102). The Royal Society: 20140891.

Miteluț, Amalia Carmen, Elisabeta Elena Popa, Mihaela Cristina Drăghici, Paul Alexandru Popescu, Vlad Ioan Popa, Oana-Crina Bujor, Violeta Alexandra Ion, and Mona Elena Popa. 2021. Latest developments in edible coatings on minimally processed fruits and vegetables: A review." *Foods* 10 (11). MDPI: 2821.

Mohamed, Heba I, Tony Kevork Sajyan, Roshan Shaalan, Rami Bejjani, Youssef Najib Sassine, and Abdul Basit. 2022. Plant-mediated copper nanoparticles for agri-ecosystem applications. In Kamel A. Abd-Elsalam (ed.), *Agri-Waste and Microbes for Production of Sustainable Nanomaterials*, 79–120. Elsevier.

Montaño-Herrera, Anay, Yair Olovaldo Santiago-Saenz, César Uriel López-Palestina, Gregorio Cadenas-Pliego, Zeus H Pinedo-Guerrero, Alma Delia Hernández-Fuentes, and José Manuel Pinedo-Espinoza. 2022. Effects of edaphic fertilization and foliar application of Se and Zn nanoparticles on yield and bioactive compounds in *Malus Domestica* L. *Horticulturae* 8 (6). MDPI: 542.

Mustafa, Fatima, and Silvana Andreescu. 2020. Nanotechnology-based approaches for food sensing and packaging applications. *RSC Advances* 10 (33). Royal Society of Chemistry: 19309–19336.

Nejatzadeh, Fatemeh. 2021. Effect of silver nanoparticles on salt tolerance of satureja hortensis l. during in vitro and in vivo germination tests. *Heliyon* 7 (2). Elsevier: e05981.

Nematian, Tahereh, Zeinab Salehi, and Alireza Shakeri. 2020. Conversion of bio-oil extracted from chlorella vulgaris micro algae to biodiesel via modified superparamagnetic nano-biocatalyst. *Renewable Energy* 146. Elsevier: 1796–1804.

Nongbet, Amilia, Awdhesh Kumar Mishra, Yugal Kishore Mohanta, Saurov Mahanta, Manjit Kumar Ray, Maryam Khan, Kwang-Hyun Baek, and Ishani Chakrabartty. 2022. Nanofertilizers: A smart and sustainable attribute to modern agriculture." *Plants* 11 (19). MDPI: 2587.

Noorbakhsh-Soltani, S M, M M Zerafat, and S Sabbaghi. 2018. A comparative study of gelatin and starch-based nano-composite films modified by nano-cellulose and chitosan for food packaging applications. *Carbohydrate Polymers* 189. Elsevier: 48–55.

Oerke, E-C, and H-W Dehne. 2004. Safeguarding production-losses in major crops and the role of crop protection. *Crop Protection* 23 (4). Elsevier: 275–285.

Oliveira, Jhones L De, Estefânia V R Campos, Taís Germano-Costa, Renata Lima, Jaqueline Franciosi Della Vechia, Sidneia Terezinha Soares, Daniel Junior de Andrade, Kelly Cristina Gonçalves, Joacir do Nascimento, and Ricardo Antonio Polanczyk. 2019. Association of zein nanoparticles with botanical compounds for effective pest control systems. *Pest Management Science* 75 (7). Wiley Online Library: 1855–1865.

Oliveira, Jhones Luiz de, Estefânia Vangelie Ramos Campos, Mansi Bakshi, P C Abhilash, and Leonardo Fernandes Fraceto. 2014. Application of nanotechnology for the encapsulation of botanical insecticides for sustainable agriculture: Prospects and promises. *Biotechnology Advances* 32 (8). Elsevier: 1550–1561.

Ouzakar, Sanaa, Nadia Skali Senhaji, Mohamed Zoubair Saidi, Mustapha El Hadri, Aicha El Baaboua, Abdelatif El Harsal, and Jamal Abrini. 2023. Antibacterial and antifungal activity of zinc oxide nanoparticles produced by phaeodactylum tricornutum culture supernatants and their potential application to extend the shelf life of sweet cherry (*Prunus avium* L.). *Biocatalysis and Agricultural Biotechnology* 49. elsevier: 102666.

Ozturk, Munir, Bengu Turkyilmaz Unal, Pedro García-Caparrós, Anum Khursheed, Alvina Gul, and Mirza Hasanuzzaman. 2021. Osmoregulation and its actions during the drought stress in plants. *Physiologia Plantarum* 172 (2). Wiley Online Library: 1321–1335.

Pimentel, David. 2009. Pesticides and pest control. In *Integrated Pest Management: Innovation-Development Process*: Volume *1*, edited by Rajinder Peshin and Ashok K Dhawan, 83–87. Dordrecht: Springer Netherlands. doi:10.1007/978-1-4020-8992-3_3.

Prakash, Chander, Sunpreet Singh, B S Pabla, and M S Uddin. 2018. Synthesis, characterization, corrosion and bioactivity investigation of nano-ha coating deposited on biodegradable Mg-Zn-Mn alloy. *Surface and Coatings Technology* 346. Elsevier: 9–18.

Prasad, Ram. n.d. Nanotechnology in the Life Sciences Series Editor. https://www.springer.com/series/15921.

Qamar, Sarmad Ahmad, Muhammad Asgher, and Muhammad Bilal. 2020. immobilization of alkaline protease from bacillus brevis using Ca-Alginate entrapment strategy for improved catalytic stability, silver recovery, and dehairing potentialities. *Catalysis Letters* 150. Springer: 3572–3583.

Qiao, Bing, Yong Liang, Ting-Jie Wang, and Yanping Jiang. 2016. Surface modification to produce hydrophobic nano-silica particles using sodium dodecyl sulfate as a modifier. *Applied Surface Science* 364. Elsevier: 103–109.

Rahemi, Majid, Saeid Reza Gharechahi, and Sahar Sedaghat. 2020. The application of nano-iron chelate and iron chelate to soil and as foliar application: Treatments against chlorosis and fruit quality in quince. *International Journal of Fruit Science* 20 (3). Taylor & Francis: 300–313.

Raliya, Ramesh, Vinod Saharan, Christian Dimkpa, and Pratim Biswas. 2017. Nanofertilizer for precision and sustainable agriculture: Current state and future perspectives. *Journal of Agricultural and Food Chemistry* 66 (26). ACS Publications: 6487–6503.

Rana, Ruhul Amin, Md Siddiqui, Milan Skalicky, Marian Brestic, Akbar Hossain, Emrul Kayesh, Marek Popov, Vaclav Hejnak, Dipali Rani Gupta, and Nur Uddin Mahmud. 2021. Prospects of nanotechnology in improving the productivity and quality of horticultural crops. *Horticulturae* 7 (10). Multidisciplinary Digital Publishing Institute: 332.

Ranjbar, Shiva, Asghar Ramezanian, and Majid Rahemi. 2020. Nano-calcium and its potential to improve 'Red delicious' apple fruit characteristics. *Horticulture, Environment, and Biotechnology* 61. Springer: 23–30.

Refaai, M M. 2014. Response of zaghloul date palms grown under minia region conditions to spraying wheat seed sprout extract and nano-boron. *Stem Cell* 5 (4): 22–28.

Reshmy, R, Eapen Philip, Ranjna Sirohi, Ayon Tarafdar, K B Arun, Aravind Madhavan, Parameswaran Binod, Mukesh Kumar Awasthi, Sunita Varjani, and George Szakacs. 2021. Nanobiocatalysts: Advancements and applications in enzyme technology. *Bioresource Technology* 337. Elsevier: 125491.

Rossi, Lorenzo, Weilan Zhang, and Xingmao Ma. 2017. Cerium oxide nanoparticles alter the salt stress tolerance of *Brassica Napus* L. by modifying the formation of root apoplastic barriers. *Environmental Pollution* 229. Elsevier: 132–138.

Sarwar, Nadeem, Saifullah, Sukhdev S Malhi, Munir Hussain Zia, Asif Naeem, Sadia Bibi, and Ghulam Farid. 2010. Role of mineral nutrition in minimizing cadmium accumulation by plants. *Journal of the Science of Food and Agriculture* 90 (6). Wiley Online Library: 925–937.

Satalkar, Priya, Bernice Simone Elger, and David M Shaw. 2016. Defining nano, nanotechnology and nanomedicine: Why should it matter? *Science and Engineering Ethics* 22. Springer: 1255–1276.

Saxena, Raghvendra, Rajesh Singh Tomar, and Manish Kumar. 2016. Exploring nanobiotechnology to mitigate abiotic stress in crop plants. *Journal of Pharmaceutical Sciences and Research* 8 (9). Journal of Pharmaceutical Sciences and Research: 974.

Seleiman, Mahmoud F, Wadei A Al-Selwey, Abdullah A Ibrahim, Mohamed Shady, and Abdullah A Alsadon. 2023. Foliar applications of ZnO and SiO_2 nanoparticles mitigate water deficit and enhance potato yield and quality traits. *Agronomy* 13 (2). MDPI: 466.

Shafiq, Fahad, Sumera Anwar, Arslan Mahmood, Muhammad Iqbal, Kaleem Ullah, Muhammad Zulqarnain, Imtiaz Haider, Muhammad Ashraf, and Lixin Zhang. 2023. Physiological effects of some engineered nanomaterials on radish (*Raphanus Sativus* L.) intercropped with pea (*Pisum Sativum* L.)." *Environmental Science and Pollution Research*. Springer, 1–14.

Shalaby, Tarek A, Emad Abd-Alkarim, Farouk El-Aidy, El-Samahy Hamed, Mohamed Sharaf-Eldin, Naglaa Taha, Hassan El-Ramady, Yousry Bayoumi, and André Rodrigues Dos Reis. 2021. Nano-selenium, silicon and H_2O_2 boost growth and productivity of cucumber under combined salinity and heat stress. *Ecotoxicology and Environmental Safety* 212. Elsevier: 111962.

Shalan, Ameer M. 2020. "Fertilization by Nano-Powder Potassium Sulfate Enhancing Production of Grapevines Cv. Crimson Seedless." *Journal of Plant Production* 11 (3). Mansoura University, Faculty of Agriculture: 207–13.

Shang, Y, M K Hasan, G J Ahammed, M Li, H Yin, and J Zhou. 2019. The use of nanotechnology in crop production and plant protection: A review. *Molecules* 24 (14). MDPI: 2558.

Shareef, Hussein J, Rashid A Al-Yahyai, Alaa El-Din K Omar, and Wan Arfiani Barus. 2020. Foliar nano-fertilization enhances fruit growth, maturity, and biochemical responses of date palm. *Canadian Journal of Plant Science* 101 (3). Canadian Science Publishing 1840 Woodward Drive, Suite 1, Ottawa, ON K2C 0P7: 299–306.

Sharma, Gaurav, Mu Naushad, Amit Kumar, Shailja Rana, Shweta Sharma, Amit Bhatnagar, Florian J Stadler, Ayman A Ghfar, and Mohammad Rizwan Khan. 2017. Efficient removal of coomassie brilliant blue R-250 dye using starch/poly (alginic acid-Cl-acrylamide) nanohydrogel. *Process Safety and Environmental Protection* 109. Elsevier: 301–310.

Sia, Paolo Di. 2017. Nanotechnology among innovation, health and risks. *Procedia-Social and Behavioral Sciences* 237. Elsevier: 1076–1080.

Sindhu, SatyavirSatyavir S, Ruchi Sharma, Swati Sindhu, and Manisha Phour. 2019. Plant nutrient management through inoculation of zinc-solubilizing bacteria for sustainable agriculture. *Biofertilizers for Sustainable Agriculture and Environment*. Springer, 173–201.

Singh, A, and S M Prasad. 2017. "Nanotechnology and Its Role in Agro-Ecosystem: A Strategic Perspective." *International Journal of Environmental Science and Technology* 14. Springer: 2277–2300.

Sinha, Rakesh Kumar, and Shiv Shankar Verma. 2021. Proteomics approach in horticultural crops for abiotic-stress tolerance. In *Stress Tolerance in Horticultural Crops*, 371–385. Elsevier, Woodhead publishing.

Song, Hao, Yusilawati Ahmad Nor, Meihua Yu, Yannan Yang, Jun Zhang, Hongwei Zhang, Chun Xu, Neena Mitter, and Chengzhong Yu. 2016. Silica nanopollens enhance adhesion for long-term bacterial inhibition. *Journal of the American Chemical Society* 138 (20). ACS Publications: 6455–6462.

Sonkaria, Sanjiv, Sung-Hoon Ahn, and Versha Khare. 2012. Nanotechnology and its impact on food and nutrition: A review. *Recent Patents on Food, Nutrition & Agriculture* 4 (1). Bentham Science Publishers: 8–18.

Subramanian, Kizhaeral S, Angamuthu Manikandan, Muthiah Thirunavukkarasu, and Christopher Sharmila Rahale. 2015. Nano-fertilizers for balanced crop nutrition. Nanotechnologies in Food and Agriculture. Springer, 69–80.

Tariq, Mohd, Shipra Choudhary, Harjeet Singh, Mohd Asif Siddiqui, Hirdesh Kumar, Asad Amir, and Neelesh Kapoor. 2021. Role of nanoparticles in abiotic stress. *Technology in Agriculture* 323. IntechOpen Rijeka, Croatia.

Tiwari, Shalini, and Charu Lata. 2018. Heavy metal stress, signaling, and tolerance due to plant-associated microbes: An overview. *Frontiers in Plant Science* 9. Frontiers Media SA: 452.

Tripathi, Durgesh Kumar, Vijay Pratap Singh, Sheo Mohan Prasad, Devendra Kumar Chauhan, and Nawal Kishore Dubey. 2015. Silicon Nanoparticles (SiNp) Alleviate Chromium (VI) Phytotoxicity in Pisum Sativum (L.) seedlings. *Plant Physiology and Biochemistry* 96. Elsevier: 189–198.

Umair Raza, Muhammad, Fozia Abasi, Muhammad Shahbaz, Maria Ehsan, Wajiha Seerat, Abida Akram, Naveed Iqbal Raja, Zia ur-Rehman Mashwani, Hammad Ul Hassan, and Jarosław Proćków. 2023. Phytomediated silver nanoparticles (AgNPs) embellish antioxidant defense system, ameliorating HLB-diseased 'kinnow' mandarin plants. *Molecules* 28 (5). MDPI: 2044.

Ureña Saborío, Hilary, Sergio Madrigal Carballo, Jorge Sandoval, José Vega Baudrit, and Alejandro Rodríguez Morales. 2017. Encapsulation of bacterial metabolic infiltrates isolated from different bacillus strains in chitosan nanoparticles as potential green chemistry-based biocontrol agents against radopholus similis. *Estados Unidos: Journal of Renewable Materials*. 5(3-4): 290–299.

Verma, Krishan K, Xiu-Peng Song, Abhishek Joshi, Vishnu D Rajput, Munna Singh, Anjney Sharma, Rajesh Kumar Singh, Dong-Mei Li, Jaya Arora, and Tatiana Minkina. 2022. Nanofertilizer possibilities for healthy soil, water, and food in future: An overview. *Frontiers in Plant Science* 13. 865048.

Vishekaii, Zohre Rohi, Ali Soleimani, Esmaeil Fallahi, Mahmood Ghasemnezhad, and Akbar Hasani. 2019. The impact of foliar application of boron nano-chelated fertilizer and boric acid on fruit yield, oil content, and quality attributes in olive (*Olea Europaea* L.). *Scientia Horticulturae* 257. Elsevier: 108689.

Wolf, Tanja, and Bettina Menne. 2007. *Environment and Health Risks from Climate Change and Variability in Italy.* World Health Organization. Regional Office for Europe.

Yadav, S K. 2010. Heavy metals toxicity in plants: An overview on the role of glutathione and phytochelatins in heavy metal stress tolerance of plants. *South African Journal of Botany* 76 (2). Elsevier: 167–179.

Yadav, S K, Subhash Babu, M K Yadav, Kalyan Singh, G S Yadav, and Suresh Pal. 2013. A review of organic farming for sustainable agriculture in Northern India. *International Journal of Agronomy* 2013. Hindawi.

Youssef, Ahmed M, Mohamed S Abdel-Aziz, and Samah M El-Sayed. 2014. Chitosan nanocomposite films based on Ag-NP and Au-NP biosynthesis by bacillus subtilis as packaging materials. *International Journal of Biological Macromolecules* 69. Elsevier: 185–191.

Zahedi, Seyed Morteza, Mostafa Abdelrahman, Marjan Sadat Hosseini, Narjes Fahadi Hoveizeh, and Lam-Son Phan Tran. 2019. Alleviation of the effect of salinity on growth and yield of strawberry by foliar spray of selenium-nanoparticles. *Environmental Pollution* 253. Elsevier: 246–258.

Zahedi, Seyed Morteza, Marjan Sadat Hosseini, Naghmeh Daneshvar Hakimi Meybodi, and Willie Peijnenburg. 2021. Mitigation of the effect of drought on growth and yield of pomegranates by foliar spraying of different sizes of selenium nanoparticles. *Journal of the Science of Food and Agriculture* 101 (12). Wiley Online Library: 5202–5213.

Zahedi, Seyed Morteza, Mahdieh Karimi, and Jaime A Teixeira da Silva. 2020. The use of nanotechnology to increase quality and yield of fruit crops. *Journal of the Science of Food and Agriculture* 100 (1). Wiley Online Library: 25–31.

Zambrano-Zaragoza, María L, David Quintanar-Guerrero, Ricardo M González-Reza, María A Cornejo-Villegas, Gerardo Leyva-Gómez, and Zaida Urbán-Morlán. 2021. Effects of UV-C and edible nano-coating as a combined strategy to preserve fresh-cut cucumber. *Polymers* 13 (21). MDPI: 3705.

Zhang, Heng, Yang Zhao, and Jian-Kang Zhu. 2020. Thriving under stress: How plants balance growth and the stress response. *Developmental Cell* 55 (5). Elsevier: 529–543.

Zhao, Jianguo, Zhaogeng Lu, Li Wang, and Biao Jin. 2020. Plant responses to heat stress: Physiology, transcription, noncoding RNAs, and epigenetics. *International Journal of Molecular Sciences* 22 (1). MDPI: 117.

Zulfiqar, Faisal, and Muhammad Ashraf. 2021. Nanoparticles potentially mediate salt stress tolerance in plants. *Plant Physiology and Biochemistry* 160. Elsevier: 257–268.

Zulfiqar, Faisal, Míriam Navarro, Muhammad Ashraf, Nudrat Aisha Akram, and Sergi Munné-Bosch. 2019. Nanofertilizer use for sustainable agriculture: Advantages and limitations. *Plant Science* 289. Elsevier: 110270.

6 Role of Digital Technologies in the Field of Horticultural Science and Technology

Tridip Boruah
Madhab Choudhury College, Barpeta

Meghali Kalita
Assam State Biodiversity Board

Sneha Hasnu
Madhab Choudhury College, Barpeta

Kaberi Saikia Das
Madhab Choudhury College, Barpeta

Rattan Singh
Guru Jambeshwar University of Science & Technology

Gulzar Ahmad Nayik
Govt. Degree College Shopian J&K

6.1 INTRODUCTION

Digital technologies are those that carry out operations or offer services by utilizing digital information as well as processing. These technologies describe and work with data using binary code, which consists of zeros and ones. In today's world, digital technologies are everywhere, impacting not just our everyday lives but also industries and communication. The advent of digital technology has completely changed the way we live, work, and interact with the external world. Digital technologies are primarily intended to swiftly and inexpensively establish connections between people. With the aid of an enormous array of digital tools and services, people are able to connect with one another. While there are many advantages to using digital devices for

DOI: 10.1201/9781003412489-6

learning, there are also threats and obstacles that come with it for individuals (Kapur 2018). Unprecedented advances have resulted from the shift from analog to digital, improving communication, streamlining tasks, and increasing access to information. The latest extended phase of humankind's socioeconomic development is represented by digital technology, which includes its ubiquitous connectivity and potent artificial intelligence (AI) (Martin 2020). Digital technology is widely acknowledged for its ability to transform lives, economies, cultures, and society. Large volumes of information are exchanged via digital networks that have been established by the internet and related information and communications technologies. Information is currently the primary element around which technologies operate, in contrast to other developments in technology. Significant turning points in the development of computing and data processing may be found throughout the intriguing, centuries-long history of digital technology. Early mathematicians created concepts of binary counting and reasoning in ancient cultures, which is where digital technology got its start.

The rapid advancement of digital technology is causing revolutionary shifts in a number of industries. Recent developments have impacted many facets of AI, communication, computing, and other fields. Some of the noteworthy advancements include 5G technology, AI and machine learning, quantum computing, extended reality, cyber securities innovation, biotechnology and digital health, green technology, etc. (Ciarli et al. 2021). All of these new developments demonstrate how digital technology is evolving and how it is changing society. Going forward, a positive and inclusive digital future will depend heavily on ethical considerations, privacy concerns, and responsible technology deployment. Global entrepreneurship and its definitions have changed as a result of the remarkable digital revolution. Giones and Bram (2017) claimed that the rapidly changing nature of our society and economy due to digitalization has left the developing area of technology entrepreneurship research unable to keep up. Their aim was to draw attention to the opportunities that new technologies and entrepreneurship offer for researchers and entrepreneurs who are eager to delve deeper into the subject. Technology entrepreneurship, digital technology entrepreneurship, and digital entrepreneurship are said to be the three distinct phenomena they suggested to conceptualize and characterize. The last 10 years have seen a surge in the use of digital technology because of its connections to digital entrepreneurship, digital economies, digital social interaction, green economics, etc. Due to this, businesses now view things differently, which has led to the digitalization of strategic policies via blockchain technology. Bai (2021) observed the advantages that the ease of implementing digital technology has brought about changes in daily life processes and methods for completing various jobs. Transparency, centralization, and access to new markets have been identified as these benefits for the companies taking into account their implications, particularly when utilizing digital technology.

Digital approaches have fundamentally changed the horticultural landscape through the incorporation of technology into numerous facets of plant management and development. Precision agriculture is a key component of contemporary horticulture. In order to collect detailed information on temperature, moisture content, and soil conditions, global positioning system-based sensors, and drones are used (Ardieu et al. 2020). By using this data, farmers may make well-informed decisions about pest management, fertilization, and irrigation, which will eventually maximize crop health and resource efficiency. This integration is demonstrated by smart irrigation systems that use sensors to track soil moisture and meteorological factors. Precise watering schedules are made possible by these devices that ensure ideal plant development while saving water (Zaagsma 2013). Furthermore, crop management practices of farmers have changed dramatically with the introduction of remote control and monitoring systems. Farmers may maintain optimal conditions for plant development by remotely monitoring and controlling environmental factors in greenhouses or controlled settings using mobile applications or web platforms. Analysis of information and predictive modeling are utilized to make use of the abundance of data that is gathered from sensors and separate sources. Farmers may increase overall productivity by using this data to improve their crop planning, disease detection, and yield prediction decisions. Significant advancements in robotics have also led to the automation of operations, including planting, harvesting, and sorting (Arts et al. 2015). This lowers labor expenses while also improving the operational effectiveness of horticultural methods. With the advent of vertical cultivation and controlled environment agriculture, production in controlled environments may now take place year-round, thanks to innovations like hydroponics and aquaponics. These techniques optimize resource efficiency and space usage. Additionally, advances in crop monitoring software and applications offer real-time data on illnesses, nutritional deficits, and plant health, enabling prompt actions to maintain crop health. Additionally, horticulture has embraced blockchain technology, especially in supply chain management. It makes it possible to trace horticultural products transparently from the farm to the market, guaranteeing their authenticity and quality. Although recently a lot of literature has surfaced in the scientific databases regarding digital tools and techniques of horticulture, the scattered characteristics of this literature are creating more confusion than providing any particular directions for facilitating research in this area. This chapter is a sincere attempt to accumulate all the relevant authentic literature and present them under a single heading to aid in literature searching and designing research hypotheses to carry out research in this field.

6.2 HISTORICAL PERSPECTIVE

6.2.1 Evolution of Digital Technologies in Horticulture

The horticultural landscape is in the throes of a revolutionary transformation, propelled by the integration of cutting-edge digital technologies. This paradigm shift marks a departure from traditional cultivation practices once reliant on manual labor and observational methods. In the not-so-distant past, the horticultural tapestry was woven with the threads of human effort and empirical insights. However, the advent of digital technologies has bestowed upon horticulture a new epoch defined by precision, efficiency, and unprecedented innovation. This transformative journey has redefined the very essence of horticulture, unlocking possibilities that were once unimaginable. From the meticulous monitoring enabled by precision farming and machine construction to the deep integration of biotechnology, engineering, and information technology in smart horticulture (Zhang et al. 2023), each stride forward has reshaped the narrative of cultivation. The historical journey of technology in horticulture commences with notable contributions from many visionaries like Budd and Buckly. In the late 19th century, Budd's pioneering work on grafting techniques laid the groundwork for advancements in plant cultivation. Buckly, in 1939, further enriched horticulture by introducing controlled environments (Nemali 2022) setting the stage for a technological integration that would redefine the field. The mid-20th century witnessed the inaugural wave of mechanization in agriculture, with tractors and machinery heralding a departure from manual labor. This era marked the early roots of digital technologies in horticulture, fostering efficiency and productivity. However, it was in the late 20th century that the integration gained momentum, aligning with the broader evolution of computing and information technologies. The real turning point materialized in the 1970s with the emergence of computers and data processing, propelling horticulture into an era of unprecedented digitalization. This narrative unfolds as a testament to the dynamic interplay between technological innovation and horticultural practices, showcasing a rich tapestry woven with contributions from pioneers across centuries.

Visionaries like Smith et al., (1978) and Bin and Johnson (1982) heralded the digital era in horticulture, envisioning the transformative potential of computers in managing vast crop datasets. Though their early systems were rudimentary, they sowed the seeds for the integration of sophisticated digital tools into horticultural practices. Visionary researchers conducted groundbreaking studies, delving into the utilization of satellite imagery to monitor crop health and evaluate vegetation indices (Jensen 1983; Rouse 1974). These developments laid the groundwork for the use of digital tools in precision agriculture. The 1990s marked a turning point with the rise of precision agriculture, a concept that leverages technology to optimize field-level management with regard to crop farming. In that era, the significance of remote sensing in monitoring the well-being of crops and assessing soil conditions was underscored (Blackmore et al. 1997). Satellite imagery and sensor technologies became integral components of precision agriculture, enabling farmers to make data-driven decisions for enhanced productivity and resource management. The ability to collect and analyze this data marked a turning point in horticultural practices. The integration of geographic information systems (GISs) and global positioning system (GPS) technologies allowed farmers to map their fields with unprecedented accuracy. This spatial data became instrumental in creating precision farming strategies, enabling farmers to tailor their approach to specific areas within a field. The ability to collect and analyze data on soil composition, weather patterns, and crop health allowed farmers to make informed decisions, optimizing resource use and improving overall productivity (Heckman et al. 2009).

As the 21st century unfolded, there was a significant surge in the development of sensor technologies tailored for horticulture. Sensors equipped with IoT capabilities became integral for real-time monitoring of environmental conditions, soil moisture, and plant health. Drones equipped with advanced imaging systems provide a bird's eye view of fields, enabling farmers to make data-driven decisions (Anderson and Gaston 2013). These devices provided farmers with actionable insights, enabling them to make data-driven decisions for irrigation, pest control, and fertilizer application. Exploration into real-time monitoring of soil moisture, temperature, and nutrient levels involved investigating the utilization of sensors (Gonzalez-Dugo et al. 2010; Liakos et al. 2018). These advancements allowed for precise irrigation and fertilization strategies, reducing waste and environmental impact. The authors explored the implementation of smart sensors for real-time monitoring of soil conditions, weather parameters, and crop health, facilitating precise interventions. The use of remote sensing for crop monitoring and the development of sophisticated IoT sensors for real-time data collection have become integral components of digital horticultural practices (Thenkabail et al. 2000; Pathan et al. 2019).

In recent years, as computing power increased and data storage became more affordable, the horticultural sector saw a surge in the use of big data analytics. Decision support systems (DSSs), powered by advanced algorithms, enabled farmers to process vast amounts of information for better decision-making. These systems incorporated data on weather forecasts, market trends, and historical crop performance to guide farmers in crop planning, risk management, and resource allocation (Srinivasan et al. 2010). The integration of robotics and automation in horticulture marked another milestone. From automated harvesting machines to robotic weeders, these technologies addressed labor shortages and increased operational efficiency. Robotic systems equipped with computer vision and machine-learning algorithms could identify ripe fruits, assess crop health, and execute precise tasks with minimal human intervention (Griepentrog et al. 2011). This amalgamation of technologies enhances the precision and efficiency of horticultural practices. Digital technologies also revolutionized plant health monitoring. Traditional methods of visually inspecting crops for diseases and pests were complemented and, in some cases, replaced by sophisticated imaging technologies. Drones equipped with multispectral cameras allow for rapid and comprehensive monitoring of large agricultural landscapes, enabling early detection of potential issues and facilitating targeted interventions (Sugiura et al. 2016). Recently, breakthroughs in genome editing technologies, such as CRISPR-Cas9, have significantly impacted horticulture. Researchers and breeders can now precisely modify plant genomes to enhance desirable traits such as disease resistance, yield, and nutritional content (Zhang et al. 2016). Biotechnological advancements have also led to the development of genetically modified crops with improved characteristics, contributing to sustainable horticultural practices. The historical journey of digital technologies in horticulture reflects a relentless pursuit of innovation to meet the demands of a growing global population. From humble beginnings in precision agriculture to the integration of advanced biotechnological tools, each phase has contributed to a more sustainable, efficient, and productive horticultural industry. As we stand at the intersection of agriculture and technology, the ongoing evolution promises a future where digital advancements continue to reshape the landscape of horticulture.

6.2.2 Impact of Previous Technological Shifts in Horticulture

Horticulture, as a crucial component of agriculture, has experienced dynamic changes spurred by technological innovations. Historical shifts in horticulture reflect not only changes in cultivation methods but also broader societal, economic, and environmental impacts. The adoption of new technologies, such as digitalization, automation, and greenhouse technology, has reshaped the operations of the horticulture industry, leading to reliable and optimized food production. These technological advancements have also addressed challenges such as climate change, fluctuating weather, and temperature, ultimately enhancing production, increasing yields, and satisfying market demands. Several challenges plague the field of horticulture, encompassing issues such as soil fertility decline, the impact of global warming, escalating land prices, water scarcity, and the missed potential of low-cost labor, all of which pose substantial threats to the sector's growth. Horticultural crops face formidable abiotic stresses, including inundation, temperature extremes, salinity, pH imbalances, and drought. The repercussions of global warming, such as reduced overall precipitation and accelerated snowmelt, contribute to severe drought conditions. These environmental changes give rise to a myriad of stress-related problems, including soil infertility, flooding, heightened evapotranspiration, low soil-oxygen levels, diminished production, nutrient leaching, and delayed planting. In tandem with rising temperatures, global warming has paradoxically contributed to unprecedented cold weather records. The prevalence of land salinity and irrigation-water salinity further compounds the challenges confronting horticultural production (Lamm et al. 2021). Seedling transplantation is a common practice in horticultural production, where the quality of seedlings directly influences the yield and overall quality of horticultural products. Unfortunately, the seedling industry faces significant challenges, including poor production due to subpar seeds and inappropriate cultivation practices. Horticultural crops are often grown in controlled environments, yet the current methods of protected cultivation rely heavily on mechanical, semi-mechanical, and manual labor, lacking smart and efficient approaches. This limitation significantly hampers the quality of seedlings produced. In the realm of protected cultivation within controlled environments, the dominance of mechanical, semi-mechanical, and manual labor underscores the need for smarter cultivation practices (Zhang et al. 2023). Furthermore, the intricate characteristics of horticultural crops necessitate increased manual labor (Sam et al. 2022). This lack of sophistication poses a substantial barrier to achieving optimal seedling quality. Concurrently, consumers express a preference for horticultural products that are not only ripe but also boast high nutritional content, exquisite flavor, and significant ornamental value. To meet these consumer demands, proper harvest procedures, transportation, and storage become paramount. Harvest timing, in particular, emerges as a crucial consideration, and the subsequent transportation and storage processes play pivotal roles in determining the ultimate quality of horticultural products. Furthermore, there is a growing focus on nutrition and health. The human dependence on horticultural crops remains substantial, serving as vital sources of protein, carbohydrates, vitamins, minerals, fiber, fat, micronutrients, and antioxidants (Gao et al. 2022). Yet, the relentless pursuit of elevated yields and profits has led to the misuse of agrochemicals, profoundly impacting both crop quality and food safety. This misuse poses a significant threat to the environment and human health throughout the entire food chain. Consequently, there is an imperative need for horticultural products to not only exhibit high nutritional quality but also to be cultivated with reduced reliance on chemical pesticides and fertilizers (Li et al. 2022).

The horticulture sector grapples with ongoing challenges posed by emerging pests and pathogens, compounded by the escalating issue of pesticide resistance. Consumer expectations of flawless fresh vegetable and fruit produce clash with the demand for food free from pesticide residues. The surge in global trading of planting material and fresh produce heightens the risk of introducing alien pests and pathogens into unfamiliar environments. Simultaneously, growers face constraints as the availability of "conventional" pesticides diminishes due to progressively stringent regulations. In light of these challenges, there is an urgent call for innovative methods to effectively combat both new and existing pests and diseases (Xu 2022). The inadequate ventilation, coupled with elevated temperatures and humidity levels, renders the environment in plant production systems harmful to human health. In the current landscape, accommodating the burgeoning global population emerges as a paramount challenge. Nonetheless, projections from the United Nations (2022) indicate a substantial growth in the global population over the next three decades, anticipating an increase of 2.0 billion from 8.0 billion in 2022 to 9.7 billion by 2050, reaching a peak of 10.4 billion in the 2080s. Consequently, there arises an imperative to augment production, achieved either directly or indirectly by expanding cultivated areas, enhancing productivity, or curbing post-harvest spoilage (the concealed harvest). In this context, horticultural crops emerge as a standout solution, surpassing other crop groups in productivity, making them pivotal in addressing the escalating demands of a population that is nearly 5.4 times higher (Tiwari et al. 2015). The comprehensive study encompasses an array of horticultural crops, including fruit trees, vegetables, flowers, aromatic plants, medicinal plants, spices, and plantation crops. Through the application of common tactics, precision farming, and advanced machinery, farms undergo a transformative shift, becoming not only more profitable but also safer and less environmentally detrimental. This synergy of innovation in agriculture not only meets the immediate challenge of food production but also ensures a sustainable and efficient approach to feeding the ever-growing global populace. Hence, this necessitates the establishment of a novel development system characterized by intelligent breeding, sophisticated cultivation practices, efficient transportation, and innovative sales strategies—all grounded in emerging technologies. Among the various concepts, the prowess of Fourth Industrial Revolution or Industry 4.0 technology manifests in its diverse applications, harnessing the synergies of cutting-edge technologies like the IoT, AI, blockchain, robotics, digital twins, and big data. The integration of Industry 4.0 with horticulture presents a promising prospect for the evolution of industrial agriculture into the next era, aptly termed Horticulture 4.0. Beyond this transformative potential, Industry 4.0 stands poised to actualize a sustainable and intelligent ecosystem, facilitating real-time management of farming activities, automation, and intelligent decision-making, as well as the instantaneous processing and analysis of data. This convergence promises not only increased efficiency but also a paradigm shift toward more informed, responsive, and sustainable practices in horticulture (Bhavana and Bhagwan 2021).

6.3 NURTURING INNOVATION: A COMPREHENSIVE LOOK AT DIGITAL TECHNOLOGIES IN HORTICULTURE

In the contemporary landscape of horticulture, the infusion of digital technologies has ushered in an era of unprecedented innovation and transformation. The symbiotic marriage of agriculture and cutting-edge digital tools presents a compelling narrative, one that transcends traditional methodologies and reshapes the very fabric of horticultural practices. This comprehensive exploration navigates through the myriad facets of digital technologies in horticulture, delving into the realms of precision farming, automated cultivation, and data-driven decision-making. From the meticulous monitoring of soil health to the orchestration of smart planting and harvesting through advanced robotics, the integration of digital technologies transcends mere mechanization, evolving into a holistic approach that harmonizes with nature while maximizing efficiency. The profound impact of these technologies extends beyond the fields, encompassing supply chain transparency, traceability, and consumer-driven demands for sustainability. As we stand at the nexus of technology and horticulture, this discourse seeks to unravel the intricate layers of this digital revolution, shedding light on its implications for productivity, environmental stewardship, and the cultivation of a future where beauty and functionality converge seamlessly in the artistry of sustainable agriculture. The emergence of Horticulture 4.0 is a contemporary and innovative development in the field, representing a concept that has gained prominence in the very recent past. Horticulture has undergone distinct evolutionary phases, each marking a significant transition in industry practices. The inception of Horticulture 1.0 denotes the traditional era reliant on manual human labor and animal assistance, utilizing rudimentary tools such as sickles and shovels. Throughout this period, labor-intensive methods dominated the landscape. The 20th century witnessed the advent of mechanization or semi-mechanization in Horticulture 2.0, revolutionizing production efficiency. Concurrently, experimentally designed breeding methodologies gained prominence. The late 20th century ushered in Horticulture 3.0, characterized by the rapid integration of biotechnology, computing, and electronic advancements. Automation equipment and molecular breeding techniques were introduced, marking a significant leap in technological applications. Around 2010, the nascent phase of Horticulture 4.0 emerged, laying the groundwork for subsequent robust development. In recent years, the vigorous advancement of emerging technologies, including AI and new-generation breeding methods, has propelled the industry further into the era of Horticulture 4.0 (Zhang et al. 2023). Within the ambit of horticultural development, various digitalized technologies are shown in Figure 6.1.

6.3.1 Indoor Vertical Farming

Indoor vertical farming has emerged as a groundbreaking solution to address the challenges posed by traditional agricultural practices, offering a sustainable alternative to crop cultivation. This innovative approach involves the cultivation of crops in vertically stacked layers, often within controlled indoor environments. By optimizing the use of space and resources, indoor vertical farming minimizes the impact of external

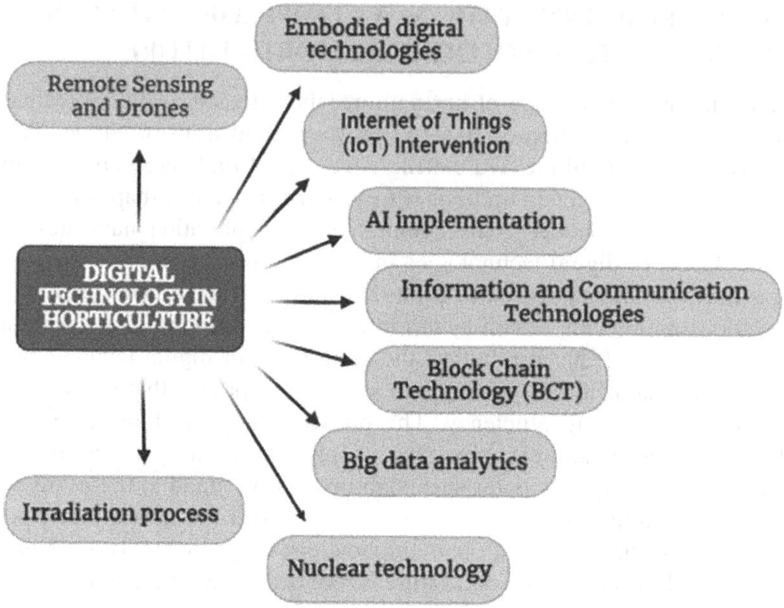

FIGURE 6.1 Digital technologies in horticulture.

environmental factors, allowing for year-round cultivation and enhanced crop yields. The utilization of advanced technologies, including hydroponics, aeroponics, and LED lighting systems, ensures precise control over factors such as nutrient delivery, humidity, and light exposure, thus fostering optimal conditions for plant growth. The rising trend of vertical farms housed within shipping containers signals a paradigm shift in modern agriculture. Employing standard 40-foot shipping containers as the vessel for these innovative farms enables seamless global transport of their produce. These containers, ingeniously transformed into self-sufficient vertical farms, boast a sophisticated setup that includes a drip irrigation system, state-of-the-art LED lighting, and vertically stacked shelving conducive to cultivating a diverse array of crops. What sets these self-contained units apart is their integration with a computerized growth management system, empowering users to remotely monitor and control all aspects of the farming process through their smartphones or computers. The adaptability of vertical farming is underscored by its accommodation within urban landscapes, utilizing vacant buildings or incorporating this forward-thinking approach into new construction projects. This ingenious method involves artificially stacking plants in three dimensions, akin to the architectural design of skyscrapers, creating an efficient use of space. This fusion of agricultural innovation and sustainable architecture not only caters to the demand for fresh produce but also presents an urban solution to feeding burgeoning populations while repurposing existing structures and minimizing the environmental impact of food production (Wamiq et al. 2022).

In the realm of vertical farming, a pivotal commonality emerges as all enterprises strategically harness one of three innovative soilless techniques—hydroponics, aeroponics, or aquaponics—to deliver essential nutrients to their cultivated plants.

Each method contributes to the sustainability and efficiency of vertical farming, addressing the challenges associated with traditional soil-based cultivation. The ensuing exploration provides a nuanced understanding of these three cultivation systems, shedding light on their unique attributes and the transformative potential they hold in reshaping contemporary agricultural practices. Hydroponics is a soilless cultivation technique that involves growing plants in a nutrient-rich, water-based solution, rather than in soil. (Khan et al. 2020). This method allows for increased efficiency in resource use, as it requires less water and land compared to traditional farming methods. Hydroponic systems can be used for various plant species and have been extensively studied for research purposes, as well as in industrial applications (Nguyen et al. 2016). Hydroponic systems provide a more homogeneous nutrient medium to the plants, with less risk of soil-borne diseases and pests compared to soil-based systems. Hydroponics can be used for a wide range of plant species, including tomatoes, tobacco, and lettuce. Hydroponic systems are favored by scientists for studying nutrient requirements, plant responses to nutrient availability, and exposure to toxic elements. Hydroponics has been used in large-scale farming for crops such as tomatoes and lettuce.

Aeroponics, a cutting-edge soilless cultivation technique, has garnered significant attention in contemporary agriculture for its efficiency and resource optimization. This innovative method involves suspending plant roots in a nutrient-rich mist, allowing for precise delivery of essential elements required for growth (Wamiq et al. 2022). The plant's exposure to an oxygen-rich environment in aeroponics enhances nutrient absorption, promoting accelerated growth rates and increased yields (Garzon et al. 2023). Aeroponic systems, in comparison to the most efficient hydroponic systems, consume up to 90% less water, establishing them as the unequivocally superior and most efficient cultivation systems for vertical farms (Wamiq et al. 2022). This method offers many benefits compared to traditional soil-based methods, including faster growth rates, higher yields, less water consumption, fewer pests, and reduced need for fertilizers. A systematic review of technology in aeroponics explored the potential of technology to enhance sustainability by optimizing resource consumption, such as water and nutrients. The review also introduced the Technology Adoption and Integration in Sustainable Agriculture Model, which can detect and address any inefficiencies, minimizing waste and maximizing resource utilization. Aquaponics, a symbiotic integration of aquaculture and hydroponics, represents a sustainable and innovative approach to agricultural cultivation. Aquaponics represents a sustainable approach to cultivating both fish and vegetables, amalgamating the fundamentals of aquaculture and hydroponics. Within aquaponic setups, fish thrive in designated tanks, and the nutrient-rich water emanating from these tanks serves as organic fertilizer for plants nurtured in a substrate-free environment. Reciprocally, cultivated plants play a vital role in cleansing the water for the fish, establishing a self-contained, mutually beneficial ecosystem (Costa et al. 2018; Rana et al. 2019).

6.3.2 Remote Sensing and Drones

Remote sensing and drone technologies have become increasingly important in horticulture, offering a range of applications for precision agriculture and smart farming.

These techniques provide valuable data on crop health, soil conditions, and environmental factors, enabling farmers to make informed decisions and optimize their agricultural practices (Thenkabail et al. 2000). Remote sensing involves the acquisition of information about an object or phenomenon without physical contact (Anderson and Gaston 2013). Multispectral and hyperspectral imaging are remote sensing techniques that capture information beyond the visible spectrum. By analyzing different wavelengths, these technologies offer detailed insights into plant health, nutrient status, and disease prevalence (Thenkabail et al. 2016). Remote sensing proves highly valuable in assessing area coverage, as well as mapping and categorizing diverse land use and land cover features such as vegetation, soil, water bodies, forests, and human activities. It facilitates the extraction of information pertaining to an image's color, texture, tone, pattern, and alignment, among other characteristics. Particularly noteworthy is the near-infrared band (>700 nm) in satellite imagery. While this band is beyond the sensitivity of the human eye, it plays a crucial role as it reflects sunlight at an elevated rate from healthy plant sources. This heightened reflection results from the band's sensitivity to leaf cell structure and leaf water content, providing nuanced insights into the health and vitality of vegetation (Wamiq et al. 2022).

In remote sensing, various sensor systems can be broadly categorized into active and passive types, each playing a crucial role in vegetation-correlated remote sensing. Passive sensors, encompassing photographic, electro-optic radiometers, and thermal imaging systems, are widely utilized for crop monitoring. While these sensors provide valuable information, a notable limitation arises from their dependency on external factors such as lighting conditions (time of day and year, latitude) and weather conditions (cloud cover). Accurate measurements of solar radiation and corrections for atmospheric effects are essential for data acquired through passive sensors. On the other hand, active sensors operate using an artificial source of energy, offering greater control over the nature (wavelength, power, duration) of the energy source in active remote sensing. Radar systems and Light Detection and Ranging (LiDAR) are examples of active sensors that do not rely on sunlight or favorable weather conditions, making them less dependent on environmental factors. This independence allows active sensors to be more targeted and dedicated to remote sensing objectives, even functioning effectively at night or in challenging conditions such as haze, clouds, or rain (Alvino and Marino 2017).

Unmanned aerial vehicles (UAVs), commonly known as drones, have emerged as powerful tools in precision horticulture. Equipped with advanced sensors, drones provide high-resolution imagery for crop monitoring, pest detection, and yield prediction. Drones, armed with LiDAR or optical sensors, facilitate the collection of spatial data with multiple variables at a remarkably high resolution. These UAVs can integrate weatherproof sensors to ensure seamless data collection irrespective of weather conditions. Additionally, satellite data from optical sources like Sentinel-2 and Landsat 8/9 (capturing images every 5 days at a resolution of 10–30 m) and radar sensors such as Sentinel-1, NISAR, and ALOS-2 PALSAR-2 offer comprehensive wall-to-wall coverage. Furthermore, high-resolution imagery from satellites like World View and Pleiades swiftly supplements information at even finer resolutions. To enhance the model's comprehensiveness, open datasets accessible through open-access repositories; national open data services containing both current and historical data on meteorological, environmental, and agricultural variables; or datasets from Copernicus can be seamlessly incorporated. The envisioned approach involves synergizing satellite images with high-resolution drone imagery, boasting resolutions of approximately 3 cm. Through the implementation of AI algorithms, this synergy aims to achieve super-resolution, yielding exceptionally detailed data across entire satellite-covered regions. Subsequently, the collected data, stemming from both project-specific endeavors and open datasets, will be channeled into rigorous data analysis processes (Gebresenbet et al. 2023). UAVs, commonly known as drones, have emerged as powerful tools in precision horticulture. Equipped with advanced sensors, drones provide high-resolution imagery for crop monitoring, pest detection, and yield prediction. Drones, armed with LiDAR or optical sensors, facilitate the collection of spatial data with multiple variables at a remarkably high resolution. These UAVs can integrate weatherproof sensors to ensure seamless data collection irrespective of weather conditions. Additionally, satellite data from optical sources like Sentinel-2 and Landsat 8/9 (capturing images every 5 days at a resolution of 10–30 m) and radar sensors such as Sentinel-1, NISAR, and ALOS-2 PALSAR-2 offer comprehensive wall-to-wall coverage. Furthermore, high-resolution imagery from satellites like World View and Pleiades swiftly supplements information at even finer resolutions. To enhance the model's comprehensiveness, open datasets accessible through open-access repositories; national open data services containing both current and historical data on meteorological, environmental, and agricultural variables; or datasets from Copernicus can be seamlessly incorporated. The envisioned approach involves synergizing satellite images with high-resolution drone imagery, boasting resolutions of approximately 3 cm. Through the implementation of AI algorithms, this synergy aims to achieve super-resolution, yielding exceptionally detailed data across entire satellite-covered regions. Subsequently, the collected data, stemming from both project-specific endeavors and open datasets, will be channeled into rigorous data analysis processes (Girma et al. 2023).

6.3.3 Embodied Digital Technologies

Embodied digital technologies in horticulture refer to the integration of digital tools and techniques into the practice of horticulture, aiming to increase efficiency, enhance precision, and improve overall productivity. These technologies not only involve the use of devices but also the embodiment of digital understanding and skills in the horticultural practice (Velden et al. 2023). Embodied digital technologies employ diverse technological strategies to enhance the efficiency of managing materials and labor. Within this realm, technologies encompass a spectrum, ranging from global positioning satellites (GPS) and GIS to sensors, DSS, and precision agriculture equipment. Each of these elements contributes to the overarching goal of optimizing processes and operations in various sectors. The use of precision farming techniques, which involve the monitoring and controlling of various environmental factors, such as soil conditions, temperature, and moisture, to optimize crop growth and reduce resource consumption (Birner et al. 2021). The functionality of the DSS involves amalgamating data from diverse sources, including sensory inputs, farmer contributions, and information on crops and animals. This aggregated data undergoes analysis and is processed through a range of models, encompassing statistical and machine-learning approaches. The primary objective is to gain a deeper understanding of potential solutions for farm-specific challenges, such as water resource management, low-input farming, grazing plan optimization, livestock performance enhancement, health and welfare improvements, increased crop yield, and overall environmental conditions. Consequently, the data earmarked for analysis necessitates efficient retrieval, storage, and organization, ensuring a streamlined decision-making process tailored to specific agricultural contexts (Gebresenbet et al. 2023). Utilizing GIS-based DSS is a common practice in agriculture. These systems often integrate data from crop simulation models to determine the optimal input levels for specific field areas, ensuring maximum yield. Within a GIS framework, essential maps encompassing soil type, nutrient levels, moisture, pH, fertility, weed and pest intensity, and other key factors are integrated. These maps, along with additional decision-making tools, are then employed to apply precise dosages of fertilizers or pesticides. The implementation of variable rate technology in the field leverages GPS for accurate location-based positioning solutions. In the context of farming management, the mapping, interpretation, and correlation of yield with the spatial and temporal variability of agronomic variables play a crucial role in shaping strategies for the upcoming season (Lingireddy et al. 2023).

6.3.4 Internet of Things (IoT) Intervention

The horticulture industry has experienced a substantial transformation due to the influence of the IoT, presenting a multitude of advantages, including heightened productivity, improved sustainability, and more efficient resource management. In recent times, there has been a notable surge in the adoption of IoT within horticulture, driven by the overarching goal of enhancing productivity, minimizing waste, and augmenting yields. The core principle of IoT lies in linking devices to the internet, facilitating the collection of data crucial for the optimization of various processes. The IoT represents an expansive network of intelligent devices capable of self-organization, information exchange, data sharing, and responsive actions based on contextual and environmental shifts. At its core, IoT architecture encompasses distinct layers, namely the perception layer, access layer, network layer, middleware layer, and application layer. The perception layer assumes a pivotal role, leveraging sensors, actuators, and identification technologies to facilitate real-time information transmission and remote control through internet connectivity. Moving to the network layer, the architecture empowers data packet transmission using embedded communication protocols. Commonly employed protocols such as Wireless Fidelity (Wi-Fi), Bluetooth, and Zigbee play a significant role in establishing wireless sensor networks that seamlessly connect with the internet, forming the foundation of the IoT ecosystem (Nuruzzaman et al. 2021). Within the realm of the IoT, sensors emerge as pivotal components, facilitating the acquisition of crucial information from the soil and environment. These sensors encompass a spectrum of measurements, including temperature, soil moisture, humidity, light intensity, pH, Nitogen, Phosphorus and Potasium (NPK), and water level. Additionally, the incorporation of a camera module enhances data collection in horticulture fields. Notably, these sensors maintain consistency whether deployed in indoor or outdoor horticultural environments. The assimilation of soil and environmental data is instrumental in creating mapped representations, empowering smart control over irrigation and fertilizer applications to optimize productivity. The soil data, in turn, fuels crop yield analyses, guiding the judicious application of fertilizers as required. All the sensor-derived information is transmitted to a computing unit, where the data-sensor processor establishes communication with the cloud server through wireless protocols and gateways. The cloud server plays a central role in visualizing the data through a graphical user interface, offering a comprehensive overview of the soil and environmental conditions (Singh et al. 2022). Moreover, IoT techniques offer a promising approach to enhancing efficiency, precision, and sustainability in horticulture, ultimately contributing to global food security and agricultural sustainability.

6.3.5 AI Implementation

The implementation of AI in horticulture has emerged as a transformative force, revolutionizing traditional agricultural practices and paving the way for enhanced efficiency, precision, and sustainability. AI is a multidisciplinary technology capable of combining machine learning, cognition, human-computer interaction, emotion recognition, data storage, and decision-making (Arya and Gangwar 2021). AI integrates machine learning, computer vision, and data analytics to process and analyze vast datasets, providing valuable insights for optimizing agricultural practices. In horticulture, AI plays a crucial role in automating tasks, predicting crop yields, detecting diseases, monitoring environmental conditions, and making informed decisions to improve overall efficiency and productivity. The hurdle faced by AI was surmounted through the evolution of processing power, accompanied by the strides in deep-learning and advanced-learning methodologies driven by extensive datasets. The continual enhancement of graphics processing units (GPUs) has not only led to the successful creation of specialized processors but has also expanded the overall computational capacity, laying the foundation for the swift advancement of AI. Within this landscape, various AI techniques such as artificial neural networks (ANNs), decision support systems (DSSs), genetic algorithms (GAs), support-vector machines, and computer vision find widespread application in agriculture. These techniques play a pivotal role in managing aspects like soil conditions, crop health, disease detection, and weed control, contributing to the optimization of agricultural practices (Zhang and Lu 2021). The implementation of AI in horticulture offers a promising approach to enhancing efficiency, precision, and sustainability in the industry. By deploying AI models and integrating advanced technologies, horticulturists can optimize agricultural practices, improve crop production, and reduce resource consumption, ultimately contributing to global food security and agricultural sustainability.

6.3.6 Information and Communication Technologies

Horticultural crops benefit significantly from the integration of information and communication technologies (ICTs), representing a pivotal component in their development. These technologies offer versatile applications throughout the agricultural value chain, encompassing production, processing, distribution, and marketing. Their implementation, coupled with sensors and the IoTs, facilitates the monitoring and control of various farming operations. Beyond this, ICTs play a crucial role in enhancing communication and collaboration among farmers, agribusinesses, and other stakeholders in the agriculture value chain. Mobile apps and social media platforms become valuable tools for disseminating vital information such as market prices, weather conditions, and other parameters influencing farming operations. The adoption of ICTs equips farmers with a comprehensive understanding of diverse facets of horticultural operations, empowering them to maintain optimal conditions with minimal effort, yielding high returns. This, in turn, fosters agribusinesses that are not only more efficient but also highly productive, economically viable, and ultimately successful (Lingireddy et al. 2023).

6.3.7 Block Chain Technology (BCT)

Blockchain, recognized as a distributed ledger technology, offers the distinct advantage of being resistant to tampering with information. Its potential lies in its ability to effectively tackle the challenge of resource allocation for transactions involving multiple untrustworthy parties within the supply chain for fresh fruit (Pass and Shi 2017). Blockchain technology emerges as a promising approach for achieving supply chain traceability in the pineapple industry. The introduced fruit-chain protocol mirrors the attributes of consistency and liveliness, akin to operating under the assumption of an honest majority of computer power (Zhang et al. 2022). While considering blockchain as a potential solution for traceability in the food chain, the investigation primarily aimed to delve deeper into understanding blockchain technology and its possible applications within the retail industry. Furthermore, the study anticipated identifying potential uses of blockchain that could benefit both the retail sector and the broader industry. Emphasizing the pivotal role of blockchain technology in the fruit retail sector and its interconnections, the research highlighted the significance of this technology in achieving traceability goals (Sharif et al. 2021). Facilitating the creation of a distributed network, blockchain technology establishes seamless connectivity among various entities within the supply chain. This enables real-time monitoring and tracking of activities from any location, ensuring the immutability and transparency of the recorded data. The integration of smart contracts between entities further empowers the establishment of digital and secure trading. Notably, blockchain serves as a crucial enabler for secure international transactions involving the export of fruits and vegetables. By employing secured hash cryptography, adherence to international standards and quality regulations for fruits and vegetables is ensured, safeguarding the integrity of the products in the global market (Singh et al. 2022). The integration of blockchain technology can enhance transaction efficiency and minimize the risk of disputes by establishing a transparent and immutable record of contracts. Beyond contract management, blockchain proves valuable in tracking policies and claims in a manner that is trustworthy and auditable. This, in turn, streamlines the delivery of insurance to farmers, ensuring effectiveness and timeliness. In the broader scope, blockchain has the potential to revolutionize transactions and supply chain management within the agriculture sector, fostering transparency, efficiency, and safety. This increased transparency not only builds trust among anonymous individuals but also addresses the challenges associated with the complex agri-food system's extensive supply chain. By offering decentralized and immutable record-keeping, blockchain not only instills confidence in the industry but also reduces the risk of fraud and related issues. Consequently, the technology provides solutions to concerns regarding food quality and safety, addressing the interests of customers, government authorities, and various stakeholders (Lingireddy et al. 2023).

6.3.8 Big Data Analytics

Big data analytics has emerged as a transformative tool in horticulture, revolutionizing the way information is processed and utilized in the cultivation of crops and management of horticultural practices. This sophisticated approach involves the collection, processing, and analysis of vast datasets generated throughout the horticultural value chain. As the volume of data accessible to field managers expands, the development of an automated mechanism for extracting operational insights becomes imperative. Big data analytics holds the potential to enhance various aspects of agricultural practices. This includes the adoption of precision agricultural techniques at the field level, the establishment of an optimized food supply chain, and the automation of the entire operation to maximize profitability. The utilization of big data analytics presents opportunities for improving efficiency, sustainability, and overall performance in the agricultural sector (Tsiolias et al. 2015). The application of big data analytics in extracting insights from vast climate datasets has brought attention to the distinctions between conventional big data techniques and the specialized methods and strategies essential for mining climate data (Rani 2017). To capture detailed ecological statistics in vegetable-growing regions, a comprehensive multi-sensor network system was implemented. Researchers utilized multidimensional information encompassing factors like environmental conditions, soil characteristics, and meteorology in vegetable fields. This multidimensional approach enabled the identification of critical elements influencing the spread of pests, leading to the development of a vegetable-pest warning system based on extensive big data (Zhang et al. 2018). The innovative nutrition-focused system for vegetable production and distribution leverages the advantages of a sophisticated big data framework. This not only provides wholesome food recommendations to end customers but also offers various predictive analyses to enhance overall system efficiency. Recognizing the significance of big data methodologies, the upcoming ICT horticulture project is set to heavily rely on them. It becomes crucial to comprehend their management and potential impact on day-to-day operations (Puri et al. 2017). Given the substantial quantities in which fruits and vegetables are produced, the sensor data generated in horticulture now qualifies as big data. This abundance of data can seamlessly be transmitted to cloud servers and accessed through wireless communication protocols, facilitating distribution box and control box connectivity. This integration underscores the pivotal role of big data in optimizing horticultural processes (Singh et al. 2022).

6.3.9 Nuclear Technology

Nuclear technology has been found to be an innovative and effective solution for addressing challenges in horticulture. While most people acknowledge the role of nuclear technology in power generation and its diversification, many remain unaware of its even greater impact in non-energy applications. Over the past two to three decades, atomic energy has significantly influenced daily life in ways not commonly realized. The substantial energy released in atomic fission finds applications beyond power generation, extending into various peaceful uses. The establishment of the International Atomic Energy Agency in 1957 marked a commitment to promoting the widespread use of radioisotopes and radiation sources across research, industry, agriculture, and medicine. Cultivation, within the realm of activities employing atomic energy for peaceful purposes, stands out as a crucial sector with profound social impact. Ionizing radiation, a form of nuclear technology aid, emerges as a sustainable alternative to numerous productivity-enhancing technologies, particularly those involving hazardous chemicals. Controlled applications of energy from gamma rays, electrons, and X-rays through ionizing radiation prove effective in enhancing the production and preservation of diverse crop products. In the field of horticulture, recent research has illuminated the potential applications of emerging nuclear energy for mutagenicity, pest control, increased agricultural yields, and improved post-harvest product quality. This underscores the continued evolution of nuclear technology beyond power generation, contributing to advancements in various sectors with positive implications for agriculture and society at large (Wamiq et al. 2022). Enhancing productivity, managing pests, and optimizing water conservation are all achievable through the application of nuclear techniques. Horticulture has effectively utilized various nuclear methods, including irradiation, isotopic and nuclear magnetic resonance, and tracer techniques, to elevate crop yield, enhance quality, and ensure safety. Irradiation emerges as a valuable tool for pest and disease control, while isotopic and nuclear magnetic resonance techniques provide insights into plant physiology and metabolism. Tracer techniques play a crucial role in studying nutrient uptake and improving water use efficiency in horticultural practices. These nuclear applications collectively contribute to advancing agricultural practices, fostering improved yields, and promoting sustainable approaches in horticulture (Chaurasia et al. 2023).

6.3.10 Irradiation Process

Irradiation stands as a pivotal procedure in horticulture, presenting a range of advantages. It serves as a phytosanitary treatment for agricultural products and food, offering a promising approach to disinfection for the control of pests and diseases. Utilizing gamma rays, X-rays, and UV light, irradiation can induce chromosomal breakages, cross-link DNA strands, delete and substitute nucleotides, and irradiate cell suspensions and pollen grains. Furthermore, the application of ionizing radiation in food irradiation contributes to food preservation by targeting pathogens and disrupting their DNA bonds. This process is instrumental in ensuring the safety and quality of fresh, pre-cut fruits, vegetables, and other minimally processed plant-based foods. Beyond food safety, irradiation plays a significant role in inducing beneficial mutations in crops through genetic modification, as well as in sterilization-based pest control and water usage management in horticulture (Chaurasia et al. 2023). The intentional irradiation of propagating material aims to bring about changes or improvements in its properties. This process involves placing the material in close proximity to a radiation source, such as Cobalt-60, for a specified duration during which the product is exposed to emitted radiation. A portion of the radiant energy reaching the product is absorbed by the material's tissues, the extent of which depends on its mass, composition, and exposure time (Vose 2013). Each product type requires a specific amount of radiant energy to achieve the desired effect, a value determined through comprehensive studies (Iaeaorg 2017). The optimal dose of ionizing radiation for inducing mutagenicity varies based on the genotype, the type of rays (gamma and X), and the processing device employed. Gamma-ray irradiation typically requires a lower dose compared to X-rays due to its superior penetration capabilities. While X-rays were commonly used in the early stages of nuclear technology in horticulture, gamma rays have now become the preferred choice for inducing mutagenesis in contemporary practices (Wamiq et al. 2022).

6.3.11 Cloud Computing

Cloud computing has emerged as a valuable tool in horticulture, particularly in the context of smart farming and indoor cultivation. It offers various benefits, including efficient management of resources, improved crop monitoring, and enhanced sustainability. Cloud computing revolutionizes the agricultural landscape by facilitating the delivery of computer services, including storage, processing, networking, analytics, and intelligence, via the Internet (the cloud), as opposed to relying on local servers or personal devices. This transformative technology empowers farmers to access and analyze data related to their crops and livestock from any location with an Internet connection, thereby enhancing efficiency, production, and sustainability in the agricultural sector. The capability of cloud computing to efficiently store and manage vast volumes of data proves instrumental in modernizing agriculture. Continuous monitoring of variables such as temperature, humidity, soil moisture, and air quality is made possible through sensors and equipment strategically placed in fields, greenhouses, and barns. The data collected is wirelessly transmitted to a centralized hub, uploaded to the cloud, and subsequently stored and analyzed using extensive cloud-based computer resources (Patil et al. 2012). This approach eliminates the need for costly servers, allowing farmers to access information remotely. Cloud computing's integration of analytics and intelligence tools enables farmers to derive valuable insights from data. Machine-learning algorithms within cloud computing contribute to farm management improvements by analyzing data on crop yields, soil moisture, and animal health. The versatility of cloud computing is further evident in the availability of numerous apps and programs that aid farmers in planning, budgeting, and forecasting. Agriculture experts also play a pivotal role by responding to farmer queries and updating the cloud service database with the latest agricultural research relevant to their expertise, thereby sharing specialized knowledge with other professionals (Gill et al. 2017). This paradigm shift in agriculture, facilitated by cloud computing, offers unparalleled convenience and mobility to farmers in data gathering and analysis. The accessibility to real-time information empowers farmers to make informed decisions, ultimately boosting efficiency, production, and longevity in the agricultural sector (Hori et al. 2010). In essence, cloud computing emerges as a game-changer, ushering in a new era of technological advancement for the benefit of farmers and the agricultural industry as a whole.

6.4 DIVERSE APPLICATIONS OF DIGITAL TECHNOLOGIES IN HORTICULTURE

Digital technologies have diverse applications in horticulture, offering solutions for crop monitoring, pest and disease management, irrigation, and post-harvest transportation. In recent years, the integration of digital technologies has ushered in a transformative era in horticulture, revolutionizing traditional agricultural practices and fostering advancements across various facets of crop cultivation and management. From precision agriculture and smart irrigation systems to blockchain-enabled traceability and genomic technologies, the diverse applications of digital technologies have become instrumental in optimizing efficiency, enhancing productivity, and promoting sustainability in horticulture. This paradigm shift not only empowers farmers with real-time data and insights but also contributes to the development of innovative solutions for challenges ranging from resource management to crop improvement. In this context, exploring the myriad ways in which digital technologies are reshaping horticulture provides a comprehensive understanding of the evolving landscape and the potential for continued growth and innovation in the field. Hydroponic farming aims to enhance crop yields, texture, and flavor while minimizing water usage. Plants grown hydroponically benefit from not requiring extensive root systems, allowing them to channel more energy into leaf and fruit development. This accelerated growth and improved defense against pests and diseases are inherent advantages of indoor hydroponic cultivation. From a sustainability standpoint, vertical farming offers the advantage of situating farms in or near densely populated areas, thereby reducing the necessity for extensive transportation and minimizing associated harmful emissions. The versatility of vertical farming extends to urban settings, making the production of fresh and locally sourced food easily accessible. Furthermore, this innovative farming method significantly reduces the land footprint required for crop cultivation when compared to traditional farming practices (Ashalatha and Rao 2023).

The application of remote sensing and drones in horticulture represents a paradigm shift in precision agriculture, offering innovative solutions to optimize crop management and enhance overall productivity. Remote sensing technologies, such as satellite imagery and aerial surveys conducted by drones, provide valuable insights into crop health, environmental conditions, and resource utilization. This transformative approach enables horticulturists to make data-driven decisions, monitor crops with unprecedented detail, and implement targeted interventions for improved yield and sustainability. By harnessing the power of these advanced technologies, the horticultural sector is poised to revolutionize traditional practices, ushering in a new era of efficiency, precision, and environmental stewardship. Farmers facing crop losses due to sudden weather changes can find relief through crop insurance, although instances of fraud do exist. To combat fraudulent claims, insurers employ advanced techniques, utilizing the red and infrared bands of satellite images along with the NDVI (normalized differential vegetation index) and crop testing. By cross-referencing this data, insurers can verify the authenticity of claims and ensure a fair and accurate assessment of losses (http://gisgeography.com). Remote sensing technologies prove invaluable in gauging soil moisture levels through the use of both active and passive sensors deployed from space. Active sensors provide high spatial resolution by illuminating the target and recording backscatter, while passive sensors measure naturally emitted electromagnetic waves with exceptional accuracy but lower resolution. A groundbreaking initiative by NASA, known as Soil Moisture Active Passive (SMAP), combines the strengths of both sensor types to deliver an optimized solution (http://gisgeography.com). In the realm of horticulture, remote sensing emerges as a powerful tool for identifying cultivated stands, determining total cultivated area, and estimating production yields to showcase the utility of remote sensing in predicting yields of various annual crops (Maja and Ehsani 2010; Usha et al. 2013). Moreover, remote sensing plays a vital role in mitigating production losses caused by pests and diseases. The early detection and identification of diseases, along with pest and nematode management, are facilitated through remote sensing's ability to detect changes in plant pigments, variations in pest-induced leaf patterns, and the identification of susceptible areas in plants (Usha et al. 2013). However, airborne multifaceted digital imaging system that assesses reflectance and canopy density under different degrees of phylloxera stress can also be detected by this technique (Johnson et al. 1996). Drones, with their robust capabilities, play a pivotal role in the fundamental task of analyzing soil and monitoring crop health. Essentially, UAVs efficiently collect and process data obtained from monitoring activities, contributing to the assessment, control, and maintenance of soil health. The application of drone technology extends beyond monitoring, as it facilitates the provision of essential nutrients to enhance soil health. Leveraging 3D mapping and data processing capabilities, drones excel in the analysis of soil health, enabling the precise spraying of fertilizers and insecticides to nurture crops and fulfill their nutritional requirements. This proactive approach results in robust and flourishing crops. Drone operators can actively oversee the application of fertilizers through drone spraying, effectively warding off insects, pests, and worms, thereby promoting the longevity of crops. Furthermore, the deployment of drone technology streamlines the labor-intensive process of sowing seeds for a diverse range of crops. An additional noteworthy advantage of utilizing drones for crop fertilization lies in their capacity to analyze, identify, and survey crops for any deficiencies. Drones contribute to the efficient planting of trees and crops, a task traditionally carried out by farmers. This technological advancement not only saves labor but also conserves fuel resources. Significantly, drones prove effective in reducing the excessive use of pesticides, insecticides, and other chemicals, contributing to sustainable and environmentally friendly agricultural practices.

The utilization of GPS-based precision farming applications has transformed various aspects of agricultural practices, encompassing tractor guidance, area mapping, soil sampling, and farm preparation, among other functions. GPS technology has empowered farmers to navigate through challenging terrain and adverse weather conditions, such as low visibility caused by rainfall, grain, fog, or darkness. The benefits of GPS extend beyond navigation, enabling farmers to optimize the use of soil enhancers and fertilizers, establish economic thresholds for pest and weed control, and strategize for sustainable organic energy sources. In the contemporary landscape, a multitude of manufacturers leverages GPS-derived products to enhance the overall efficiency of their agribusiness operations. GPS receivers play a crucial role in collecting location data for mapping area boundaries, roads, irrigation systems and identifying specific areas with plant issues like weeds or diseases. The precision offered by GPS technology is instrumental in creating highly accurate farm maps and detailing estates, roads, and even coastal areas. Farm owners harness the power of GPS for targeted site surveys, allowing them to gather soil samples and monitor crop health across different seasons. Moreover, GPS facilitates pilots in providing farmers with precise aerial maps. As GPS technology continues to evolve, manufacturers and agricultural service providers anticipate further enhancements, including the steady upgrade of GPS satellites with second and third city signals, marking a significant stride toward more advanced and accurate agricultural practices (Wamiq et al. 2022).

In horticulture, the IoT takes center stage in the monitoring of pests and diseases that pose significant threats to plant health and have detrimental effects on the overall horticultural output. The implementation of an IoT system is strategically designed to reduce the frequency of pesticide and fungicide usage while also offering predictive insights into pest occurrences. Additionally, the synergy of IoT and cloud computing proves invaluable in the precise identification of fruits, leveraging three key characteristics—color, shape, and texture. For farmers involved in vegetable cultivation, a spectrum of challenges, ranging from seed-related issues to pest and disease management, commodity pricing, and product marketing, necessitates innovative solutions. The integration of IoT, cloud computing, and mobile technology facilitates an interactive and information-rich platform, enabling farmers to access valuable insights and engage in collaborative processes related to vegetable production through online platforms and the concept of the mobile cloud (Singh et al. 2022). By employing this approach, there is a reduction in the feature vector's dimensions. Consequently, the amalgamated and normalized features of the image result in improved classification accuracy, requiring less training data (Rajasekar and Sharmila 2019). Monitoring the supply chain in real-time provides stakeholders with valuable insights into perishable food items, enabling better price management and timely actions to maintain quality standards (John et al. 2021). For farmers engaged in vegetable cultivation, a spectrum of challenges, encompassing seed-related issues, pest and disease management, commodity costs, and product marketing, necessitates innovative solutions. Leveraging the internet and the concept of mobile cloud computing, farmers can access information and participate in interactive discussions regarding vegetable production through mobile learning platforms (Shwetapadma and Sethy 2017).

In horticulture, the application of AI algorithms is versatile, encompassing various functionalities such as fruit detection, identification of pests and diseases, weed detection, plant stress detection, and yield prediction. These tasks are achieved through the utilization of a spectroscopy-and-camera system (Kingsley et al. 2020). The utilization of AI in the drying process of fruits and vegetables is gaining significant traction due to its ability to enhance the quality of dried products when integrated with an effective physical field. Utilizing an IoT-driven tool enables the identification of whether a climacteric fruit has undergone artificial ripening. Machine-learning algorithms play a crucial role in distinguishing between fruits ripened artificially and those ripened naturally (Singh et al. 2022). Capturing images of various crops under white/UV-A light serves as a method to determine the optimal stage of fruit maturity. Farmers can establish distinct maturity grades based on the crop/fruit category, organizing them into separate stacks before market distribution, particularly for highly perishable horticultural crops. Harvesting at the proper stage of maturity enhances post-harvest shelf life. The use of high-definition images from airborne systems, such as drones or copters, allows for real-time assessments during cultivation. By creating a field map and identifying areas requiring water, fertilizer, or pesticides, resource optimization is significantly improved. Additionally, smart irrigation systems, based on the IoT, offer automation in the irrigation process by analyzing soil moisture status and weather parameters. This technology, driven by AI, leverages historical weather patterns, soil quality, and crop types to automate irrigation, alleviating the labor-intensive nature of the process. Automated irrigation systems, employing real-time devices, maintain desired soil conditions to enhance water use efficiency and overall returns. Given that nearly 70% of the world's freshwater is used in irrigation, automation becomes a pivotal tool in addressing water management challenges, aligning with the goal of achieving more crops per drop. In recent years, there has been a growing trend in using image processing for fruit grading. Grading, a crucial step in the post-harvest process, involves categorizing fruits based on disease severity, defects, and contamination. Manual grading is both time-consuming and unreliable, emphasizing the need for automated systems. An effective method is the application of automatic image processing techniques for sorting and grading fruits (Gill et al. 2017).

Integration of blockchain technology into the fruit supply chain optimizes post-harvest and inventory management, leading to increased operational efficiency and reduced losses. The implementation of QR codes on fruits ensures end-to-end traceability, providing consumers with a transparent and trustworthy narrative about the product's journey. Collaborative efforts on a unified platform guarantee fair pricing for producers, fostering confidence and transparency among stakeholders. Additionally, blockchain finds application in monitoring the pre-harvest process for both yield and quality. Post-harvest management involves vigilant oversight of critical phases to prevent losses and enhance output, adherence to sustainable practices through monitored procedures, and the maintenance of unalterable digital records that accurately reflect information, meeting essential legal requirements (Singh et al. 2022). Immutable and distributed, blockchain technology serves as a digital ledger, ensuring a continuous stream of trustworthy data. With the proliferation of interconnected devices gathering information in smart agriculture, the complexity of data security rises. To meet this challenge, there is a crucial need to establish a system that seamlessly integrates IoT with blockchain technology. This integration facilitates a decentralized, transparent, and reliable data storage service, ensuring the automated and optimized management of agriculture systems (Gebresenbet et al. 2023). Enhancing transaction efficiency and minimizing the risk of disputes, blockchain technology establishes a clear and unchangeable record of contracts. Furthermore, its applications extend to monitoring policies and claims in a manner that is both reliable and auditable. This contributes to the prompt and reliable delivery of insurance to farmers. In summary, the potential of blockchain technology lies in its ability to bring transparency, efficiency, and safety to transactions and supply chain management within the agriculture sector, thereby enhancing trustworthiness among stakeholders (Lingireddy et al. 2023). Various digital technologies and their applications in the recent times is shown in Table 6.1.

Employing sensors and automated tools for data collection, coupled with the application of suitable analytical tools, holds the potential to significantly reduce

TABLE 6.1
Digital Technologies and Their Applications in the Recent Times

Sl. No.	Name of the Technology	Application(s)	Reference
1	Digital twin technology	Technology in greenhouse horticulture Digital representations of the (near) real-time state and real-life physical objects' behavior. Predictive capabilities, e.g., for growth simulation and yield prediction.	Ariesen-Verschuur et al. (2022)
2	Drones for pest and disease identification and management	Determine the "hotspots" for pests and illnesses in standing crops and then precisely release helpful insects into the atmosphere.	Kaur and Kumar (2022)
3	Robotics	Utilized for horticultural tasks like weeding, crop scouting, irrigation, microspraying, selective harvesting, sorting, and packaging, as well as for parks, golf courses, polyhouses, and other applications.	Kaur and Kumar (2022)
4	Digital scouting applications	Offer biological solutions for horticulture and a platform for disease registration monitoring.	Kaur and Kumar (2022)
5	Digital image processing	Employed with suitable optics and imaging equipment to gather data on characteristics such as floral size, shape, color, and defects in the surface.	Reddy and Samanta (2013)

(Continued)

TABLE 6.1 (*Continued*)
Digital Technologies and Their Applications in the Recent Times

Sl. No.	Name of the Technology	Application(s)	Reference
6	"SICANTIK" mobile application	A digital marketing platform designed to help farmers assess market potential, create and implement marketing strategies, and collaborate with the Agriculture Office to determine needs-based guidance, opportunities for marketing, and the growth of agricultural entrepreneurial endeavors.	Luckyardi et al. (2022)
7	Remote sensing	Used to record and handle information about plot boundaries, crops, the quality of the crops, agricultural volume, etc.	Ashalatha and Rao (2023)
8	Deep-learning technology	Identification of crop varieties, yield assessment, quality assessment, identification of stress phenotypes, growth tracking, and further tasks.	Yang and Xu (2021)
9	Artificial intelligence (AI)	Used in smart agriculture with deep-learning and machine-learning technology. It makes quick decisions possible for farmers, such as automating and optimizing the management of water and soil, detecting crop diseases, using pesticides, and managing livestock, among other things.	Gebresenbet et al. (2023)
10	Big data, cloud and edge computing	Cloud computing offers services like cloud storage and data access. Edge computing is a useful tool for connecting with cloud and IoT apps and processing data. Employed in smart agriculture, such as for meteorological forecasting and gathering agricultural data.	Gebresenbet et al. (2023)
11	IoT	Uses smart technologies, such as Bluetooth and wireless sensor networks, to connect several devices over the internet so that users may control the linked objects efficiently.	Gebresenbet et al. (2023)
12	BlockChain technology	A distributed digital ledger system that is utilized in smart agriculture to make agri-food supply networks more transparent, for instance.	Gebresenbet et al. (2023)

(*Continued*)

TABLE 6.1 (*Continued*)
Digital Technologies and Their Applications in the Recent Times

Sl. No.	Name of the Technology	Application(s)	Reference
13	Cyber physical systems (CPS)	Constituted by the machines' networking and interaction, as well as the inventories and operational resources utilized. When their surroundings change, these components respond by communicating with one another independently and taking self-directed action.	Ludwig-Ohm et al. (2023)
14	Digital insect traps	A digital support system for pest management, as well as automated pest detection and counting.	Ludwig-Ohm et al. (2023)

the time required for technology generation. Big data analytics enables swift and comprehensive assessment of large-scale field crosses, replacing the laborious and subjective ratings traditionally employed by plant breeders. This approach is applicable in both breeding nurseries and field experiments, utilizing sensors, drones, or satellite imagery to gather spatial and temporal information. Monitoring crop growth throughout a season yields valuable insights. Big data finds diverse applications in high throughput sequencing, genome assembly, proteomics, and metagenomics. Agricultural researchers continually generate genome data for crucial plants, animals, and microbes. The development of infrastructure, data-sharing policies, and parallel/grid computing is imperative for creating de novo assemblies and identifying genes associated with various traits. Initiatives like the studies conducted by Indian Council of Agricultural Research (ICAR) on database creation and genome/proteome sequence mining through machine-learning algorithms are underway. Establishing genomic data warehousing, workflow parallelization, SNP resources for animal species, and genomic resources for various crops/animals, along with other transcriptome resources, should be prioritized. Big data analytics plays a vital role in Mega Environment Mapping of Crops, while the application of image recognition for assessing the quality of different crops emerges as another promising area for big data analytics (http://naas.org.in/Policy%20Papers/policy%20101.pdf).

Harnessing ionizing radiation, such as gamma or neutron irradiation, proves instrumental in inducing mutations in plant breeding. This process leads to the creation of novel crop varieties that exhibit heightened resistance to pests and diseases, along with increased adaptability to challenging climatic conditions. The Sterile Insect Technique involves utilizing radiation to sterilize insects like fruit flies, releasing them into the environment to mate with wild counterparts. This strategic approach effectively reduces pest populations, safeguarding horticultural crops.

The application of irradiation contributes to enhancing food safety and prolonging the shelf life of horticultural produce. Its efficacy lies in eradicating microorganisms and pests responsible for illness and rot, leaving no residual radiation. Nuclear techniques play a crucial role in studying and managing soil and water resources. This involves utilizing isotopic tracers to monitor water movement in soil and employing nuclear techniques for effective water resource management in horticultural production. In crop breeding, nuclear technology is leveraged to develop improved varieties that better withstand the impacts of climate change, thereby bolstering food security. Additionally, nuclear technology serves as a valuable tool in controlling and preventing trans boundary plant diseases, ultimately enhancing overall crop production (https://www.fao.org/fao-stories/article/en/c/1390726/).

Together with this in horticulture, the utilization of cloud computing involves gathering and analyzing data pertaining to crop growth, soil conditions, and environmental factors. This data can be securely stored in the cloud, granting farmers remote access for informed decision-making in crop management. Weather data for specific regions and forecasts for designated time periods can also find a home in cloud storage, aiding farmers in making timely and crop-centric decisions. Cloud computing further extends its utility by automating diverse horticultural processes like irrigation and fertilization, resulting in time savings and heightened productivity. Take smart irrigation systems, for instance, which can be programmed to automatically water plants based on real-time soil moisture levels, minimizing the necessity for manual intervention. Moreover, cloud computing opens avenues for farmers to directly sell their produce to consumers, bypassing intermediaries and ensuring farmers receive fair prices for their products (Choudhary et al. 2016).

6.5 CURRENT UNDERSTANDING AND FUTURE PROSPECTIVE

A variety of tools, including GPS guidance, control systems, sensors, robots, drones, autonomous vehicles, variable rate technology, GPS-based soil sampling, and automated hardware, are used in this farming method, along with data analysis and information technology. Another potential development includes the smart agriculture, sometimes referred to as digital farming, which is the application of cutting-edge technologies and data-driven solutions to improve a range of agricultural elements. Smart agricultural systems are made possible by the integration of smart sensors and IoT devices. In addition to automating irrigation and providing real-time data on crop health, these systems can monitor and regulate the surrounding environment. Therefore, a more sustainable and fruitful future for the farming sector depends on smart agriculture, which is essential in tackling the problems of contemporary agriculture. A schematic representation of future prospective associated with digital technologies in horticulture is displayed in Figure 6.2.

Controlled environment agriculture and vertical farming also heavily rely on digital technologies. These techniques entail cultivating crops in regulated settings, frequently inside or in stacks of three. These installations require automated lighting, temperature management, and nutrient delivery systems in order to maximize productivity and achieve year-round growth. Given the current state of technology and the growing acceptance of automation in agriculture, robotics in horticulture

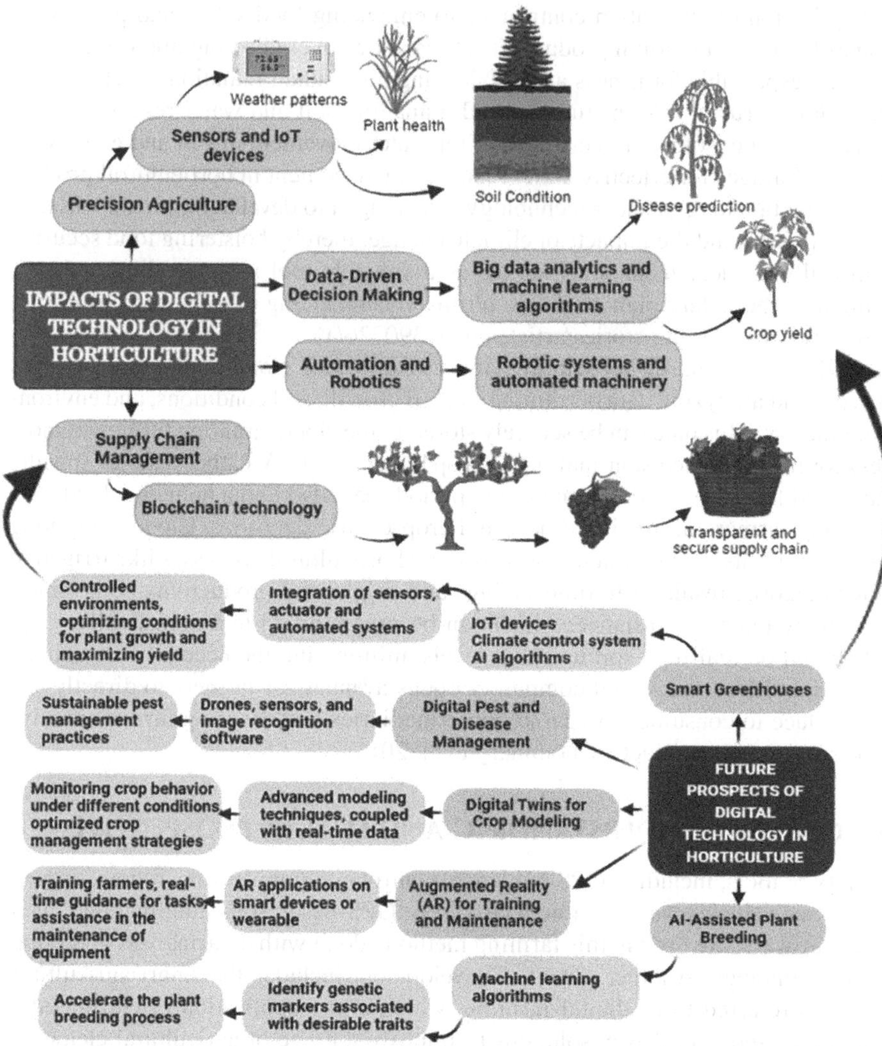

FIGURE 6.2 A schematic representation of future prospective associated with digital technologies in horticulture.

has a promising future. In horticulture, robotics is being used more and more for chores like planting, harvesting, and weeding. Robotic arms and autonomous cars can explore fields and carry out repetitive activities, eliminating the need for manual labor and helping areas that are experiencing a labor shortage. The horticultural sector can benefit greatly from blockchain technology, which offers transparency, traceability, and efficiency in a number of activities. The horticulture supply chain can benefit from increased traceability and transparency, thanks to blockchain technology. Blockchain records and verifies every transaction in a secure, decentralized manner, guaranteeing the accuracy of data pertaining to the cultivation, handling, and distribution of agricultural goods. Crop development techniques have been greatly expanded by genomic technology, which allows scientists and breeders to comprehend and modify the genetic composition of crops to achieve improved qualities, including yield, disease resistance, nutritional content, and environmental adaptation. Genetics and bioinformatics along with digital technologies can improve crops by making it possible to analyze plant genomes. With the help of this knowledge, new cultivars with enhanced qualities, including drought resistance, disease resistance, and nutritional value, can be created. These tools may mimic a variety of situations, assisting people in learning critical skills such as disease detection and handling crops.

The control of diseases and pests has been impacted by the expanding development of digital technologies, and this trend is expected to continue. In order to enable timely action, advanced monitoring technologies such as sensors, drones, and image recognition software can identify early indicators of diseases and insect infestations. This supports environmentally friendly methods of controlling pests. Greenhouse with smart digital technology has a high significance in the near future. AI algorithms, climate control systems, and IoT devices can all be used in greenhouses to create an environment that is both responsive and effective. By combining automated systems, sensors, and actuators, greenhouses may be made into controlled environments that optimize growth conditions for plants and maximize productivity. Training programs for farmers are crucial to the growth of both agriculture and horticulture. In these situations, farmers and laborers can be trained using augmented reality programs on smart devices and wearable, which offer real-time task assistance. It can also help with infrastructure and machinery upkeep. Horticultural production, sustainability, and resistance to economic and environmental difficulties are predicted to increase with the further evolution of digital technology integration. Regional differences may exist in adoption rates, but generally, the trend points toward a more efficient and technologically sophisticated horticultural industry.

6.6 CONCLUSION

The transformative impact of digital technologies on the field of horticulture science and technology is undeniable. As we navigate the ever-evolving landscape of innovation, the role played by these technologies becomes increasingly pivotal in shaping the future of horticulture. From precision agriculture and real-time monitoring to the integration of blockchain and AI, the chapters of this book have woven a narrative of possibilities and advancements. Digital technologies have not only revolutionized the way we cultivate, monitor, and manage horticultural crops but have also fostered sustainable practices, increased efficiency, and heightened productivity. The synergy between traditional horticultural wisdom and cutting-edge digital tools has paved the way for a more resilient and adaptive agricultural sector. As we reflect on the multifaceted contributions discussed in this book, it becomes evident that the journey doesn't end here. The intersection of digital technologies with horticulture science is an ongoing exploration, offering a spectrum of opportunities for further research, development, and implementation. The collaborative efforts of scientists, technologists, and practitioners in the horticulture community will continue to be the driving force behind innovations that address challenges and propel the field into new realms of discovery. In this dynamic landscape, embracing and harnessing the potential of digital technologies is not merely a choice but a necessity. It is a journey of continuous learning, adaptation, and integration, ensuring that the fruits of our labor benefit not only the present generation but also those that follow. The role of digital technologies in horticulture science and technology is an unfolding story, rich with promise, and its chapters will be written by the collective endeavors of the horticultural community worldwide.

REFERENCES

Alvino, A., and Marino, S. (2017). Remote sensing for irrigation of horticultural crops. *Horticulturae* 3(2):40.

Anderson, K., and Gaston, K. J. (2013). Lightweight unmanned aerial vehicles will revolutionize spatial ecology. *Frontiers in Ecology and the Environment*, 11(3):138–146.

Ardieu, H., Daly, D., Esteban-Lauzán, J., Hall, J., Miller, G. (2020). The birth of digital-A brief history of digital technologies. In: *Deliberately Digital. Future of Business and Finance.* Springer, Cham. https://doi.org/10.1007/978-3-030-37955-1_1

Ariesen-Verschuur, A., Verdouw, C., Tekinerdogan, B. (2022). Digital twins in greenhouse horticulture: A review. *Computers and Electronics in Agriculture*, 199:107183.

Arts, K., van der Wal, R., and Adams, W. M. (2015). Digital technology and the conservation of nature. *Ambio*, 44 (Suppl 4), 661–673. https://doi.org/10.1007/s13280-015-0705-1

Arya, P.S., and Gangwar, M. (2021). A proposed architecture: Detecting freshness of vegetables using Internet of Things (IoT) & deep learning prediction algorithm. *Advances in Computing, Communication Control and Networking*, ICAC3N:718–723.

Ashalatha, D., and Rao, S. (2023). Role of digital technology in agriculture and allied services. *International Journal of Creative Research Thoughts*, 11(4):627–633.

Bai, H. (2021). Role of digital technology in transforming organizational competencies influencing green economy: Moderating role of product knowledge hiding. *Frontiers in Psychology*, 12:792550. https://doi.org/10.3389/fpsyg.2021.792550.

Bhavana, H., and Bhagwan, A. (2021). Review on: Role of robotics in horticulture. *Journal of Pharmacognosy and Phytochemistry*, 10(1S):306–309.

Birner, R., Daum, T., and Pray, C. (2021). Who drives the digital revolution in agriculture? A review of supply-side trends, players and challenges. *Applied Economic Perspectives and Policy*, 43(4): 1260–1285.

Bin, F. and Johnson, N.F. (1982). Some new species of Telenomus (Hym., Scelionidae) egg-parasitoids of tropical pyralid pests (Lep., Pyralidae). *Redia*, 65: 229–252.

Blackmore, S., Godwin, R. J., Fountas, S., and Humphries, A. (1997). Precision agriculture: A new approach to crop management.

Chaurasia, J., Tripathi, V. K., and Tripathi, R. (2023). Nuclear technology in horticulture: Boosting productivity, controlling pests, conserving water. *Asian Journal of Agricultural and Horticultural Research*, 10(4): 178–182.

Choudhary, S. K., Jadoun, R. S., and Hardwari, L. M.. (2016). Role of cloud computing technology in agriculture fields. *Computer Engineering and Intelligent Systems* 7.

Ciarli, T., Kenney, M., Massini, S., and Piscitello, L. (2021) Digital technologies, innovation, and skills: Emerging trajectories and challenges, *Research Policy*, 50(7):104289. https://doi.org/10.1016/j.respol.2021.104289.

Costa, R., Rana, M. M., Oshita, S., and Kondo, T. (2018) Development of sustainable aquaponics system for green lettuce and *Nile tilapia* cultivation. *Sustainability*, 10(11):4221.

Gao, T., Liu, X., Tan, K., Zhang, D., Zhu, B., Ma, F., and Li, C. (2022). Introducing melatonin to the horticultural industry: Physiological roles, potential applications, and challenges. *Horticulture Research*, 9: p.uhac094.

Garzón, J., Montes, L., Garzón, J., and Lampropoulos, G. (2023). Systematic review of technology in aeroponics: Introducing the technology adoption and integration in sustainable agriculture model. *Agronomy* 13(10):2517.

Gebresenbet, G., Bosona, T.,Patterson, D., Persson, H., Fischer, B., Mandaluniz, N., Chirici, G., Zacepins, A., Komasilovs, V., Pitulac, T., and Nasirahmadi, A. (2023), A concept for application of integrated digital technologies to enhance future smart agricultural systems, *Smart Agricultural Technology*, 5(100255):1–12. https://doi.org/10.1016/j.atech.2023.100255.

Gill, S.S., Chana, I., and Buyya, R. (2017). IoT based agriculture as a cloud and big data service: The beginning of digital India. *Journal of Organizational and End User Computing (JOEUC)*, 29(4):1–23.

Giones, F., and Brem, A. (2017). Digital technology entrepreneurship: A definition and research agenda. *Technology Innovation Management Review*, 7(5):44–51.

Gonzalez-Dugo, V., Mateos, L., Testi, L., Orgaz, F., and Fereres, E. (2010). An automated system for continuous field measurements of canopy fluorescence, transpiration, and photosynthesis during canopy closure and steady state. *Transactions of the ASABE*, 53(4):1263–1277.

Griepentrog, H. W., Zhang, Q., and Zhang, M. (2011). Developments in weed control. *Advances in Agronomy*, 113:73–141.

Girma, D., Mebratu, A. and Belete, F. (2023). Response of potato (*Solanum tuberosum* L.) varieties to blended NPSB fertilizer rates on tuber yield and quality parameters in Gummer district, Southern Ethiopia. *The Scientific Temper*, 14(03): 680-691.

Heckman, J., Welacky, T., Lee, E. H., and Onclin, C. (2009). Precision agriculture technologies positively affecting the environment. *Computers and Electronics in Agriculture*, 67(1–2): 1–8.

Hori, M., Kawashima, E. and Yamazaki, T. (2010). Application of cloud computing to agriculture and prospects in other fields. *Fujitsu Science & Technical Journal*, 46(4): 446–454.

Policy papers regarding regulatory guildelines for data sharing among countries. http://naas.org.in/page.php?pageid=61 Last Opened in 4th April, 2024.

Regulatory guidelines for USA and European unions regarding nuclear energy. https://www.fao.org/fao-stories/article/en/c/1390726/ (FAO Policy) – Last Opened in 4th April, 2024.

Jensen, J. R. (1983). Remote Sensing of the Environment: An Earth Resource Perspective. Pearson Education India, Bangalore

John, K., Haw, S., Su, N., and Mulaafer, S. (2021). Kiwi fruit IoT shelf life estimation during transportation with cloud computing. In *Proceedings of the 2021 IEEE International Conference on Artificial Intelligence in Engineering and Technology (IICAIET)*, Kota Kinabalu, Malaysia, 13–15 September.

Johnson, L., Lobitz, B., Armstrong, R., Baldy, R., Weber, E., DeBenedictis, J., and Bosch, D. (1996). Airborne imaging aids vineyard canopy evaluation. *California Agriculture*, 50:14–18.

Kapur, R. (2018). Significance of digital technology. *International Journal of Transformations in Business Management*, 8 (I), 20–33.

Kaur, J., and Kumar, N. (2022). *Role and Potential of Digital Horticulture for Rural Development: Adoption & Enhancement, Rural India and Sustainable Development.* Vats Publication, New Delhi 1, pp 126–140.

Khan, S., Purohit, A., and Vadsaria, N. (2020). Hydroponics: Current and future state of the art in farming. *Journal of Plant Nutrition*, 44(10): 1515–1538.

Kingsley, J., Isong, I.A., Kebonye, N.M., Ayito, E.O., Agyeman, P.C., and Afu, S.M. (2020). Using machine learning algorithms to estimate soil organic carbon variability with environmental variables and soil nutrient indicators in an alluvial soil. *Land*, 9: 487.

Lamm, K., Powell, A., and Lombardini, L. (2021). Identifying critical issues in the horticulture industry: A Delphi analysis during the COVID-19 pandemic. *Horticulturae*, 7: 416.

Li, C., Zhu, H., Li, C., Qian, H., Yao, W., and Guo, Y. (2021). The present situation of pesticide residues in China and their removal and transformation during food processing. *Food Chemistry*, 354:129552.

Liakos, K.G., Busato, P., Moshou, D., Pearson, S., and Bochtis, D. (2018). Machine learning in agriculture: A review. *Sensors*. 18(8): 2674.

Lingireddy, H., Bhargava, K., Abhishek, E., Barla Madhusudhan, C., and Sri, B.G. (2023). An overview of digital technologies in agriculture and their applications. Biological Forum. 15(2): 750–757.

Luckyardi, S., Hurriyati, R., Dirgantari, D. P. (2022), Mobile application design for digital marketing and management of horticultural. *Journal of Engineering Science and Technology*, 17(1): 0064–0074.

Ludwig-Ohm, S., Hildner, P., Isaac, M., Derksmeyer, W., Schattenberg, J. (2023), The contribution of Horticulture 4.0 innovations to more sustainable horticulture, *Procedia Computer Science*, 217, 465–477

Maja, J.M., and Ehsani, R. (2010). Development of a yield monitoring system for citrus mechanical harvesting machines. *Precision Agriculture Journal* 11(5): 475–487.

Martin, H. (2020). Digital technology and social change: The digital transformation of society from a historical perspective. *Dialogues in Clinical Neuroscience*, 22(2), 189–194, https://doi.org/10.31887/DCNS.2020.22.2/mhilbert

Mohammad Nuruzzaman, B., Rahman, M., Billah, M., and Saha, D. (2021). Internet of things (IoT): A review of its enabling technologies in healthcare applications, standards protocols, security, and market opportunities. *IEEE Internet of Things Journal*, 8: 10474–10498.

Nemali, K. (2022). History of controlled environment horticulture. *Greenhouses*, 57(2):239–246.

Nguyen, H.T.T., Dang, D.T., Van Pham, C. and Bertin, P. (2016). QTL mapping for nitrogen use efficiency and related physiological and agronomical traits during the vegetative phase in rice under hydroponics. *Euphytica*, 212: 473-500.

Pass, R., and Shi, E. (2017). Fruitchains: A fair blockchain. In *Proceedings of the ACM symposium on principles of distributed computing*. Salerno, Italy pp. 315–324.

Pathan, A.K., Jia, K., and Zhang, S. (2019). Internet of Things (IoT) in Agriculture: System architecture and implementation. *IEEE Access*, 7:129551–129557.

Patil, V.C., Al-Gaadi, K.A., Biradar, D.P., and Rangaswamy, M. (2012). Internet of things (Iot) and cloud computing for agriculture: An overview. Proceedings of agro-informatics and precision agriculture (AIPA 2012), India 292: 296.

Puri, V., Nayyar, A., and Raja, L. (2017). Agriculture drones: A modern breakthrough in precision agriculture. *Journal of Statistics and Management Systems*, 20(4): 507–518.

Rajasekar, L., and Sharmila, D. (2019) Performance analysis of soft computing techniques for the automatic classification of fruits dataset. *Soft Computing*, 23: 2773–2788.

Rana, M.M., Oshita, S., Fujioka, Y., Takemura, A., and Osada, T. (2019). Effects of stocking density of Nile tilapia (*Oreochromis niloticus*) on the water quality and performance of tomatoes in an aquaponics system. *Aquaculture Reports*, 13: 100173.

Rani, A.S. (2017). The impact of data analytics in crop management based on weather conditions. *International Journal of Engineering Technology Science and Research*, 4(5): 299–308.

Reddy, K. R., and Samanta, D. (2013), Application of digital image processing for horticulture, *International Journal of Engineering Sciences & Emerging Technologies*, 6(2): 228–232.

Rouse, J. W. (1974). *Monitoring Vegetation Systems in the Great Plains with ERTS*. NASA Special Publication, Washington, D.C. USA 351.

Sam, S., Mira, L., and Kai, S. (2022). The impact of digitalization and automation on horticultural employees-A systematic literature review and field study. *Journal of Rural Studies*, 95: 560–569.

Sharif, K.I.M., Hassan, M.G., Miraz, M.H., Zulkifly, E., Udin, Z.M., and Omar, M. (2021). Factors affecting blockchain in fruit retail market: An unveiling myth of blockchain. In *Intelligent Manufacturing and Mechatronics: Proceedings of SympoSIMM 2020*. pp. 1095–1106. Perlis, Malaysia.

Shwetapadma, P., and Sethy, P.K. (2017). Post-harvest grading of carica papaya fruit using image segmentation and soft computing. *International Journal of Advanced Research in Computer Science*, 8: 7.

Singh, R., Singh, R., Gehlot, A., Akram, S.V., Priyadarshi, N., and Twala, B. (2022). Horticulture 4.0: Adoption of industry 4.0 technologies in horticulture for meeting sustainable farming. *Applied Sciences*, 12(24): 12557.

Smith, O.S., Lower, R.L. and Moll, R.H. (1978). Estimates of Heritabilities and Variance Components in Pickling Cucumber. *Journal of the American Society for Horticultural Science*, 103(2): 222–225.

Srinivasan, A., Zhang, M., and Shannon, D. (2010). Agricultural water management: Concepts, issues, and applications in research and practice. *Advances in Agronomy*, 109: 3–44.

Sugiura, R., Nemoto, M., Saito, T., and Maki, H. (2016). Development of a multispectral camera system for detecting diseases in field crops. *Precision Agriculture* 17(2): 139–153.

Thenkabail, P.S., Lyon, J.G., and Huete, A. (2000). Hyperspectral Remote Sensing of Vegetation. *Photogrammetric Engineering & Remote Sensing* (TSI). 80(8): 695–723.

Thenkabail, P.S., Smith, R.B., Pauw, E.D., and Hyperspectral, C. (2016). Hyperspectral vegetation indices and their relationships with agricultural crop characteristics. *Remote Sensing of Environment*, 158: 257–277.

Tiwari, A., Singh, A.K., Singh, S., Singh, B., and Pal, A.K. (2015). University-Industry based partnership for horticultural development in India-needs and challanges. *International Conference on Issues and Challenges in Doctoral Research*, Noida, India 25:21.

Tsiolias, D., Keramydas, C., Iakovou, E., and Vlachos, D. (2015). Big Data and agricultural supply chains: Opportunities for increased food security. *Operational Excellence in Logistics and Supply Chains: Optimization Methods, Data-driven Approaches and Security Insights. Proceedings of the Hamburg International Conference of Logistics (HICL)*, 22:331–354. Hamburg, Germany.

Usha, K., and Singh, B. (2013). Potential applications of remote sensing in horticulture-A review. *Scientia Horticulturae*, 153:71–83.

Velden, D.V.D., Klerkx, L., Dessein, J., and Debruyne, L. (2023). *Cyborg Farmers: Embodied Understandings of Precision Agriculture. Sociologia Ruralis*, 64 (1): 3–21.

Vose, P.B. (2013). *Introduction to Nuclear Techniques in Agronomy and Plant Biology: Pergamon International Library of Science, Technology, Engineering and Social Studies*. Elsevier, Amsterdam, The Netherlands.

Wamiq, M., Kumar, M., Kaushik, K., and Kumar, A. (2022). Modern technologies in horticulture. *Modern Concepts in Farming*, 123.

Xu, X. (2022). Major challenges facing the commercial horticulture. *Frontiers in Horticulture*, 1:980159.

Yang, B., Xu, Y. (2021), Applications of deep-learning approaches in horticultural research: A review. *Horticulture Research*, 8(123): 1–31

Zaagsma, G. (2013). On digital history, *BMGN—Low Countries Historical Review*, 128(4): 3–29.

Zhang, C., and Lu, Y. (2021). Study on artificial intelligence: The state of the art and future prospects. *Journal of Industrial Information Integration*, 23:100224.

Zhang, C., Cai, J., Xiao, D., Ye, Y., and Chehelamirani, M. (2018). Research on vegetable pest warning system based on multidimensional big data. *Insects*, 9(2):66.

Zhang, M., Han, Y., Li, D., Xu, S., and Huang, Y. (2023). Smart horticulture as an emerging interdisciplinary field combining novel solutions: Past development, current challenges, and future perspectives. *Horticultural Plant Journal*, 9(4): 1–29

Zhang, Y., Bai, Y., Wu, G., Zou, S., Chen, Y., Gao, C., and Zhao, Y. (2016). Simultaneous modification of three homoeologs of TaEDR1 by genome editing enhances powdery mildew resistance in wheat. *The Plant Journal*, 88(4):574–583.

Zhang, Y., Chen, L., Battino, M., Farag, M.A., Xiao, J., Simal-Gandara, J., Gao, H., and Jiang, W. (2022). Blockchain: An emerging novel technology to upgrade the current fresh fruit supply chain. *Trends in Food Science & Technology*, 124:1–12.

7 Strategies for Adding Value to Food Byproducts (Recovery and Valorization)

Shamas Murtaza
Jiangsu University
MNS-University of Agriculture

Nighat Raza, Muhammad Shehbaz, Umar Farooq, Samara Qaisar alvi, Hammad Naeem, and Abdul Mannan Shakir
MNS-University of Agriculture

Saima Rafique
University of Poonch

Sanabil
Jiangsu University
University of Central Punjab

Iram Hafiz
University of Sargodha

Bismillah Mubeen and Yong Kun Ma
Jiangsu University

7.1 INTRODUCTION

The agri-food industry sector makes a significant contribution to the global economy by providing vast market opportunities, supporting local economies and creating job opportunities. The agri-food sector has been affected by globalization and free trade policies. The food supply chain has expanded massively worldwide, and new and useful novelties are constantly being introduced. In the international market,

DOI: 10.1201/9781003412489-7

novel food products are being introduced because of tremendous globalization. Significant worldwide waste and/or byproduct production occurs at several levels of the agri-food chain. The geopolitical, economical and environmental conditions of a place can be seen as being reflected in inefficient and unsustainable waste and/or byproduct management.

Effective agri-food waste/byproduct upcycling can ensure the sustainable future of the entire food production and transport chain in addition to making a significant contribution to local food security. In today's global environment (e.g., recession, economic insecurity, food insecurity, pandemics like COVID-19), reducing food sector waste is essential, as is effectively using byproducts to make valuable products. In this scenario, an eco-conscious system based on the concepts of "circular economy" and "zero waste generation" aims to effectively increase the value of food industry waste/byproducts, thereby contributing to economic improvement while minimizing negative environmental impact. "Turning waste into wealth" and "tasting waste" have become buzzwords that emphasize the need of repurposing waste and byproducts. Several novel methodologies and practical models for reducing waste and maximizing the use of byproducts in agri-food production and supply chains have been proposed.

The recently presented concept of circular economy is an auspicious framework for increasing resource efficiency. Reducing waste generation while maximizing its utilization can benefit the economy and can dig out innovative business options. The circular economy perception principally aims at sustainable food production and consumption with lesser emission of greenhouse gas as well as food waste management. Furthermore, the Circular Economy Action Plan (CEAP) employs a novel approach to reduce food waste production. To support the regional economy, the program relies heavily on material reduction, reuse, recovery and recycling, in addition to improving resource efficiency. A waste-free economy and a circular society built on sustainable biomaterials are urgently needed to address the current global crisis. However, for upcycling of industrial waste, environmentally sustainable technologies must be used (Bhat, 2021). This chapter identifies and discusses some technologies that are anticipated to have an impression on the value addition of agri-food waste and byproducts.

7.2 GLOBAL SCENARIO OF FOOD WASTAGE

The food industry has made tremendous progress over the last decade, making it one of the world's fastest growing industries. The industry's rapid growth has brought with it a number of challenges, the most notable of which are food security, safety and food waste management. According to Baysal and Ulku (2022), almost one-third of the food produced each year is wasted or lost. The food loss can be defined as the reduction in the amount of edible food during the production, post-harvest and processing stages of the food chain, primarily in developing countries (FAO, 2023). Food waste is the disposal of edible food at retail and consumer levels, primarily in developed countries. Food waste represents lost opportunities to improve food security and has a high environmental cost. According to estimates, this has a minimum value of 1.6 billion tons, with a carbon footprint of 3.3 billion tons of CO_2 (FAO,

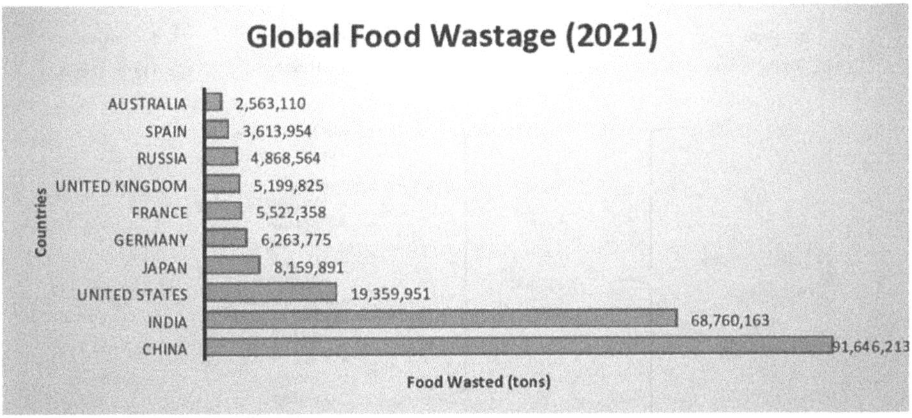

FIGURE 7.1 Global scenario of food wastage of the year 2021.

2022). However, the processing sector is responsible for a significant portion of these wastes.

In terms of food categories, tubers, roots and oilseed crops had the highest post-harvest to distribution losses (25%), followed by fruits and vegetables (22%), animal products and meat (12%) and pulses and cereals (9%) (FAO, 2023). A recent study in Europe found that vegetables and fruits have shown the highest total food waste, with similar amounts of food waste generated during primary production and consumption (Caldeira et al., 2019). Figure 7.1 depicts the global scenario of food wastage of the year 2021.

7.3 APPLICATIONS OF FOOD BYPRODUCTS

Non-edible portions of agricultural goods may contain significant quantities of phytonutrients and bioactive substances that, if extracted, might add significantly to the value of a wide range of products, including nutraceuticals, therapeutics, food additives and cosmetics (Nayak and Bhushan, 2019). Pectin, sugars, fibers, polysaccharides, lipids, proteins, polyphenols, essential minerals, vitamins, fatty acids, pigments and anthocyanins are among the value-added products extracted from fruit and vegetable byproducts or waste (Socas-Rodrguez et al., 2021). Figure 7.2 demonstrates the ways to isolate agri-food waste byproducts and their potential applications.

7.4 VALORIZATION OF FOOD BYPRODUCTS

Food waste is a byproduct or leftover from the conversion of raw materials into higher-value items (Trigo et al., 2022). Food waste has historically been burned or dumped in landfills, causing soil and food contamination as well as air and water pollution. The European Union (EU) promotes reducing food waste and creating new uses for food byproducts in order to address these problems. According to these guidelines, food waste from several agri-food industries (fruits, vegetables, meat, marine items and aquaculture, beverages and sugar, etc.) represents an intriguing and

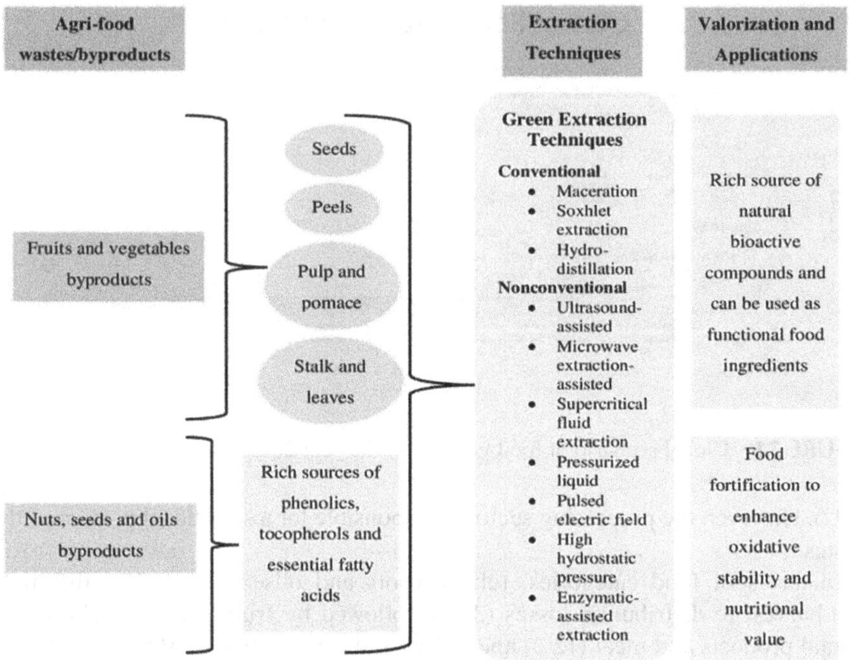

FIGURE 7.2 Extraction and valorization of agri-food byproducts and their potential food application.

less expensive potential source of functional or bioactive compounds. Food byproducts can also be used in nutraceuticals and pharmaceuticals, or they can be converted into livestock feed (Wunder et al., 2018).

Despite these promising prospects, adding value to food byproducts is difficult for quite a few reasons: (i) traditional extraction processes use non-food-grade solvents; (ii) technical difficulty in large-scale production; (iii) traditional extraction processes are inefficient and expensive in terms of energy; and (iv) varying stability of value-added components during processing. The idea of biofuels and value-added byproducts arose from the context of integrated and combined extraction or recovery processes for the conversion of biomass into energy and various products, respectively (Mazzutti and Salvador, 2020). Food waste, on the other hand, may contain pathogens, organic residues, high moisture content, and physical and chemical contaminants that are hazardous to one's health. Legislation, rules and additional scientific research are therefore essential. In addition, value-added byproducts' quality and safety must be verified, just as it is with other goods used for human or animal use (Socas-Rodrguez et al., 2021).

7.4.1 Valorization of Fruit and Vegetable Byproducts

Fruit production has increased due to changes in the eating habits of the people and an increasing population (FAO, 2017), reaching nearly 900 million tons in 2017.

A great number of byproducts and waste are unavoidable depending on geographical location, consumption and population boom. According to FAO (2016), one-third of the food that is being produced is lost or wasted, counting the tropical fruits. The total amount of food wasted or lost each year is projected to be around US$940 billion. Food waste in Europe has reached 95 kg/year, compared to 115 kg in North America. Food losses in Africa and Asia range from 6 to 11 kg/year (FAO, 2015). In beverage industry, fruits after beverage production generate a significant amount of waste in the form of pomace. In most cases, pomace, a byproduct of fruit processing, contains far more valuable bioactive compounds than juice itself. Because of their chemical abundance and heterogeneity, abundantly occurring pomace may represent an interesting natural biological resource. The issue of converting fruit processing waste into value-added products has been extensively researched. Berries and apples, for example, have been proposed as additives in baked goods and dairy formulations to boost their natural antioxidant and fiber content. Furthermore, the existence of natural colors as well as the volatile compounds can improve the final product's sensory attributes.

Tropical fruit byproducts contain a high concentration of bioactive compounds, primarily carotenoids and phenolic compounds. To obtain these compounds, they must be extracted from the matrix, and various extraction methods, such as conventional and unconventional methods, are being investigated. Figure 7.3 depicts a flowchart showing the extraction of bioactive compounds from tropical fruit waste and/or byproducts and their potential uses.

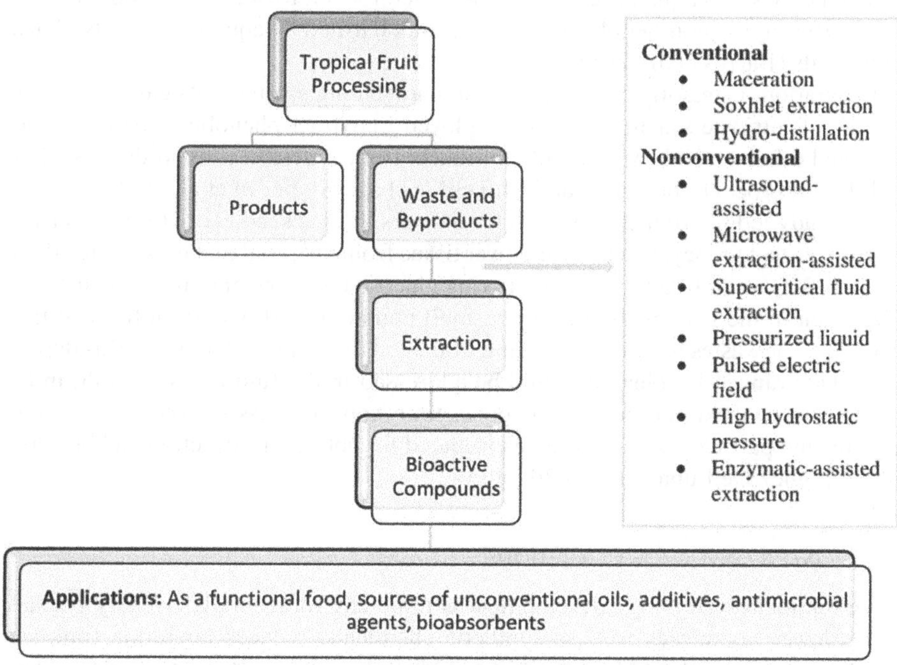

FIGURE 7.3 General scheme for the extraction of bioactive compounds from tropical fruit byproducts and wastes along with their applications.

7.4.2 Conventional Extraction Methods and Their Health Effects

Traditional techniques rely on heat treatment, solvent extraction or a combination of the two. Hydro-distillation, soxhlet extraction and maceration are the most commonly used solid-liquid extraction or liquid–liquid methods. It is well established that bioactive substances attained from tropical fruit byproducts and vegetables as depicted in Figure 7.3 using conventional extraction methods have a wide range of health benefits. The extract has numerous biological activities like the scavenging of active substances that may lead to severe health related issues which provide significant value as tropical fruit industry byproducts and wastes.

By isolating different natural bioactive element, primarily phenolic compounds, which find applications in the medical and food industries, tropical fruit byproducts and vegetables and wastes can be given additional economic benefits (value added). Some of the studies listed below emphasize the significance of bioactive compounds in tropical fruit waste or byproducts. Methanol and ethanol are the most frequently used solvents to extract important secondary metabolites from vegetables' byproducts such as phenolic acids and flavonoids (Albuquerque et al., 2018).

The inhibitory activity of rambutan seed ethanol extract and its constituents (ethyl acetate, hexane, water and butanol) on glucosidase was estimated and according to high-performance liquid chromatography (HPLC), the key phenolic compounds present in the extract were rutin, diosmin, corylatin, ellagic acid and quercetin. These researchers came to the conclusion that rambutan bark has the potential to be a natural antimicrobial agent in various foods.

Acai berry seeds aqueous extracts were tested for cell toxicity in various carcinomas. Furthermore, non-neoplastic cells were not harmed by aqueous extracts of acai berry seeds (Barros et al., 2015).

Maceration, is majorly employed extraction techniques due to its ease of use and absence of intricate machinery. It is employed to extract phenolic acids and flavonoids and other secondary metabolic products from vegetables' byproducts such as cauliflower, broccoli, bok choy and Chinese cabbage (Bahorun et al., 2004).

In a study led by Man et al. (2016) for 6 weeks, lychee seed extract treatment was found to protect kidney, pancreas and liver tissue from potential diabetes-related damage after 30 days of treatment by improving glucose tolerance and insulin resistance.

As seen in the studies listed, only a small number of extracts from tropical fruit byproducts or wastes were extracted and evaluated for biological activity; this depicts that a wide range of molecules could be addressed in the future. As a result, in the use of traditional extraction methodologies, time, solvent, mass-to-volume ratio, temperature and particle size needs to be evaluated for optimum extraction and bioactivity (Rodr'guez De Luna et al., 2020).

7.4.3 Nonconventional Methods

Unconventional technology development is more environmentally friendly because it uses fewer organic as well as synthetic chemicals, reduces operation time and optimizes compound yield and quality. Although some of these techniques have been proposed for nearly 50 years, they are still regarded as novel. Microwave and ultrasound-assisted extraction and also supercritical fluids are the most commonly used unconventional techniques for the extraction of bioactive compounds.

7.4.3.1 Ultrasound-Assisted Extraction (UAE)

This denotes the usage of ultrasound frequencies ranging from 20 to 2000 kHz. These waves cause cavitation, which includes the formation and growth of gas bubbles and also their collapse. At this stage, energy is generated uncontrollably in all directions of the medium; this enormous release of energy leads to extreme variations in pressure and temperature within the bubble. The ultrasonic extraction mechanism involves diffusion (Azmir et al., 2013). The extraction of phenolic compounds from mango peel was studied using ethanol and ultrasound-assisted extraction (UAE). Routine extraction by using a methanol: water mixture in a ratio of 8:2, v/v with constant stirring for 5 minutes (67%) yielded the highest total phenolic content (TPC), according to the experimental design; the authors concluded that ultrasound-assisted extraction (UAE) had no effect on the phenolic extracted compounds (Guandalini et al., 2019).

7.4.3.2 Microwave-Assisted Extraction (MAE)

This is also a method of extracting compounds from a variety of materials into fluids employing the microwave energy, and it has been recognized for extracting phenolic compounds from tropical fruit byproducts and also their wastes. Solvents with similar polarities to the compound to be extracted are commonly utilized to extract isoflavones, carotenoids and flavonoids; water, methanol or ethanol, are used to extract aglycones and flavonoid glycosides. MAE was tested on phenolic compounds in mango peel by Ordon ez-Torres et al. (2020). Vallejo-Castillo et al. (2020) identified optimal conditions of MAE to maximize total flavonoid contents (TFC) and TPC extraction efficiency from the rind of papaya.

MAE was optimized for compounds which were valorized by yellow and red pitaya peel (Ferreres et al., 2017). Therefore, deduction can be made that microwaves can speed up the process of extraction and increase the amount of bioactive compounds while decreasing solvent volume. Furthermore, byproducts such as tropical fruit peels can be repurposed to recover compounds, primarily anthocyanins and tannins while avoiding thermal degradation. Jia et al. (2017) separated apigenin, quercetin, catechin and various other flavonoids from black rice, along with p-coumaric acid, gallic acid, ferulic acid, syringic acid, chlorogenic acid and ethanol was utilized as a solvent. In a similar way Rodr'guez-Pe'rez et al. (2016) employed a hydroalcoholic solvent for extracting kaempferol, quercetin, isorhamnetin-3-O-glucoside, vitexin and various other flavonoids from *Moringa oleifera* leaves. Ho et al. (2015) also employed this method of extraction for the isolation of lycopene from the skin of tomato.

7.4.3.3 Supercritical Fluid Extraction (SFE)

Shorter processing times and more yield of the bioactive compounds are two advantages of this method. CO_2 is regarded as an ideal SFE solvent since it can resist pressure and also temperature fluctuations, and the gas is changed into the form of a supercritical fluid (Capuzzo et al., 2013). SFE was employed for phenolic compound recovery from the seeds of papaya using CO_2 and its combination with ethanol (CO_2-EtOH). This technique was used to derive lycopene from tomato waste, flavonoids from the leaves of spearmint, polyphenols from grape byproducts, trans-resveratrol, catechins, rutin, epicatechin and quercetin (Panja, 2018) Using SFE, Sa'nchez-Mesa et al. (2020) investigated gallic acid, mangiferin, quercetin and ellagic acid in mango peel. RP-UHPLC-DAD was used to characterize carotenoid profiles. Hence, SFE is a novel technique for recovering bioactive compounds from tropical fruit agri-industrial wastes, as well as a value-added alternative.

7.4.3.4 Pressurized Liquid Extraction (PLE)

A combination of the high pressure and temperature, uses less solvent, allowing for faster extraction. The technology creates conditions that allow for high solubility and mass transfer rates, as well as lowering the solvent's viscosity and surface tension, increasing the level of extraction of the bioactive compounds (Khan et al., 2019). In this respect, Figueroa et al. (2018) used ethanol to optimize the adsorption of phenolic compounds from avocado peel. Arau'jo et al. (2020) employed PLE for the extraction of phenolic compounds from avocado peel and high concentrations of flavonols catechins and phenolic acids were found. At room temperature, a pulsed electric field (PEF) is a procedure that uses high-intensity pulses of electric fields that are of short-duration. PEF disrupts cell membranes, enhancing extraction (Lopez et al., 2009). In this regard, water extraction followed by PEF application resulted in a 400% extraction of phenolics from mango peel (Parniakov et al., 2016).

7.4.3.5 High Hydrostatic Pressure (HHP) Extraction

High hydrostatic pressure (HHP) operates between pressure ranging from 100 to 1000 MPa; this pressure increases mass transfer rates, bioactive compound diffusion and cell permeability. In this context, phenolic compounds from the peel of longan was attained, TPC and TFC extraction yields from papaya seeds was assessed and carotenoid present in two varieties of passion fruit was assessed to increase from 9% to 27% by using high hydrostatic pressure (HHP) (Briones-Labarca et al., 2015; De Ancos et al., 2000; Prasad et al., 2010).

7.4.3.6 Enzymatic-Assisted Extraction (EAE)

In this, enzymatic treatments have been used to release and increase the release of bound compounds. Important factors in obtaining bioactive compounds included particle size of tropical fruit byproducts, enzyme concentration, enzyme composition, hydrolysis time and solid-to-water ratio. EAE may be employed to overcome the limitations of other extraction procedures. This procedure is regarded as environmentally friendly and aids in the reduction of environmental concerns. Pectinases, cellulases and hemicellulases are the most commonly used enzymes to extract bioactive compounds from the byproducts of tropical fruits (Le and Le, 2012; Mehmood et al., 2016). The above technique was applied to derive different substances like carotenoids from peppers (Nath et al., 2016) and squash (Ghosh and Biswas, 2016), mignonette from celery (*Apium graveolens*) leaves by employing apigenin and pectinase (Zhang et al., 2011) and multiple anthocyanins and polyphenols were extracted from grape skins via a combination of cellulase and pectinase (Maier et al., 2008). Also, alkaloids, flavonoids, coumarins, phenolic acids, terpenoids and lectins have been isolated using the EAE process (Kumar et al., 2017).

7.5 FERMENTATION

Secondary metabolites produced by fermenting microorganisms can be used to obtain bioactive compounds (Petkova and Antova, 2015). Studies on liquid cultures have revealed that these compounds begin to be produced when growth is hampered by nutrient depletion, such as phosphate, carbon or nitrogen sources (Martins et al., 2011; Mosunova et al., 2020).

Green coconut shells are high in ferulic acid, which can be converted into vanillin using *Phanerochaete chrysosporium* under SSF (solid-state fermentation) conditions (dos Santos Barbosa et al., 2008). Obtention of the bio-actives by fermenting technique is a strategic approach to produce extracts of high-quality while avoiding toxicity from the extractants, particularly organic solvents. Due to the lack of toxic solvents, the bioactive elements derived from byproducts of tropical fruits have excellent value because they can be used directly in the production of functional foods. Furthermore, this process is fairly profitable for the food industry and is cost-effective due to the lower cost and richness of the phenolic compounds attained from the byproducts got from tropical fruits.

7.5.1 Sources of Unconventional Oils

A large number of seeds are generated by the tropical fruit processing industry which can be used to produce unconventional vegetable oils with high levels of bioactive compounds and unsaturated fatty acids. Oils extracted from the seeds of honeydew melon (*Cucumis melo*), watermelon (*Citrullus lanatus*) and pomegranate (*Punica granatum*) were studied for their fatty acid (FA) and phytosterol composition. The recovered oil is primarily composed of phenolic compounds, vitamin E, phytosterols as well as FA, with a high concentration of 4.25%–64.2% linoleic acid. Punicic acid (86.2%) is abundant in pomegranate seed oil (Gornas and Rudzinska, 2016). Watermelon seed oil contains more and oleic acid (17%) and linoleic acid (55%).

7.5.1.1 As Colorants

Fruit byproducts are a good source of natural pigments like carotenoids and anthocyanins, which are the major natural coloring substances (Sigurdson et al., 2017); they have also been linked to health benefits. Some of the additives now in use are covered by regulation (EC) No. 133/2008 and may be found in tropical fruit byproducts. Anthocyanins have been found to be abundant in banana bracts and acerola cherries (Yaemchuen et al., 2018). Furthermore, the tropical fruit guajillo contains a higher concentration of acylated anthocyanins. As these natural colors are less intense and less stable in color, they may impart flavor or odor to food so their use is limited (Faustino et al., 2019).

7.5.1.2 As Texturizing Agents

Thickeners, stabilizers and emulsifiers are all considered as texturizing agents and they can be used to change the texture and mouthfeel of foods (Saha and Bhattacharya, 2010). Thickeners can be used to change or increase the viscosity of food without altering their properties and emulsifiers hold oil and water together in an emulsion (Faustino et al., 2019). Citrus seeds, powders derived from the peel of pineapple and passion fruit peels have been deliberated as texturizing agents owing to their pectin and dietary fiber contents.

7.5.1.3 As Antimicrobial Agents

Phenolic compounds derived from byproducts of various tropical fruits have been found to exert antibacterial activity in diverse ways like (i) microbial cell death by influencing acidification of the cytoplasm by causing the diffusion of the dissociated phenolic acids, (ii) altering the cell membranes of the microbes, increasing cellular aggregation and permeability and also causing the deprivation of cell nutrients, and (iii) forming complexes with extracellular and soluble protein (Fidelis et al., 2020; Miller and Zachary, 2017; Paz et al., 2015; Santos and de Aquino Santana, 2019). Phenolic compounds derived from fruits and their byproducts can be the innovative sources for antimicrobial agents that may propose new profitable scenarios for the food industry. Regrettably, the evidence concerning the inhibition of the bacteria present in food is limited, so it is needed to pay immediate consideration on this aspect to benefit and assist the food industry.

7.6 POTENTIAL BENEFITS OF FRUIT AND VEGETABLE BYPRODUCTS

7.6.1 Bioactive Compounds of Fruit Byproducts as Potential Prebiotics

A substrate that is selectively utilized by the microorganisms of the host and that benefits the health is known as prebiotic. This is a well-accepted definition as it incapacitates the apprehensions that are related to the exclusiveness of oligosaccharides that are fermentable and are known as prebiotic substrates. Biologically active compounds like carotenoids, flavonoids and peptides benefits the host by intestinal putrefaction, peroxidation, tumorigenesis, mutagenesis suppression and immunomodulation (Vieira da Silva et al., 2016). Prebiotics and probiotics work well together to improve the balance of the microbiota in the gut by inhibiting the proliferation of the undesirable bacteria and enhancing the growth of the beneficial ones. Probiotics are live microorganisms that provide health benefits to the host when consumed in sufficient amounts (Fraga et al., 2019). In order to understand the mechanism of the new compounds in the modulation of the intestinal microbiota, an increase interest is evoked in the use and sustainability of fruit byproducts (Banerjee et al., 2017).

Fruit pieces like husks, peels, seeds and bagasse are removed during the process of manufacturing fruit juices, which results in the loss of up to 30% of the fruit mass (Esparza et al., 2020; Medeiros et al., 2019). Byproducts derived from the temperate, subtropical and tropical fruits may contain various bioactive compounds such as melatonin, carotenoids, dietary fiber and phenolic compounds (Mateos-Aparicio et al., 2020). Phytochemicals are isolated from fruit byproducts and incorporated into natural matrices to make innovative products enriched with dietary fibers (Gil-Sanchez et al., 2018).

On the basis of the prebiotic characteristics associated with the gut microbiota, the fruit byproducts studies can be classified into three different niches: (i) Those in which different extracts and high concentration of phytochemicals of byproducts were tested for both in vivo and in vitro activity; (ii) Those in which the byproducts were dehydrated or converted into fine powders were used for modulating the gut microbiota, keeping in view the synergistic association between the gut microbiota and the chemical constituents of the byproducts; (iii) Those in which fine powders of concentrated extracts are employed as functional ingredients in foods like fermented foods and tests their ability of enhancing the nutritional content of the food as well as also tests the viability of the probiotic strains (Espin et al., 2017).

The waste that is left after the fruit juice processing is considered as a propitious source of bio-actives and so, the waste or byproducts should be explored more as an attractive and innovative food technological approach that may have economic and environmental benefit (Esparza et al., 2020). The fruit byproducts have proved evidence of their bio-efficiencies as anti-inflammatory and antioxidant and also as prebiotics that promotes and benefits the human health.

The incorporation of fruit byproducts in the formulation of functional foods, such as bakery products, can provide health benefits (Sudha et al., 2016). Cakes made with various percentages of guava peel flour (70%, 50% and 30%) revealed that the addition of flour resulted in a high amount of phenolic compounds, fiber and beta-carotene, as well as a lower content of carbohydrates and lipids was also reported. As a result, guava peel flour can be used to partially replace wheat flour in the production of cakes, increasing their nutritional value while maintaining their organoleptic characteristics (Bertagnolli et al., 2014). Using dietary fiber and phenolic compounds derived from tropical fruit byproducts may be an effective way to boost the fiber content and antioxidant potential of bakery products.

7.6.2 Functional Compounds of Fruit and Vegetable Byproducts and Their Impact on Gut Microbiota

Epidemiological studies have shown promising evidence for the protective role of vegetables and fruits when consumed regularly in the normal diet against many chronic diseases like CVDs, obesity, diabetes and cancer. 70 kg of fresh fruits/year/person is the minimum recommendation by the health authorities for the consumption of fruits. Overall health can be improved and also the risk of non-communicable diseases can be reduced by consuming more than 400g of fresh vegetables and fruits per day (WHO, 2019). Few industrialized nations consume the recommended amount, however many western nations should boost their consumption because it is below that amount. In 2017, about 3.7 million deaths globally were attributed to insufficient intake of vegetable and fruits (WHO, 2019). This alarming situation have aroused great interest in the fruit industry to propose innovative fruit-based commodities like juices, sorbets, mix beverages, jellies, jams, ice-creams, yogurts and others (Rico et al., 2020).

The wasted fruit and vegetables parts are concentrated with various primary metabolites like sugars and insoluble and soluble fibers, and secondary metabolites such as phytochemicals (carotenoids, polyphenols, glycosylates and phytosterols) (Rico et al., 2020). Byproducts of berry have 25%–50% procyanidins (Struck et al., 2016). Secondary compounds of the plants are produced as a result of their metabolism (primary metabolism), they act as protective agents against UV-B radiation, microbiological predators and herbivores (Zaynab et al., 2018). These secondary metabolites have aroused great interest nutritional interest because of their role in the risk reduction of several chronic diseases like type 2 diabetes, cancer and hypertension (Bohn et al., 2015). Fruit byproducts are a great source of natural pigments and dietary fibers, supporting the international recommendations for fiber intake, maintains the natural color of food and provides sustainability of the agri-industry (de Souza et al., 2019; Medeiros et al., 2019).

Intensive research has addressed the various metabolic pathways that contribute to the release of biologically active compounds from byproducts via the human gut (Farag et al., 2020). Fruit seeds, peels and residues can provide energy to gut microbiota, resulting in a number of bioactive compounds with increased activity or/and bioavailability (Zhang et al., 2018).

An active and diverse population of microorganisms, mostly anaerobes, reside in symbiotic partnerships on the luminal mucosal surface of the human intestine in their natural habitat (Li et al., 2016). Synthesis of nutrients, their digestion and also absorption, immune regulation and pathogen control are all important functions of the gut microbiota (Russell et al., 2011). Gut microbiota may improve the digestion and absorption of food, affect appetite (Perry et al., 2016; Wang et al., 2017) and serve as a barrier to pathogens via environmental acidification, antimicrobial compounds production and competitive exclusion (Lin and Zhang, 2017). As a result, the gut microbiota is critical to the immune system and intestinal mucosa (Li et al., 2016). Dysbiosis (a condition in which there is an imbalance in the microorganisms residing in the human gut) raises the risk of many diseases, including respiratory diseases (Brown et al., 2017), liver disease (Goel et al., 2014), problems with metabolism associated with gastrointestinal disorders (Parker et al., 2018), psychological issues and autoimmune disorders.

The microbial population of the gut is dynamic, that increases in diversity and number in the gut significantly up to 1012 CFU/mL (Duen et al., 2015). The microbial population of the human gut has been divided into distinct categories as a result of molecular biology advancements (Ruminococcus, Bacteroides and Prevotella), suggesting there is a balanced. Microbial symbiotic gut enterotype population that may respond differently to drug and dietary intake. Large food particles and potentially harmful parasites, bacteria and yeasts are prevented from passing through the digestive mucosa by the quantity and diversity of these enterotypes, along with diet and other lifestyle choices (Conlon and Bird, 2015). Eating habits can cause up to 57% of changes in the gut microbiota (Clark and Mach, 2016). Tight junctions are compromised in an unhealthy gut, allowing xenobiotic compounds, undigested or the remnants of food particles and pathogens to pass through the gut and stimulate the immune response, resulting in inflammatory diseases and food allergies (Michielan and DInca, 2015). The microbiota of the gut is an essential "organ" that impacts the host's biology and affects every part of the body (Mehta et al., 2017).

The intimate relationship between the microbiota of the gut and bio-actives derived from fruit and its byproducts may shed light on the importance of secondary metabolites derived from these rejected fractions. Their true biological activity must be further studied, and new approaches for recovering, to concentrate and employ these new compounds and matrices as food, prebiotic ingredients and dietary supplements needs to be discovered (Kowalska et al., 2017).

7.6.3 Impact of Fruit and Vegetable Byproducts–Derived Phenolics and Dietary Fibers on Human Health

Phenolics and dietary fibers are the components of fruit, that when consumed, not only contribute to human nutrition but also improve human health due to antioxidant and their anti-inflammatory effects. They also improve the beneficial microbial population of the gut by stimulating the activity and growth of beneficial microbes (de-Souza et al., 2019). The effects of these bio-actives derived from fruit byproducts have been thoroughly researched, particularly their effects on human metabolic disorders like inflammatory bowel disease, obesity, metabolic syndrome and various other diseases (Coman et al., 2020). Given to the wellness potential of some bioactive compounds found in fruit remains, researchers have been looking into their prebiotic potential in vivo or/and in vitro (Quiros-Sauceda et al., 2018). These investigations found chemicals in fruit residues can stimulate the development of bacteria that are advantageous to humans, producing a prebiotic impact. A complex is referred to as "A dietary fibre concentrate that contains a substantial quantity of natural antioxidants that are linked to non-digestible compounds" when phenols and dietary fiber are combined (Beres et al., 2016). When bound phenolic compounds are combined with fiber from food during intestinal digestion, their antioxidant effects are enhanced compared to free phenols. Once the complex has been liberated from human digestive enzymes, it is being metabolized by the microbes of the gut, which can help to regulate the gut health through minimizing dysbiosis and thus optimize human health in an antioxidant condition (Gustavo et al., 2016).

7.6.4 Impact of Fruit and Vegetable Byproducts on Beneficial Microorganisms and Folate Production

The application of fruit byproducts as potent prebiotic ingredients that promote the activity and growth of beneficial microbes has been extensively researched in recent years (Vieira et al., 2017). Until 2017, prebiotics were only defined in terms of the beneficial effects of dietary fiber, which included ingredients like inulin and fructo-oligosaccharides. These prebiotic ingredients are well-known for their ability to promote the growth of beneficial microbial species such as Bifidobacteria and lactic acid bacteria (*Streptococcus thermophilus*, *Lactobacillus*, etc.). These microbial organisms are widely used in the food sector, particularly in the manufacture of fermented dairy products, as well as by pharmaceutical companies in the manufacturing of biological supplements.

As previously described, the International Scientific Association for Probiotics and Prebiotics published a scientific acceptance document in 2017 establishing a new scope and definition for prebiotics (Gibson et al., 2017). Given that other substrates than dietary fiber may benefit health. The influence of fruit byproducts and substances obtained from these matrices (including phenolic compounds) can be employed to improve the activity and growth of beneficial microorganisms in prebiotics. Fruit byproducts or compounds with significant nutritional value obtained from fruits can be investigated as prebiotics, with possibilities for use in the production of novel functional foods featuring probiotics. As fruit byproducts are acidic in nature and can contain antimicrobial components, they may impair the viability and growth of certain probiotic and starter cultures (de Oliveira et al., 2020). Fruit waste products, on the other hand, could assist in encouraging the growth of advantageous and commensal microorganisms in the gastrointestinal microbiota.

During the process of fermentation of soy beverages, byproducts of passion fruit were found to trigger the growth of *Lactobacillus reuteri RC-14.* The microbes could not grow in the absence of the fruit byproduct or when the residue was combined with fructo-oligosaccharides, implying that phytochemicals in the byproducts of passion fruit have an antimicrobial impact during fermentation. Additional investigations have looked into the effect of fruit byproducts on the growth of probiotic and starter cultures in order to add value to these byproducts, which are discarded in the environment or frequently being utilized as animal feed. Strains of *Bifidobacterium*, *Lactobacillus* and *Streptococcus thermophilus* were stimulated by byproducts of mango (*Mangifera indica*), orange (*Citrus sinensis*), acerola cherry (*Malpighia emarginata*) and passion fruit (*Passiflora edulis f. Flavicarpa*). Nevertheless, along with acerola cherries, these fruit byproducts promoted the growth of *Clostridium perfringens* and *E. coli* (Vieira et al., 2019). Grape (*Vitis vinifera*), known for its functional and nutritional value, has been devoted to the evaluation of bioactive compounds, particularly phenolic compounds such as tannins, phenolic acids and flavonoids, all of which possess substantial antioxidant properties (Beres et al., 2017).

Lactobacillus spp. growth was observed in a fermented beverage study. When grape byproducts were added, activation increased when compared to un-supplemented beverages. Lactobacilli are bacteriostatic against *E. coli*, *L. monocytogenes* and *Salmonella subsp. enterica serovar Enteritidis*, and remained viable during storage (Dias et al., 2018).

According to Albuquerque and his colleagues (2016), the orange byproducts stimulated folate synthesis in the majority of investigated microbial cultures, particularly the probiotics *Bifidobacterium animalis subsp. Lactic Acid BB-12* and *Lactobacillus acidophilus LA-5.* As none of these microorganisms had de novo folate synthesis genes in their genomes, it was hypothesized that the byproducts of the fruit may have generated new pathways of metabolism in these probiotics, that might be contributing to folate synthesis

The value addition of vegetable byproducts has resulted in the isolation of multiple bioactive substances that have potential medicinal value. These biologically active substances could be less expensive as well as more efficient than bioactive compounds obtained from plants having medical benefits. This could help contribute to the long-term viability of medicinal plants and protect them from being extinct (Barbulova et al., 2015). While some vegetables, such as pepper and bitter melon, contain particular bioactive components. But, most vegetables as well as their waste and/or byproducts, share common biological active compounds like as anthocyanins, polyphenols and carotenoids. Table 7.2 lists bioactive compounds derived from plant byproducts.

7.7 VALORIZATION OF NUT BYPRODUCTS

The material discarded during the process of peeling of tree nuts and edible peanuts like pistachios, almonds, hazelnuts and pecans, is referred to as nut byproducts. The valorization of byproducts is of interest because high-value-added substances derived from waste products may play an important part in the long-term sustainability of food systems. Byproducts can be incorporated into the circular economy, an establishing field that includes waste-reduction methods such as effective disposal of waste to avoid energy and biological losses, and contribute to the generation of downstream components, ensuring food safety and considering their future usage (Campos et al., 2020). More than half of the weight produced by the nut business today is made up of byproducts, which means that producers must dispose of a sizable amount of solid waste. Nearly 50 million tons of byproducts, largely made from cashews, almonds and walnuts, were created in 2018 from the 97.6 million tons of nuts produced (FAO, 2020). The largest contributors to such a large amount of waste generated are the United States, Australia, Singapore and India, and so efforts must be made to find a suitable way to utilize this waste that can be provide us valuable compounds. Furthermore, given the large footprint of some of these nuts, growing awareness of their environmental impact has led to an increased interest in their byproducts. For example, raw peanuts, pistachios, hazelnuts and almonds are estimated to yield 0.62, 0.53, 0.40 and 0.23 kg CO_2 eq/kg, respectively (Volpe et al., 2015).

Currently, there aren't many effective methods for dealing with these byproducts, and some even point to potential for higher-fidelity purposes. They have the potential to be used as natural additives that can improve the quality and/or shelf life of a wide variety of foods, as well as functional food ingredients and/or nutraceuticals to protect the health of individuals. They are a great source of dietary fibers and phenolic compounds (Vazquez-Flores, 2018). The nutraceutical prospective of phytochemicals derived from nut byproducts, how they are used in the food industry and their physiological impacts on human well-being are discussed.

7.7.1 Extraction and Valorization of Nut Byproducts

As previously stated, nut processing produces byproducts that are rich in bioactive molecules. This section provides information on the functional and nutritional value of nut byproducts, as well as their positive health effects and potential uses in the future.

7.7.2 Pistachio Byproducts

Pistachios consist of a supple shell encased in a husk that shields the nut kernel. Before the ripening process starts, pistachios are quick picked and processed because the fleshy shell emits tannins that could discolor the shell. Green shells are separated from hard woody shells during pistachio processing to obtain the major byproduct, also called as green shells; pistachio production also produces hard shells and leaves as secondary byproducts. However, the outer green shell is the most visible, accounting for roughly 60% of pistachios (Arjeh et al., 2020). Thus, from 2014 to 2018, the Latin American pistachio industry produced approximately 144 tons of green shell byproduct. Green shells are susceptible to microbial contamination due to their high water content (70%).

7.7.2.1 Chemical Composition of Pistachio Byproducts

Protein, ash, lipids and carbohydrates make up the majority of pistachio green hulls (PGH) (13%, 14%, 20% and 40%, respectively) (Hamed et al., 2020). Polysaccharide isolation from such byproduct carbohydrates is of particular interest. However, the yields of extraction are lower in comparison to other plant polysaccharides. Carbohydrates (80%) are high in PGH's polysaccharide portion, primarily composed of α-xylopyranose, α-arabinofuranose, α-glucopyranose, β-galacturonic acid and β-fructofuranose, while, low in fat and protein (0.04% and 1.80%, respectively) (Akbari-Alavijeh et al., 2018). Pectin is also present in PGH. This substance underwent acid extraction, which left it with high levels of galacturonic acid and low levels of methoxyl pectin. It has a wide range of uses in both the food and pharmaceutical industries as a result of its high purity and ability to produce gels without the addition of sugar (Kazemi et al., 2019).

7.7.2.2 Extraction of Pistachio Byproducts

The added value of PGH is primarily due to the fact that it contains bioactive substances that have significant antioxidant activities, primarily polysaccharide and phenolic compounds, which can be reused as byproducts of mass production. Different extraction techniques and solvents have been analyzed, resulting in increased compound yield (Arjeh et al., 2020). Because the polarity of the PGH compounds is implied by their variability, the solvent used will depend on the type of compound to be extracted, by employing a polar solvent such as water (Tabaraki and Ghadiri, 2016) or by also utilizing a nonpolar organic solvent such as acetone or ethyl acetate and hexane. Greater matrix release is possible with the use of various extraction methods, including supercritical fluid, enzymatic, ultrasonic, gamma-ray irradiation and MAE (Abolhasani et al., 2018).

7.7.2.3 Valorization of Pistachio Byproducts

The TPC of PGH aqueous extract was approximately 197 mg gallic acid equivalent (GAE)/g dry weight (DW), which was a lot more than reported for other fruits and vegetables that are thought to be good sources of such compounds. The major phenolic compounds found in PGH were tannins, flavonoids and phenolic acids; though, phenolic compound distribution varied depending on maturity, environmental conditions and variety (Garavand et al., 2017).

The major flavonoid, quercetin, is generally present as glycosylated quercetin-O-hexoside or quercetin-3-O-rutinoside, depending on pistachio variety and its ripening state (Barreca et al. People, 2016). Anthocyanins, such as anthocyanin-3-O-galactoside, have been reported to exist in glycosylated forms. Phloroglucinol, galloylshikimic acid, theogalline, galloyl-O-hexoside, pyrogallol and catechin were all found (Lalegani et al., 2018). Anacardic acids are natural compounds that occurs in the form of groups, of the Anacardiaceae family. PGH consists of 11 different anacardic acids, which are thought to be the main components found in this byproduct (Erşan et al., 2016). PGH also contains polar compounds such as fatty acids, carotenoids, phytosterols, tocopherols, chlorophyll and essential oils.

7.7.2.4 Health Benefits of Pistachio Byproducts

PGH has been positively linked to a variety of health benefits due to its nutritional content. PGH extracts, both nonpolar and polar, elicited responses that were anti-inflammatory in vitro. In cultures being treated with the oxidant tert-butyl hydroperoxide, PGH methanolic and aqueous extracts, as well as essential oils produced from PGH, enhance cell viability, having cryoprotective effects on human lymphocytes in vitro (Smeriglio et al., 2017). This effect was linked to the presence of monoterpene derivatives or anthocyanin-3-O-galactoside in essential oils, specifically 3-pinene, 4-pinene, α-pinene and D-limonene. Along with the priorily stated impacts, the methanol-based extract of PGH exhibits an anti-melanogenic influence on SKMEL-3 melanoma cells from humans in vitro and is regarded as an effective alternative for the medical management of hyperpigmentation of the skin disorders like melanoma or melasma and may enhance the tyrosinase's inhibitory activity (Sarkhail et al., 2017).

7.7.3 Hazelnut Byproducts

Commercial nuts are those with the hazelnut kernel inside of a hard shell. The hazelnut is best eaten raw (with the skin on) or roasted (without the skin on) and is used in the confectionary business (Bottone et al., 2019). The skin, hard shells, husk and green leafy parts, which are usually discarded during harvesting and processing, are examples of hazelnut byproducts and waste. Skin accounted for 2.5%, nuts for 37.5%–47.5% and hard shells for 50%–60% of the total (Balart et al., 2018). Thus, Latin American countries produced 1642 tons of skin and 36,344 tons of hard shells between the years 2015 and 2018 (FAOSTAT, 2020). Plant-derived foods contain a diverse range of phytochemicals with known biological activities. With continued research on the hazelnut kernel, there has been increased interest in hazelnut byproducts in recent years (Bottone et al., 2019). These waste products could be utilized as useful food ingredients.

7.7.3.1 Chemical Composition of Hazelnut Byproducts

Carbohydrates (6 g/100 g), protein (8.26 g/100 g) and fat (16 g/100 g) are the most abundant components of hazelnuts. The ash content and moisture content of hazelnut skin is low (1.98 g/100 g) (5.8 g/100 g), respectively (Zarie et al., 2018). Tocopherols (226–593 μg/g) are found in skin, with α–tocopherol being the most abundant isomer (Ta and Go kmen, 2015). The fatty acids with the highest concentrations were oleic and linoleic (Ivanovic et al., 2020). Dietary fiber, which accounts for up to 73% of the carbohydrate portion, is of particular concern. Lignin, arabinose and uronic acid (212 mg/L) are also present while, mannose and galactose are present in trace amounts (Zari et al., 2018). Calcium and magnesium were the most abundant minerals in terms of elemental content (85.8 and 113.8 mg/100 g, respectively), followed by potassium, phosphorus and sodium (15.9, 52.6 and 60 mg/100 g, respectively). Other elements (chromium, antimony, selenium, cadmium and iron) as well as trace elements (zinc, barium and copper) have also been identified (Ivanovic et al., 2020).

7.7.3.2 Extraction and Valorization of Hazelnut Byproducts

The skin of hazelnut consists of about 1413 mg GAE/g total phenolic compounds with catechin up to 550 mg (CE)/g of the total flavonoids. As in accordance to the methods of extraction and quantification, total phenolic compounds are 168–378 times higher in roasted and raw nuts, and also considerably higher than in green tea, coffee, cinnamon and dark chocolate, which are some of the most antioxidant-rich foods (Ivanovic et al., 2020). Gallic acid, along with epicatechin, catechin and protocatechuic acid, are the most abundant phenolic compounds found (Krol et al., 2020).

Hydroxycinnamic acid derivatives like caffeic, p-coumaric, sinapinic acids and ferulic have been identified in the skin (Bottone et al., 2019). A green process was used to extract phenolic compounds from the shells of hazelnut (Pe'rez-Armada et al., 2019). Sinapic acids, coumaric, ferulic and caffeic are present in the shell (Shataer et al., 2020).

7.7.3.3 Health Benefits of Hazelnut Byproducts

Since phytochemicals found in hazelnuts are linked to astonishing health benefits (Amaral and Oliveira, 2016), the health benefits of hazelnut skins and shells are limited. In high-fat-fed hamster's ratio diet, ingestion of a fiber-rich skin extract (Fiberox) showed that the lipid profile of plasma was improved and reduced the deoxycholic/lithocholic bile feces (Caimari et al., 2015).

The bioavailability of hazelnut skin polyphenols in beverages was found to be high (27%). On the other hand, immediate ingestion of beverages did not result in statistically significant changes in endothelial/vascular function, despite the fact that numerous manufacturers of particular secondary metabolites were involved (Mocciaro et al., 2019).

7.7.4 Almond Byproducts

Almonds produces plenty of byproducts, including shell, skin and hull (70%–85% of the total amount of almond fruit), which have limited industrial use but exhibit great functional properties and contain a large number of bio-actives which, in industrial applications, can be capitalized. The residue that made up the most of the total fresh weight was green hull, followed by kernel and shell (52%, 15% and 33%, respectively) (Prgomet et al., 2017).

7.7.4.1 Chemical Composition of Almond Byproducts

The chemical makeup of almond skin may vary conditional to the industrial processing techniques (natural peeling or blanching). Lipids (24.2 and 22.2 g/100g natural skin and blanched, respectively) and total dietary fiber (45.1 and 47.5 g/100g natural skin and blanched, respectively) were abundant in almond skin. Additionally, the skin has large levels of total phenolic compounds, which account for 60%–80% of the total polyphenol content of nuts (Prgomet et al., 2017). The endocarp is made up of lignocellulose arrangements that are primarily made up of hemicellulose (19.3%–29%), lignin (20.4%–50.7%) and cellulose (29.8%–50.7%).

7.7.4.2 Characterization and Valorization of Almond Byproducts

Numerous studies have shown that the polyphenol content of nuts is predominantly concentrated in the outer layer or skin; hence, the bulk of this substance is lost during the processing of the nut. Regardless of the type of almonds or how they are processed (raw or roasted), the most prevalent polyphenols in almonds and their derivatives are flavonoids, hydrolyzable tannins and pro-anthocyanidins (61–162 mg/100g of almonds).

Almonds skin polyphenols are composed of relatively low molecular weight polymer compounds, primarily hydroxycinnamic acids (such as ferulic acid, sinapic acid, coumaric acid and caffeic acid), hydroxybenzoic acids (such as protocatechuic acid, vanillic acid and p-hydroxybenzoic acid), some flavonoid groups (aglycones, flavonol glycosides, stilbenes and dihydroflavonols) and aldehydes (Galali et al., 2020). Prodelphinidins of types A and B, progeranidins of types A and B, and pro-anthocyanidins of types A and B are the primary classes of high molecular weight polymers. Seventy percent of the skin's total flavonoid content is made up of isorhamnetin (3-O-rutinoside or 3-O-glucoside). Some phenolic substances (derivatives of chlorogenic acid) were only discovered in hulls. Protocatechuic acid, one of the principal benzoic acid derivatives present in hull, exhibits strong antioxidant effects as well as an inhibitory effect on a number of enzymes. The abundance of flavonoids suggests that they act as phytoalexins, insulating the skin from a variety of stressors (Prgomet et al., 2017).

7.7.4.3 Health Benefits of Almond Byproducts

Byproducts from almonds have interesting health benefits, according to extracts. An oil-in-water emulsion containing 2% blanched almond aqueous extract successfully reduced UV-induced erythema in humans by 50%. Vanillic acid, catechin, p-hydroxybenzoic acid and protocatechin isolated from almond skin were found to have high 2, 2-diphenyl-1-picrylhydrazyl (DPPH) scavenging action.

Skin of almond also demonstrated prebiotic potential by raising *Clostridium Eubacterium rectalum* and *Bifidobacterium* populations while decreasing *Clostridium histolyticum* populations. Changes in bacterial activity of enzymes, like enhanced fecal β-galactosidase action and reduced fecal azoreductase, nitroreductase and β-glucuronidase and actions, also show that almond skin polyphenols benefit the microbiota. Blanched and natural polyphenols derived from almonds' skin exhibited antiviral effects by inhibiting infection caused by herpes simplex virus type 1 (HSV-1) and preventing HSV-1 particle production and spread to neighboring cells. The phenolic compound quercetin was the most prevalent in the extract, proving that virucidal polyphenols could be made from almond skin extracts. Almond hulls contain terpenoids with anti-HIV effects.

7.7.5 Cashew Nut Byproducts

This tree produces a type of cashew known as pseudo fruit, which accounts for 90% of the fruit's weight (Talasila and Shaik, 2015). According to Mohod et al. (2011), the cashew seed, which makes up around 67% of the total weight of the cashew seed and has a double shell attached to the bottom, is considered to be the genuine fruit. During the ripening process, the nuts and apples fall to the ground; the nuts are then separated from the apples for further processing. Nearly 25% of the weight of the cashew nut is made up of liquid, which is regarded as a byproduct with little additional value and makes up the cashew nut shell liquid (CNSL) (Mgaya et al., 2019).

Bagasse is also produced during processing and is regarded as a byproduct. Regardless of these potential uses, just 6% of the product is used, resulting in a large amount of byproducts after the nuts are separated. Because of very little post-harvest life correlated with low industrial application capacity and the lack of economically feasible raw material preservation methods, this byproduct is wasted.

7.7.5.1 Chemical Composition of Cashew Nut Byproducts

Cashew apples are high in fiber, minerals (zinc, sodium, copper, potassium, iron, magnesium, phosphorus and calcium), vitamins and amino acids (Salehi et al., 2019). Moreover, the fruit was found to be an excellent source of low-cost dietary fiber (35%), primarily in the form of insoluble fiber (27%). This byproduct's protein (9.9% dry basis) is primarily made up of the amino acids like aspartic acid, alanine, serine, glutamic acid, proline and threonine. It contains organic acids like lactic, acetic, citric, tartaric and malic acids, as well as reducing sugars like glucose and fructose (6.29.8 g/100 g) (Batista et al., 2018).

7.7.5.2 Characterization and Valorization of Cashew Nut Byproducts

In addition to its nutritional value, cashew apples include a lot of physiologically active compounds, with a concentration of 65–88 mg GAE per 100 g dry basis of total phenolic components (Guedes-Oliveda et al., 2016). According to Batista et al. (2018), the main phenolic compounds included in this fruit are flavonoids, hydroxycinnamic acids, salicylic acid, rutin, 3–4-dihydroxybenzoic acid, naringenin, syringic acid, myricetin and catechin. Cashew apples are an excellent source of natural organic coloring agents, the main carotenoid being β-carotene. It has also been reported that ζ-carotene, α-carotene, cis-β-carotene, lutein, cryptoxanthin, violaxanthin, zeaxanthin and aureoxanthin are present (Talasila and Shaik, 2015). Contrarily, cashew nut shell liquid (CNSL) is a mixture of phenolic compounds with aliphatic side chains, with cardanol (5%), cardanol (18%) and anacardic acid (70%) making up the majority of the mixture (Malhotra et al., 2017).

7.7.5.3 Health Benefits of Cashew Nut Byproducts

The apple juice of cashew has antimicrobial activity against *Helicobacter pylori*, a bacterium that causes stomach problems and has even been linked to stomach cancer development. It also has antitumor, urease and lipoxygenase inhibitory properties (Talasila and Shaik, 2015). The presence of phenolic components such anacardic acid, as well as high quantities of tannins and vitamin C, is connected to these health advantages, which may account for its high antioxidant capacity. Cashew apple byproducts supplemented in rat diets protected against the effects of a dyslipidemia, improved lipid metabolism and gut health (Batista et al., 2018).

7.7.6 Walnut Byproducts

The most widely cultivated tree nuts in the world, walnuts (*Juglans regia* L.), are grown in every continent. Commercial cultivation takes place in the United States, northern Africa, southern Europe, western South America and eastern Asia (Mart et al., 2010). From 2015 to 2018, approximately 655,392 tons of walnuts were produced. Central America accounts for approximately 87% of Latin American nut production, with walnut production increasing by 4%–16% per year (FAO, 2020). Due to their natural health benefits, a large number of byproducts, including broken seed, leaves, shells and husks byproducts, have the potential to be high value-added commodities after extraction of the edible part (kernel) using processing operations due to enormous health benefits (Surmen and Demirba, 2003).

7.7.6.1 Characterization and Valorization of Walnut Byproducts

Walnut kernels or seeds make up 30%–40% of the total weight of the nut, and seed oil makes up 52%–70%. Compared to other nuts, such as almonds and hazelnuts, walnuts exhibit lower quantities of the antioxidant -tocopherol (Pereira et al., 2008). The majority of phenolics in walnuts are found in the skin. The walnut seed coat's average TPC was 12.7 times higher than that of cereals (Trandafir, Cosmulescu et al., 2016). As a result, the majority of the antioxidants are found in the skin of the walnut.

Even though individual walnut constituents have antioxidant activity, scientific data on the antioxidant effects of bioactive substances in walnut skin is still limited. As a result, assessing these properties remains an important task, particularly in the search for new sources of natural antioxidants, functional foods and nutraceuticals. The husk, or green shell that surrounds the walnut, is an agricultural waste that is extensively being employed in traditional medicine for dermatological conditions and alleviating pain. Efficient shell utilization may be a critical issue, as the plentiful supply of phytochemicals will bring attention to the significance of walnut production and recommend uses for a byproduct of mass production (Jahanban-Esfahlan et al., 2019).

7.8 VALORIZATION OF OIL BYPRODUCTS

Waste accounts for 36% of total usable oilseeds in the oilseeds industry, the majority of which is generated during the process of processing and production. The pressed cake that results from oil extraction can be processed into flour, dried and formed into meal and hulls. These leftovers of the oil industry are significant because they include high-value compounds, and their extraction can make them a rich source of bioactive molecules that can be used in both practical applications and academic study. The proportion of macromolecules present in the byproducts is determined by the method of extraction and the raw material being employed (Sagar et al., 2018). The oil industry produces a large number of byproducts; that, with proper treatment, can be used in the formulation of foods. Fiber, protein and amino acid contents along with antioxidants are frequently used to characterize the composition of oil seeds and their byproducts obtained from the oil industry. Hempseed flour and cake, sunflower oil cakes, grape seed and walnut flour are rich sources of protein, amino acids and crude fiber. High content of protein is present in flaxseed and rapeseed meal (Multescu et al., 2022)

Furthermore, considerable supplies of phenolic compounds with antioxidant qualities can be found in byproducts of the oil industry like tocopherols, polyphenols, phenolic acids and flavonoids Multescu et al. (2022) examined the ability of 14 different vegetable byproducts of the oil industry (groats, meals and flour) to act as fat-soluble antioxidants and found that these products may be major sources of phenolic compounds. Consumers' demand for foods containing high levels of health-promoting biomolecules is determined by the health concerns of the consumers. Vegetable oils, that produce byproducts from the processing of oil, offer the high levels of polyunsaturated fatty acids (PUFAs) required for a balanced and nutritious diet. The best-known PUFAs are omega-3 and omega-6, which are essential fatty acids ALA (Alpha Linolenic Acid) and LA (Linolenic Acid), and both can be found in vegetable oils and seeds, but ALA is in lower concentrations.

These edible oil industry byproducts would be efficient in terms of the circular economy (Sharma et al., 2022). To reduce their environmental impact and improve the economic value of plant resources, alternative approaches for the productive utilization of such byproducts or/and wastes must be developed.

7.8.1 Valorization of Olive Oil Byproducts (OOBL)

The increased production of olive oil has fueled global recognition of the health benefits associated with its use in the context of a Mediterranean diet, boosting the sector's economic significance, particularly in the countries that produce more (Stillitano et al., 2017).

The olive oil industry, on the other hand, generates a large amount of unwanted OOBP such as olive stones, olive leaves, olive oil wastewater (OOWW) and olive oil pomace (OOP). The first OOBL obtained during olive pretreatment is olive leaf, according to OOBP. When compared to other common OOBPs, olive pomace is the solid residue left over from the production of olive oil.

Most of the compounds in olive drupes are phenolic in nature. A significant amount of phenolic compounds (approximately 98%) remain in OOBL and just two percent of these bioactive substances are released into olive oil (El-Abbassi et al., 2014).

7.8.2 Olive Oil Byproducts and Valorization

Two different types of residues are created during the production of olive oil, whether it is continuous or discontinuous: (i) a liquid called olive oil effluent or wastewater, which is created in three steps, and (ii) a solid called olive oil residue, which is also known as olive oil pomace (OOP) or prina (Naziri et al., 2014). Olive leaves are also the first waste gathered during the pretreatment stage of olive oil production, which is a standard feature of all methods for extracting olive oil.

7.8.2.1 Olive Leaves

Leaves of olives are a mixture of leaves and twig that accumulates during olive tree pruning and harvesting, as well as during the cleaning process of olives before processing in the olive industry and olive groves. Long ago, this byproduct has been employed for a variety of purposes, including food preservation and health promotion. Several studies over the last decade have linked the beneficial attributes of leaves of olives to their high content of bioactive compounds. The chemical composition of olive leaves was similar to that of lignocellulosic material, with a high percentage of hemicellulose (3.8%), cellulose (5.7%), lignin (39.6%) and extractives (36.52%). Contreras and his colleagues (2020) also found crude protein in unmilled and milled leaves, with values ranging between 9.34% and 8.10%, respectively. These researchers also discovered trace amounts of mannitol. The composition of olive leaf extract can vary in quality and quantity due to factors such as cultivars, cultivation conditions, analytical methods and extraction techniques (Romero-Garca et al., 2014).

The updated literature focuses on the use of olive leaf as an excellent source of antioxidant. Based on their molecular structure, phenolic compounds in olive leaf are classified as simple phenols, phenolic acids, flavonoids, secoiridoids and derivatives of cinnamic acid (Dini et al., 2020).

7.8.2.2 Olive Oil Pomace (OOP)

Olive pomace, which makes up 35%–40% of the total mass of olives processed at the mill, is the most important solid byproduct of the extraction of olive oil. Depending on the oil recovery extraction technology utilized, this biological waste is made up of olive skin, pulp and crushed olive stones as well as varied amounts of water (Chanioti and Tzia, 2018). Depending on the olive type, culture conditions and manufacturing method, this residue's composition varies slightly, but overall, it has a comparable makeup (Rodrigues et al., 2015). There are currently two technologies for producing olive oil: two phase and three phase. Two-phase separation systems yield moist semi-solid pomace (also known as alperujo) with a higher percentage of moisture (50%–70%) than three-phase systems (40%–54%) or conventionally pressed pomace plants (22%–25%), which are difficult to handle and store (Nunes et al., 2018).

A utilized solid known as orujillo or olive cake (dry or extracted) can be produced by further processing the olive pomace in order to collect residual oil (pomace oil). Olive pomace is also a rich source of added value substances such proteins, carbs and phenolics in addition to water (Bermez-Oria et al., 2019). Mineral (4.4%–6%) and oil (7.5%–14%) make up its composition. According to Rodrigues et al. (2015), it mostly consists of lignocellulose, which includes lignin (30.0%–41.6%) and polysaccharides from the cell wall (35.3%–49.0%) as hemicellulose, cellulose and pectin polymers. According to Rodrigues et al. (2015), the presence of squalene in olive pomace is a great source of a variety of phytochemicals, including phenolic compounds, peptides, quercetin and tocopherols.

7.8.2.3 Olive Oil Wastewater (OOWW)

In general, OOWW is an organic residue that contains high levels of phenolic compounds and other important nutrients. Phosphorus (P), nitrogen (N), magnesium (Mg) and potassium (K) are also abundant in this highly conductive liquid. It may contain organic acids and sugars in addition to its high water content (83%–92%), phenolic compounds and minerals (Dermeche et al. 2013). Khoufi et al. (2008) described the liquid–liquid extraction of a phenolic-rich extract from OOWW using ethyl acetate as the extraction solvent. Extraction procedure was implemented as a pretreatment for electrocoagulation to reduce the toxicity of the wastewater exhausted fractions for anaerobic digestion. The main compound present in the OOWW extract, as determined by gas chromatography-mass spectrometry (GC-MS), was hydroxytyrosol (66.5%). Ferulic acid, p-coumaric acid, caffeic acid and tyrosol were also found in trace amounts.

The authors claim that OOWW is a fantastic source of free radicals. Furthermore, Ghanbari et al. (2012) came to the conclusion that the high concentration of phenolic acids and phenolic compounds in OOWW's extracts was principally responsible for their phytotoxic and antibacterial properties.

7.8.2.4 Health Benefits of Olive Oil Byproducts

A number of investigations have documented the antioxidant qualities of OOBL (olive leaf, olive pomace and OOWW) using various assays like the ABTS, FRAP, TEAC, DPPH and ORAC assays. Generally, phenolic compounds like hydroxytyrosol, oleuropein, oleuropein aglycone and the derivatives thereof were found to be significantly associated with the antioxidant properties of OOBL. OOBL have anti-inflammatory effects, although less studied than those focused on antioxidant activity. Extracts of phenolic compounds derived from OOWW and olive pomace were shown to inhibit interleukin-8 production, the pro-inflammatory cytokine present in humans' adenocarcinoma of the colorectal (Caco-2) cells and also reduced the production of NO in lipopolysaccharide macrophages (Di-Nunzio et al., 2018).

Hydroxytyrosol has been shown to inhibit the proliferation and growth of cancer cells, and also induces the process of apoptosis in various in vivo models and cancer cell lines by stimulating cell cycle arrest and regulating the gene expression and pathways of different genes associated with the progression of tumor (Robles-Almazan et al., 2018).

According to several studies (Guex et al., 2019), olive leaf, pomace and peel are the main ingredients with anti-obesity and anti-diabetic properties. These biological effects are attributed primarily to phenolic compounds, particularly oleuropein and its derivatives. In a study, oleuropein prevented cardiac decline in rats suffering from acute myocardial infarction through minimizing the oxidative damage and enhancing the production of pro-inflammatory cytokines (Janahmadi et al., 2015).

7.8.2.5 Application of Olive Oil Byproducts in the Food Industry

Phenolic compounds such as oleuropein, hydroxytyrosol and tyrosol and also the dietary fibers derived from the byproducts of olive oil, exhibit significant biological characteristics (particularly the antioxidant property) and have been scientifically proven to be utilized as preservatives and food additives in many different kinds of food products, such as dairy products, in the oil industry, as well as in the meat industry. Hydroxytyrosol has been indicated to provide an advantage over the rest of the phenolic compounds in that it remains active for a longer period of time, whereas the concentration of other phenolic compounds decreases with time.

The extracted and purified bioactive compounds from the OOBL can be used to improve the stability and functional properties of other foods. Some studies in this field have found that incorporating byproducts' phenolic components (hydroxytyrosol and oleuropein) into olive oil can improve its antioxidant activity (Arajo et al., 2015). In the feed supplementation of animals, olive leaves are used as a food supplement in poultry feed, in order to produce eggs high in long-chain ω-3 fatty acids.

Furthermore, olive byproducts are used in the meat, milk and wine industries, providing some novel food applications. Because of its antimicrobial characteristics, hydroxytyrosol has been employed as an alternative for sulfur dioxide in the process of winemaking. A preventive effect of the phenolic compounds obtained from the olive byproducts in the Maillard reaction was investigated in dairy products. Troise et al. (2014) added phenolic powdered ingredients from OOWW to raw milk prior to ultra-pasteurization, inhibiting the formation of off-flavor compounds during heat treatment and thus improving organoleptic and nutritional properties. Moreover, phenolic compounds' extracts of olive pomace were added to the raw milk as an additional functional ingredient to enhance its properties (Aliakbarian et al., 2015).

7.9 VALORIZATION OF OILSEED MEALS

According to the report of USDA (2021), global production of oilseeds in the year 2019/2020 was 580.69 million metric tons and thus resulting in a significant quantity of oilseed meal, the oilseed's industry byproduct. There are two types of oilseed meal: inedible and edible. The inedible meal derived from linseed has toxic compounds and should only be used as a fertilizer (Gupta et al., 2018). Edible meals, contrary to inedible meals, like soybeans, can be obtained for both human and animal consumption owing to their substantial amount of nutrients and lack of toxins.

Since the majority of oilseeds contain at least 15% fat, this results in a sizeable amount of oilseed meal that is wasted once the oil is extracted. These resources are usually overlooked and seen as waste. The disposal of such waste materials poses health and environmental concerns due to their susceptibility to microbial contamination and economic difficulties. These byproducts are high in valuable substances like phytochemicals, antioxidants, dietary fiber and protein that have beneficial health effects. These underutilized leftovers may be utilized in food preparations as functional ingredients (Ben Jeddou et al., 2017).

Functional food components owing wellness-promoting ability can be created through value-added processes. The incorporation of functional ingredients into food industries (for example, bakeries) is gaining popularity because it reduces the likelihood of chronic diseases while also provide essential nutrients (Eswaran et al., 2013). Oilseed byproducts are a major concern today, and they are in high demand due to the possible health advantages they offer. They produce a large number of byproducts like chia seed powder, peanut meal, soybean okara, kenaf seed meal, hemp seed powder and sesame seed meal (Prakash et al., 2018).

Protein hydrolysates, isolates and concentrates can all be made from oilseed meal. They also provide antioxidants and act as a substrate for the production of antibiotics, enzymes, amino acids, vitamins, bioactive compounds, flavors, surfactants and pigments. In addition, meals of edible oilseed can be added to the diets of undernourished people by including it into nutritional supplements, bakery items and infant food commodities.

7.9.1 Soybean Meals

Soybean is the world's most important oilseed and one of the most widely produced. In 2019, the global production of oilseeds was 580.69 million metric tons, with soybeans contributing to 58% of that total production of oilseed (USDA, 2021). Despite thousands of years of soybean cultivation in Asia, the largest exporters are the United States, Brazil and Argentina (Arrutia et al., 2020). After oil extraction, there is an abundance of soybean meal. In 2019, 244.49 million metric tons of soybean meal was produced, accounting for more than 70% of the world's production of oilseed meal (USDA, 2021).

Soybean meal is high in protein, with concentrations ranging from 33.04% to 56.27% (Rosset et al., 2014), and has long been utilized as fertilizer and animal feed. Additionally, the world's rapidly growing population has raised the need for dietary protein, leading to an increase in efforts to investigate soybean meal as an alternative source of protein. Lujan-Rhenals and colleagues (2014) reported that treatment of soybean meal with dilute acid improved fermentable sugars and protein contents and activity in the soybean meal. They concluded that hydrolysis for 1–4 hours at an acid concentration of 0.9%–1.9% produced a final product with increased protein content as well as sugar contents.

7.9.2 Sesame Seed Meals

One of the oldest and most significant oilseeds grown in Africa, Asia and South America is sesamum. Commonly sesame seeds are used as sesame oil and as raw seeds in baked goods or salads (Saatchi et al., 2019). Sesame contains significant amounts of minerals like zinc and magnesium, carbohydrates, proteins and fats (Karnika and Naik, 2014).

Nantarat et al. (2020) purified and characterized the isolated sesaminol diglucoside (SDG) present in the sesame meal, derived from sesame powder, and observed anti-collagenase, anti-hyaluronidase and antioxidant properties. They discovered that SDG has a high reducing potential as well as iron-reducing antioxidant capacity.

Idowu and colleagues (2021) investigated the composition of amino acids, enzyme inhibitory and antioxidant properties of the products of sesame meal protein, which included protein components (like globulin, albumin, prolamin and glutenin) and their hydrolysates and isolates. Sesame meal powder demonstrated the highest radical scavenging capacity (91.8%).

7.9.3 Potential Food Application of Oilseed Meals

Functional foods have been successfully invented for nearly every food group, such as infant market products, beverage, bakery, dairy and confectionary (Siro et al., 2008). Likewise, food scientists are becoming more intrigued in employing underutilized oilseed meals in foods as functional components.

7.9.3.1 Bakery Products

Bakery items are thought to represent the biggest group of gluten-free options compared to other functional food commodities; thus, edible oilseed meals high in both fiber and protein are suitable for inclusion in supplementary products such as biscuits, bread and cakes. Moreover, meals in the form of nutritional supplements and infant products can be added to the diets of malnourished populations.

Cookies are regarded as one of the most popular bakery snacks and are thought to be an effective means for adding nutrients and value to the diet. The nutritional value of traditional commercial biscuits was improved by replacing wheat flour with a variety of oilseed flours such as flax, sesame, chia seeds and poppy, and differences in dietary value and acceptance by consumers were evaluated by Martinez and colleagues (2021). The oilseed meal-replaced biscuits were higher in fiber and protein, and the carbohydrate content was reduced than the control biscuits. Based on consumer evaluations, flax and sesame cookies outperformed conventional wheat cookies across all study parameters. Bread is another popular bakery product. The inclusion of valuable and nutritious food byproducts like oilseed meal into bread has grown in popularity.

7.9.3.2 Beverage Products

The various dietary characteristics of oilseed meals as well as their functional attributes have sparked considerable interest in their potential use as additional components in a variety of food products, such as beverages. Potential use of flaxseed meal for the making of novel fermented beverages that are comparable to kefir was investigated by Lopusiewicz et al. (2019). Due to fermentation, the beverage has significant antioxidant activity. Because of the functional and nutritional properties imparted to flaxseed meal beverages, the use of kefir grains has evoked effective industrial potential. As a result, these beverages might serve as an innovative non-dairy beverage with beneficial microbial communities, particularly for vegan and lactose-intolerant consumers.

Enzymatic hydrolysis of hemp seed meal yields two fractions: a sediment and a liquid that has distinct organoleptic, nutritional as well as anti-nutritional properties. In comparison to liquids, the sediment fraction contains more vital nutrients (fats, proteins, ash, cannabinoids and pro-anthocyanidins) and has a greater potential for the formulation of various products. Yet, due to its low solubility, the precipitate's application is limited to solid foods (for example, bread, crackers and biscuits). The liquid fraction, on the other hand, is much more soluble in comparison to the sediment phase, allowing it to be added as a supplement for improving the dietary properties of beverages, like smoothies, and other foods that are liquid (Pap et al., 2020). Surprisingly, the few studies on the use of oilseed meal in beverage products have prompted more research in this area.

7.10 CONCLUSION AND FUTURE REMARKS

As demonstrated in this chapter, management of food wastage is a major issue around the world, and new techniques and approaches are being created and implemented in order to minimize its impact on the environment while increasing the value of byproducts derived from agri-food products. Despite the increasing number of safety concerns in this area of study, little research involves safety evaluation investigations including microbiological contaminant assays and physicochemical quality. The worldwide demand for nutritious foods from various sources like fruits and vegetables must satisfy the food requirement while also providing bioactive compounds and other nutrients for a balanced diet. But because of their taste and properties, tropical fruit byproducts such as peels, pits and seeds cannot be eaten directly as food. Tropical fruit byproducts are a rich source of essential bioactive compounds, and converting them into nutritious alternatives remains an important endeavor for food scientists as well as for the food sector.

The findings indicated that assessing the added value of wellness-promoting components or the incorporation of byproducts to create novel products with enhanced functionality is insufficient to encourage their production and also their application. Significant efforts are needed to be done toward developing suitable approaches that enable the sensitive and selective detection of any potentially dangerous substances, as well as effective processes that guarantee appropriate processing without jeopardizing the final product's safety. Furthermore, in most cases, the absence of particular legislation governing the safety and suitability of innovative products to ensure consumer safety renders evaluating the results very complicated and challenging. The dearth of regulation jeopardizes the monetization of many potentially useful agri-food industry byproducts. This situation highlights the need for specific regulations to determine the added value and safety of food byproducts.

In addition, new strategies for demonstrating the safety of the final products obtained from these byproducts are required in this field of study. New international approaches, such as "Foodomics" in conjunction with novel cellular models for examining toxicity without employing animals, may additionally have an intriguing part in this regard, as they can provide data regarding the compound/s causing toxicity as well as the expression levels by employing various techniques like metabolomics, proteomics and/or transcriptomics. Finally, it will be critical in the future to explore the role played by food waste products to the exposome, as they will undoubtedly be coupled with various other environmental exposures that may have an impact on the well-being of humans.

REFERENCES

Abolhasani, A., M. Barzegar and M.A. Sahari. 2018. Effect of gamma irradiation on the extraction yield, antioxidant, and antityrosinase activities of pistachio green hull extract. *Radiation Physics and Chemistry*. 144: 373–378.

Agidew, M.G., A.A. Dubale, M. Atlabachew and W. Abebe. 2021. Fatty acid composition, total phenolic contents and antioxidant activity of white and black sesame seed varieties from different localities of Ethiopia. *Chemical and Biological. Technologies in Agriculture*. 8: 1–10.

Akbari-Alavijeh, S., S. Soleimanian-Zad, M. Sheikh-Zeinoddin and S. Hashmi. 2018. Pistachio hull water-soluble polysaccharides as a novel prebiotic agent. *International Journal of Biological Macromolecules*. 107: 808–816.

Albuquerque, B.R., M.A. Prieto, J.A. Vazquez, M.F. Barreiro, L. Barros and I.C. Ferreira. 2018. Recovery of bioactive compounds from Arbutus unedo L. fruits: Comparative optimization study of maceration/microwave/ultrasound extraction techniques. *Food Research International*. 109: 455–471.

Albuquerque, M.A.C., D. de-Yamacita, R. Bedani, J.G. LeBlanc and S.M. I. Saad. 2019. Influence of passion fruit by-product and fruc- tooligosaccharides on the viability of Streptococcus thermophilus TH-4 and Lactobacillus rhamnosus LGG in folate bio-enriched fer- mented soy products and their effect on probiotic survival and folate bio-accessibility under in vitro simulated gastrointestinal conditions. *International Journal of Food Microbiology*. 292: 126–136.

Albuquerque, M.A.C., R. de-Benadi, A.D. Silva Vieira, J.G.L. LeBlanc and S.M. Isay Saad. 2016. Supplementation with fruit and okara by-products and amaranth flour increases the folate production by starter and probiotic cultures. *International Journal of Food Microbiology*. 236: 26–32.

Aliakbarian, B., M. Casale, M. Paini, A.A. Casazza, S. Lanteri and P. Perego. 2015. Production of a novel fermented milk fortified with natural antioxidants and its analysis by NIR spectroscopy. *Lebensmittel-Wissenschaft and Technologie*. 62(1): 376–383.

Amaral, J.S. and M.B.P.P. Oliveira. 2016. Bioactive compounds of hazelnuts as health promoters. *Natural Bioactive Compounds from Fruits and Vegetables*. 2: 154–178.

Amofa-Diatuo, T., D.M. Anang, F.J. Barba and B.K. Tiwari. 2017. Development of new apple beverages rich in isothiocyanates by using extracts obtained from ultrasound-treated cauliflower by-products: Evaluation of physical properties and consumer acceptance. *Journal of Food Composition and Analysis*. 61: 73–81.

Ancuța, P. and A. Sonia. 2020. Oil press-cakes and meals valorization through circular economy approaches: A review. *Applied Sciences*. 10(21): 7432.

Araujo, M., F. Pimentel, R.C. Alves and M.B.P.P. Oliveira. 2015. Phenolic compounds from olive mill wastes: Health effects, analytical approach and application as food antioxidants. *Trends in Food Science and Technology*. 45(2): 200–211.

Araujo, R.G., R.M. Rodriguez-Jasso, H.A. Ruiz, M. Govea-Salas, M.E. Pintado and C.N. Aguilar. 2020. Process optimization of microwave-assisted extraction of bioactive molecules from avocado seeds. *Industrial Crops and Products*. 154: 112623.

Ares, A.M., J. Bernal, M.J. Nozal, C. Turner and M. Plaza. 2015. Fast determination of intact glucosinolates in broccoli leaf by pressurized liquid extraction and ultra high-performance liquid chromatography coupled to quadrupole time-of-flight mass spectrometry. *Food Research International*. 76: 498–505.

Arjeh, E., H.R. Akhavan, M. Barzegar and A. Carbonell-Barrachina. 2020. Bio-active compounds and functional properties of pistachio hull: A review. *Trends in Food Science and Technology*. 97: 55–64.

Arora, S. and P. Itankar. 2018. Extraction, isolation and identification of flavonoid from Chenopodium album aerial parts. *Journal of Traditional and Complementary Medicine*. 8(4): 476–482.

Arrutia, F., E. Binner, P. Williams and K.W. Waldron. 2020. Oilseeds beyond oil: Press cakes and meals supplying global protein requirements. *Trends in Food Science and Technology* 100: 88–102.

Azmir, J., I.S.M. Zaidul, M.M. Rahman, M.K. Sharif, A. Mohamed, F. Sahena, A.K.M. Omar. 2013. Techniques for extraction of bioactive compounds from plant materials: A review. *Journal of Food Engineering*. 117(4): 426–436.

Bahorun, T., A. Luximon-Ramma, A. Crozier and O.I. Aruoma. 2004. Total phenol, flavonoid, proanthocyanidin and vitamin C levels and antioxidant activities of Mauritian vegetables. *Journal of the Science of Food and Agriculture*. 84(12): 1553–1561.

Balart, J.F., D. Garcia-Sanoguera, R. Balart, T. Boronat and L. SanchezNacher. 2018. Manufacturing and properties of biobased thermoplastic composites from poly (lactid acid) and hazelnut shell wastes. *Polymer Composites*. 39(3): 848–857.

Ballesteros-Vivas, D., G. Alvarez-Rivera, S.J. Morantes, A. del Pilar Sanchez-Camargo, E. Ibanez, F. Parada-Alfonso and A. Cifuentes. 2019. An integrated approach for the valorization of mango seed kernel: Efficient extraction solvent selection, phytochemical profiling and antiproliferative activity assessment. *Food Research International*. 126: 108616.

Banerjee, J., R. Singh, R. Vijayaraghavan, D. MacFarlane, A. Patti and A. Arora. 2017. Bioactives from fruit processing wastes: Green approaches to valuable chemicals. *Food Chemistry*. 225: 10–22.

Barbulova, A., G. Colucci and F. Apone. 2015. New trends in cosmetics: By-products of plant origin and their potential use as cosmetic active ingredients. *Cosmetics*. 2(2): 82–92.

Barreca, D., G. Lagana, U. Leuzzi, A. Smeriglio, D. Trombetta and E. Bellocco. 2016. Evaluation of the nutraceutical, antioxidant and cytoprotective properties of ripe pistachio (*Pistacia vera L*., variety bronte) hulls. *Food Chemistry*. 196: 493–502.

Barreira, J.C., A.A. Arraibi and I.C. Ferreira. 2019. Bioactive and functional compounds in apple pomace from juice and cider manufacturing: Potential use in dermal formulations. *Trends in Food Science and Technology*. 90: 76–87.

Barros, L., R.C. Calhelha, M.J.R.P. Queiroz, C. Santos-Buelga, E.A. Santos, W.C.B. Regis and I.C.F.R. Ferreira. 2015. The powerful in vitro bioactivity of Euterpe oleracea Mart. seeds and related phenolic compounds. *Industrial Crops and Products*. 76: 318–322.

Batista, K.S., A.F. Alves, M. Dos Santos Lima, L.A. Da Silva, P.P. Lins and D. Gomes. 2018. Beneficial effects of consumption of acerola, cashew or guava processing by-products on intestinal health and lipid metabolism in dyslipidaemic female wistar rats. *British Journal of Nutrition*. 119(1): 30–41.

Baysal, S. Su, and M. Ali Ülkü. 2022. Food loss and waste: A sustainable supply chain perspective. In Book: *Disruptive technologies and eco-innovation for sustainable development* edited by Ulas Akkucuk, IGI Global Publishing, Pennsylvania, USA, 90–108.

Ben-Jeddou, K., F. Bouaziz, S. Zouari-Ellouzi, F. Chaari, S. E. Chaabouni, R. Ellouz Ghorbel and O. Nouri-Ellouz. 2017. Improvement of texture and sensory properties of cakes by addition of potato peel powder with high level of dietary fiber and protein. *Food Chemistry*. 217: 668–677.

Ben-Othman, S., I. Joudu and R. Bhat. 2020. Bioactives from agri-food wastes: Present insights and future challenges. *Molecules*. 25(3): 510.

Ben-Romdhane, M., A. Haddar, I. Ghazala, K. Ben Jeddou, C. Boisset Helbert and S. Ellouz-Chaabouni. 2017. Optimization of poly- saccharides extraction from watermelon rinds: Structure, functional and biological activities. *Food Chemistry*. 216: 355–364.

Beres, C., G.N.S. Costa, I. Cabezudo, N.K. Silva-James, A.S.C. da, Teles, A.P.G. Cruz and S.P. Freitas. 2017. Towards integral utilization of grape pomace from winemaking process: A review. *Waste Management*, 68: 581–594.

Beres, C.F., F. Simas-Tosin, I.P. Cabezudo, S. Freitas, M. Iacomini, C. Mellinger-Silva and L.M.C. Cabral. 2016. Antioxidant dietary fibre recovery from Brazilian Pinot noir grape pomace. *Food Chemistry*. 201: 145–152.

Bermudez-Oria, A., G. Rodriguez-Gutierrez, M. Alaiz, J. Vioque, J. Giron-Calle and J. Fernandez-Bola. 2019. Polyphenols associated to pectic polysaccharides account for most of the antiproliferative and antioxidant activities in olive extracts. *Journal of Functional Foods*. 62: 103530.

Bertagnolli, S.M.M., M.L.R. Silveira, A.O. Fogac, L. Umann and N.G. Penna. 2014. Bioactive compounds and acceptance of cookies made with Guava peel flour. *Food Science and Technology*. 34(2): 303–308.

Bhat, R. 2021. Sustainability challenges in the valorization of agri-food wastes and byproducts. In *Valorization of Agri-Food Wastes and Byproducts* (pp. 1–27). Academic Press, London, UK

Bohn, T.J., G. McDougall, A. Alegria, M. Alminger, E. Arrigoni, A.M. Aura and C. Santos. 2015. Mind the gap-deficits in our knowledge of aspects impacting the bioavailability of phytochemicals and their metabolites: A position paper focusing on carotenoids and polyphenols. *Molecular Nutrition and Food Research*. 59: 1307–1323.

Bottone, A., A. Cerulli, G. Durso, M. Masullo, P. Montoro, A. Napolitano and S. Piacente. 2019. Plant specialized metabolites in hazelnut (*Corylus avellana*) kernel and by-products: An update on chemistry, biological activity, and analytical aspects. *Planta Medicine*. 85(11–12): 840–855.

Briones-Labarca, V., M. Plaza-Morales, C. Giovagnoli-Vicuna and F. Jamett. 2015. High hydrostatic pressure and ultrasound extractions of antioxidant compounds, sulforaphane and fatty acids from Chilean papaya (Vasconcellea pubescens) seeds: Effects of extraction conditions and methods. *LWT - Food Science and Technology*. 60(1): 525–534.

Brown, D. 2003. FDA considers health cailms for nuts. *Journal of the American Dietetic Association*. 103(4): 426–427.

Brown, R.L., R.P. Sequeira and T.B. Clarke. 2017. The microbiota protects against respiratory infection via GM-CSF signalling. *Nature Communications*. 8(1): 1512.

Caimari, A., F. Puiggros, M. Suarez, A. Crescenti, S. Laos, J.A. Ruiz and L. Arola. 2015. The intake of a hazelnut skin extract improves the plasma lipid profile and reduces the lithocholic/ deoxycholic bile acid faecal ratio, a risk factor for colon cancer, in hamsters fed a high-fat diet. *Food Chemistry*. 167: 138–144.

Caldeira, C., V. De Laurentiis, S. Corrado, F. van Holsteijn and S. Sala. 2019. Quantification of food waste per product group along the food supply chain in the European Union: A mass flow analysis. *Resources, Conservation and Recycling*. 149: 479–488.

Campos, D.A., R. Gomez-Garcia, A.A. Vilas-Boas, A.R. Madureira and M.M. Pintado. 2020. Management of fruit industrial by-products: A case study on circular economy approach. *Molecules*. 25(2): 320.

Capuzzo, A., M.E. Maffei and A. Occhipinti. 2013. Supercritical fluid extraction of plant flavors and fragrances. *Molecules*. 18(6): 7194–7238.

Castro-Vargas, H.I., D.B. Vivas, J.O. Barbosa, S.J.M. Medina, F.A. Gutierrez and F. Parada-Alfonso. 2019. Bioactive phenolic compounds from the agroindustrial waste of Colombian mango cultivars 'sugar mango' and 'tommy atkins'-An alternative for their use and valorization. *Antioxidants*. 8(2): 41.

Castro-Vazquez, L., M.E. Alanon, V. Rodriguez-Robledo, M.S. Perez- Coello, I. Hermosin-Gutierrez, M.C. Diaz-Maroto and M.D.M. Arroyo-Jimenez. 2016. Bioactive flavonoids, antioxidant behaviour, and cytoprotective effects of dried grapefruit peels (Citrus paradisi macf.). Oxidative Medicine and Cellular Longevity, 8915729. https://doi.org/10.1155/2016/8915729

Chanioti, S. and C. Tzia. 2018. Extraction of phenolic compounds from olive pomace by using natural deep eutectic solvents and innovative extraction techniques. *Innovative Food Science and Emerging Technologies*. 48: 228–239.

Chavez-Gonzalez, M.L., L. Sepulveda, D.K. Verma, H.A. Luna-Garcia, L.V. Rodriguez Duran, A. Ilina and C.N. Aguilar. 2020. Conventional and emerging extraction processes of flavonoids. *Processes*. 8(4): 434.

Clark, A. and N. Mach. 2016. Exercise-induced stress behavior, gut-microbiota-brain axis and diet: A systematic review for athletes. *Journal of the International Society of Sports Nutrition*. 13(1): 43.

Coman, V., B.E. Teleky, L. Mitrea, G.A. Martau, K. Szabo, L.F. Calinoiu and D.C. Vodnar. 2020. Bioactive potential of fruit and vegetable wastes. *Advances in Food and Nutrition Research. Academic Press Inc*. 91: 157–225.

Conlon, M.A. and A.R. Bird. 2015. The impact of diet and lifestyle on gut microbiota and human health. *Nutrients*. 7(1): 17–44.

Contreras, M., M. del, A. Lama-Munoz, F. Espínola, M. Moya, I. Romero and E. Castro. 2020. Valorization of olive mill leaves through ultrasound-assisted extraction. *Food Chemistry*. 314: 126218.

Cujic, N., K. Savikin, T. Jankovic, D. Pljevljakusic, G. Zdunic and S. Ibric. 2016. Optimization of polyphenols extraction from dried chokeberry using maceration as traditional technique. *Food Chemistry*. 194: 135–142.

De-Ancos, B., E. Gonzalez and M.P. Cano. 2000. Effect of high- pressure treatment on the carotenoid composition and the radical scavenging activity of persimmon fruit purees. *Journal of Agricultural and Food Chemistry*. 48(8): 3542–3548.

De-Andrade Lima, M., D. Charalampopoulos and A. Chatzifragkou. 2018. Optimisation and modelling of supercritical CO2 extraction process of carotenoids from carrot peels. *The Journal of Supercritical Fluids*. 133: 94–102.

Del-Pilar Sanchez-Camargo, A., L.F. Gutierrez, S.M. Vargas, H.A. Martinez-Correa, F. Parada-Alfonso and C.E. Narvaez-Cuenca. 2019. Valorisation of mango peel: Proximate composition, supercritical fluid extraction of carotenoids, and application as an antioxidant additive for an edible oil. *The Journal of Supercritical Fluids*. 152: 104574.

De-Oliveira, S.D., C.M. Arau´jo, G. Borges, M. Lima, V.B. Viera and E.F. Garcia. 2020. Improvement in physicochemical characteristics, bioactive compounds and antioxidant activity of acerola (Malpighia emarginata D. C.) and guava (Psidium guajava L.) fruit by-products fermented with potentially probiotic lactobacilli. LWT. 134: 110200.

Dermeche, S., M. Nadour, C. Larroche, F. Moulti-Mati and P. Michaud. 2013. Olive mill wastes: Biochemical characterizations and valorization strategies edited by foxit reader copyright (C) by Foxit Software Company, 2005-2008 for evaluation only. *Process Biochemistry*. 48(10): 1532–1552.

De-Souza, C.B., M. Jonathan, S.M.I. Saad, H.A. Schols and K. Venema. 2019. Degradation of fibres from fruit by-products allows selective modulation of the gut bacteria in an in vitro model of the proximal colon. *Journal of Functional Foods*. 57: 275–285.

Dias, J.F., B.D. Simbras, C. Beres, K.O. dos-Santos, L.M.C. Cabral and M.A.L. Miguel. 2018. Acid lactic bacteria as a bio-preservant for grape pomace beverage. *Frontiers in Sustainable Food Systems*. 2: 58.

Dini, I., G. Graziani, F.L. Fedele, A. Sicari, F. Vinale, L. Castaldo and A. Ritieni. 2020. Effects of trichoderma biostimulation on the phenolic profile of extra-virgin olive oil and olive oil by-products. *Antioxidants*. 9(4): 1–12.

Di-Nunzio, M., G. Picone, F. Pasini, M.F. Caboni, A. Gianotti, A. Bordoni, A. 2018. Olive oil industry by-products. Effects of a polyphenol-rich extract on the metabolome and response to inflammation in cultured intestinal cell. *Food Research International.* 113: 392–400.

Dos-Santos Barbosa, E., D. Perrone, A.L. Amaral Vendramini and S.G. Ferreira Leite. 2008. Vanillin production by Phanerochaete chrysosporium grown on green coconut agroindustrial husk in solid state fermentation. *BioResources.* 3(4): 1042–1050.

Duenas, M., I. Munoz-Gonzalez, C. Cueva, A. Jimenez-Giron, F. Sanchez-Patan, C. Santos-Buelga and B. Bartolome. 2015. A survey of modulation of gut microbiota by dietary polyphenols. *BioMed Research International.* 2015: 850902.

Dulf, F.V., D.C. Vodnar and C. Socaciu. 2016. Effects of solid-state fermentation with two filamentous fungi on the total phenolic contents, flavonoids, antioxidant activities and lipid fractions of plum fruit (*Prunus domestica L.) byproducts. Food chemistry.* 209: 27–36.

El-Abbassi, A., H. Kiai, J. Raiti and A. Hafidi. 2014. Cloud point extraction of phenolic compounds from pretreated olive mill wastewater. *Journal of Environmental Chemical Engineering.* 2(3): 1480–1486.

Encalada, A.M.I., C.D. Perez, S.K. Flores, L. Rossetti, E.N. Fissore and A.M. Rojas. 2019. Antioxidant pectin enriched fractions obtained from discarded carrots (*Daucus carota* L.) by ultrasound-enzyme assisted extraction. Food Chemistry. 289: 453–460.

Ersan, S., O. Guclu Ustundag, R. Carle and R.M. Schweiggert. 2016. Identification of phenolic compounds in red and green pistachio (*Pistacia vera* L.) hulls (exo-and mesocarp) by HPLC-DAD-ESI-(HR)-MS. *Journal of Agricultural and Food Chemistry.* 64(26): 5334–5344.

Espin, J.C., A. Gonzalez-Sarrias and F.A. Tomas-Barberan. 2017. The gut microbiota: A key factor in the therapeutic effects of (poly) phenols. *Biochemical pharmacology.* 139: 82–93.

Esposito, T., F. Sansone, S. Franceschelli, P. del-Gaudio, P. Picerno, R.P. Aquino and T. Mencherini. 2017. Hazelnut (*Corylus avellana* L.) shells extract: Phenolic composition, antioxidant effect and cytotoxic activity on human cancer cell lines. *International Journal of Molecular Sciences.* 18(2): 392.

Eswaran, S., J. Muir and W.D. Chey. 2013. Fiber and functional gastrointestinal disorders. *The American Journal of Gastroenterology.* 108(5): 718–727. FAO (2022). Food wastage footprint impacts on natural resources. Available online: https://www.fao.org/news/story/en/item/196402/icode/ (accessed on 20 March 2023).

FAO. 2023. Sustainability pathways, Food wastage FootprintAvailable online: https://www.fao.org/nr/sustainability/food-loss-and-waste/en/ (accessed on 20 March 2024).

FAO. 2015. Food loss and waste facts. Save food: Global initiative on food loss and waste reduction. Available online: https://www.fao.org/3/i7657e/i7657e.pdf (accessed on 20 March 2024).

FAO. 2017. Food losses and waste in the context of sustainable food systems. https://www.fao.org/3/a-i3901e.pdf

FAO FAOSTAT Agricultural Data. (2020). Available online: https://www.fao.org/faostat/en/#data. *QC* (accessed on 26 March 2021).

Farag, M.A., A. Abdelwareth, I.E. Sallam, M. el-Shorbagi, N. Jehmlich, K. Fritz-Wallace and M. Bergen. 2020. Metabolomics reveals impact of seven functional foods on metabolic pathways in a gut microbiota model. *Journal of Advanced Research.* 23: 47–59.

Farzaneh, V. and I.S. Carvalho. 2017. Modelling of microwave assisted extraction (MAE) of anthocyanins (TMA). *Journal of Applied Research on Medicinal and Aromatic Plants.* 6:92–100.

Fathalizadeh, J., V. Bagheri, H. Khorramdelazad, M.K. Arababadi, A. Jafarzadeh and M.R. Mirzaei. 2015. Induction of apoptosis by pistachio (*Pistacia vera* L.) hull extract and its molecular mechanisms of action in human hepatoma cell line hepg2. *Molecular and Cellular Biology*. 61(7): 128–134.

Faustino, M., M. Veiga, P. Sousa, E.M. Costa, S. Silva and M. Pintado. 2019. Agro-food byproducts as a new source of natural food additives. Molecules (Basel, Switzerland). 24(6): 1056.

Fernandes, F.A.N., T.V. Fonteles, S. Rodrigues, E.S. de-Brito, and B.K. Tiwari. 2020. Ultrasound-assisted extraction of anthocyanins and phenolics from jabuticaba (*Myrciaria cauliflora*) peel: Kinetics and mathematical modeling. *Journal of Food Science and Technology*. 57(6): 2321–2328.

Ferreira, A.S., C. Nunes, A. Castro, P. Ferreira and M.A. Coimbra. 2014. Influence of grape pomace extract incorporation on chitosan films properties. *Carbohydrate Polymers*. 113: 490–499.

Ferreres, F., C. Grosso, A. Gil-Izquierdo, P. Valentao, A.T. Mota and P.B. Andrade. 2017. Optimization of the recovery of high-value compounds from pitaya fruit by-products using microwave-assisted extraction. *Food Chemistry*. 230: 463–474.

Fidelis, M., M.A.V. do-Carmo, T.M. da-Cruz, L. Azevedo, T. Myoda, M. Miranda Furtado and D. Granato. 2020. Camu-camu seed (Myrciaria dubia) from side stream to an antioxidant, antihyper- glycemic, antiproliferative, antimicrobial, antihemolytic, anti-inflammatory, and antihypertensive ingredient. *Food Chemistry*. 310: 125909.

Figueroa, J.G., I. Borras-Linares, J. Lozano-Sanchez, R. Quirantes-Pine and A. Segura Carretero. 2018. Optimization of drying process and pressurized liquid extraction for recovery of bioactive com- pounds from avocado peel by-product. *Electrophoresis*. 39(15): 1908–1916.

Fraga, C.G., K.D. Croft, D.O. Kennedy and F.A. Tomas-Barbera. 2019. The effects of polyphenols and other bioactives on human health. *Food and Function*. 10(2): 514–528.

Friedman, M., N. Kozukue, H.J. Kim, S.H. Choi and M. Mizuno. 2017. Glycoalkaloid, phenolic, and flavonoid content and antioxidative activities of conventional nonorganic and organic potato peel powders from commercial gold, red, and Russet potatoes. *Journal of Food Composition and Analysis*. 62: 69–75.

Fu, Y.J., W. Liu, Y.G. Zu, M.H. Tong, S.M. Li, M.M. Yan and H. Luo. 2008. Enzyme assisted extraction of luteolin and apigenin from pigeonpea (*Cajanus* cajan (L.) Millsp.) leaves. *Food Chemistry*. 111(2): 508–512.

Galali, Y., Z.A. Omar and S.M. Sajadi. 2020. Biologically active components in by-products of food processing. *Food Science and Nutrition*. 8(7): 3004–3022.

Garavand, F., A. Madadlou and S. Moini. 2017. Determination of phenolic profile and antioxidant activity of pistachio hull using high performance liquid chromatography diode array detector electro spray ionization mass spectrometry as affected by ultrasound and microwave. *International Journal of Food Properties*. 20: 19–29.

Ghanbari, R., F. Anwar, K.M. Alkharfy, A.H. Gilani and N. Saari. 2012. Valuable nutrients and functional bioactives in different parts of olive (*Olea europaea L.): A Review. International Journal of Molecular Sciences*. 13(3): 3291–3340.

Ghazala, I., A. Sila, F. Frikha, D. Driss, S. Ellouz-Chaabouni and A. Haddar. 2015. Antioxidant and antimicrobial properties of water-soluble polysaccharide extracted from carrot peels by-products. *Journal of Food Science and Technology*. 52(11): 6953–6965.

Ghendov, M.A., E. Cristea, A. Patras, R. Sturza, S. Padureanu, O. Deseatnicova, N. Turculet, O. Boestean and M. Niculaua. 2020. Potential application of *Hippophae rhamnoides* in wheat bread production. *Molecules*. 25(6): 1272.

Ghosh, D. and P.K. Biswas. 2016. Enzyme-aided extraction of carotenoids from pumpkin tissues. *Indian Chemical Engineer*. 58(1): 111.

Gibson, G.R., R. Hutkins, M.E. Sanders, S.L. Prescott, R.A. Reimer, S.J. Salminen and G. Reid. 2017. Expert consensus document: The international scientific association for probiotics and prebiotics (ISAPP) consensus statement on the definition and scope of prebiotics. *Nature Reviews Gastroenterology and Hepatology*. 14(8): 491–502.

Gil-Sanchez, I., A. Esteban-Fernandez, D. Gonzalez de Llano, M. Sanz Buenhombre, A. Guadarrana, N. Salazar and M.V. Moreno-Arribas. 2018. Supplementation with grape pomace in healthy women: Changes in biochemical parameters, gut microbiota and related metabolic biomarkers. *Journal of Functional Foods*. 45: 34–46.

Goel, A., M. Gupta and R. Aggarwal. 2014. Gut microbiota and liver disease. *Journal of Gastroenterology and Hepatology*. 29: 1139–1148.

Gonzales, G.B., K. Raes, S. Coelus, K. Struijs, G. Smagghe and J. Van-Camp. 2014. Ultra high pressure liquid chromatography electrospray ionization time of flight ion mobility high-definition mass spectrometry for the rapid identification and structural characterization of flavonoid glycosides from cauliflower waste. *Journal of Chromatography A*. 1323: 39–48.

Gornas, P. and M. Rudzinska. 2016. Seeds recovered from industry by products of nine fruit species with a high potential utility as a source of unconventional oil for biodiesel and cosmetic and pharmaceutical sectors. *Industrial Crops and Products*. 83: 329–338.

Gozaydin, G. and A. Yuksel. 2017. Valorization of hazelnut shell waste in hot compressed water. *Fuel Processing Technology*. 166: 96–106.

Guandalini, B.B.V., N.P. Rodrigues and L.D.F. Marczak. 2019. Sequential extraction of phenolics and pectin from mango peel assisted by ultrasound. *Food Research International*. 119: 455–461.

Guedes-Oliveira, J.M., R.L. Salgado, B.R. Costa-Lima, J.G. Oliveira and C.A. Conte-Junior. 2016. Washed cashew apple fiber (*Anacardium occidentale* L.) as fat replacer in chicken patties. *LWT-Food Science and Technology*. 71: 268–273.

Guex, C.G., F.Z. Reginato, P.R. de-Jesus, J.C. Brondani, G.H.H. Lopes and L.F. Bauermann. 2019. Antidiabetic effects of *Olea europaea* L. leaves in diabetic rats induced by high fat diet and low dose streptozotocin. *Journal of Ethnopharmacology*. 235: 1–7.

Gupta, A., R. Sharma, S. Sharma and B. Singh. 2018. Oilseed as potential functional food ingredient. Trends and prospects in food technology, processing and preservation, 1st Ed.; Prodyut Kumar, P., Mahawar, M. K., Abobatta, W., Panja, P., Eds. pp 25–58. Urbana, IL: Today and Tomorrow's Printers and Publishers.

Gustavo, V.R., Q.S. Ana, M.M. Gil, A.Z.J. Fernando, A.G. Humberto, R.S.R. Maribel, G.A.G Adolfo. 2016. Effect of dietary fiber on the bioaccessibility of phenolic compounds of mango, papaya and pineapple fruits by an in vitro digestion model. *Food Science and Technology*. 36: 188–194.

Hamed, M., H. Bougatef, W. Karoud, F. Krichen, A. Haddar and A. Bougatef. 2020. Polysaccharides extracted from pistachio external hull: Characterization, antioxidant activity and potential application on meat as preservative. *Industrial Crops and Products*. 148: 112315.

Hernandez-Brenes, C., P.A. Ramos-Parra, D.A. Jacobo-Velazquez, R. Villarreal-Lara and R.I. La-Garza. 2013. High hydrostatic pressure processing as a strategy to increase carotenoid contents of tropical fruits. *ACS Symposium Series*. 1129: 2942.

Ho, K.K.H.Y., M.G. Ferruzzi, A.M. Liceaga and M.F. Gonzalez. 2015. Microwave-assisted extraction of lycopene in tomato peels: Effect of extraction conditions on all trans and cis isomer yields. *LWT-Food Science and Technology*. 62(1): 160–168.

Idowu, A.O., A.A.N. Famuwagun, T. Fagbemi and R.E. Aluko. 2021. Antioxidant and enzyme inhibitory properties of sesame seed protein fractions and their isolate and hydrolyzate. *International Journal of Food Properties*. 24(1): 780–795.

Ivanovic, S., N. Avramovi, S. Trifunovi, M. Novakovi, V. Tesevi and B. Mandi. 2020. Chemical composition, total phenols and flavonoids contents and antioxidant activity as nutritive potential of roasted hazelnut skins (*Corylus avellana* L.). *Foods (Basel, Switzerland).* 9(4): 430.

Jahanban-Esfahlan, A., A. Ostadrahimi and M. Tabibiazar. 2019. A comparative review on the extraction, antioxidant content and antioxidant potential of different parts. *Molecules.* 24: 2133.

Janahmadi, Z., A.A. Nekooeian, A.R. Moaref and M. Emamghoreishi. 2015. Oleuropein offers cardioprotection in rats with acute myocardial infarction. *Cardiovascular Toxicology.* 15(1): 61–68.

Jia, Y.P., L. Sun, H.S. Yu, L.P. Liang, W. Li, H.J.L. Ding. 2017. The pharmacological effects of lutein and zeaxanthin on visual disorders and cognition diseases. *Molecules.* 22(4): 610.

Karnika, P. and S.N. Naik. 2014. Bioactive constituents as a potential agent in sesame for functional and nutritional application. *Journal of Bioresource Engineering and Technology.* 1: 48–66.

Kazemi, M., F. Khodaiyan, M. Labbafi, S.S. Hosseini and M. Hojjati. 2019. Pistachio green hull pectin: optimization of microwave assisted extraction and evaluation of its physicochemical, structural and functional properties. *Food Chemistry.* 271: 663–672.

Khan, D.T., N.T. Dang, S.H. Jabarov, T.G. Naghiyev, R.M. Rzayev, T.Q. Nguyen, L.V.T. Son. 2019. Study on luminescent properties of Tb3 1 and Sm3 1 co-doped CaSiO3 phosphors for white light emitting diodes. *Materials Research Express.* 7(1): 016507

Khoufi, S., F. Aloui and S. Sayadi. 2008. Extraction of antioxidants from olive mill wastewater and electro-coagulation of exhausted fraction to reduce its toxicity on anaerobic digestion. *Journal of Hazardous Materials.* 151(2–3): 531–539.

Kowalska, H., K. Czajkowska, J. Cichowska and A. Lenart. 2017. What's new in biopotential of fruit and vegetable by-products applied in the food processing industry. *Trends in Food Science and Technology.* 67: 150–159.

Krol, K., M. Gantner, A. Piotrowska and E. Hallmann. 2020. Effect of climate and roasting on polyphenols and tocopherols in the kernels and skin of six hazelnut cultivars (*Corylus avellana* L.). *Agriculture.* 10(2): 36.

Kumar, K., A.N. Yadav, V. Kumar, P. Vyas, P. and H.S. Dhaliwal. 2017. Food waste: A potential bioresource for extraction of nutraceuticals and bioactive compounds. *Bioresources and Bioprocessing.* 4(1). 1–14.

Kunarto, B. and E. Yuniarti Sani. 2018. Antioxidant activity of extract from ultrasonic-assisted extraction of durian peels. *Journal of Applied Food Technology.* 5(2): 25–29

Lalegani, S., H.A. Gavlighi, M.H. Azizi and R.A. Sarteshnizi. 2018. Inhibitory activity of phenolic rich pistachio green hull extract enriched pasta on key type 2 diabetes relevant enzymes and glycemic index. *Food Research International.* 105: 94–101.

Lasta, H.F.B., L. Lentz, N. Mezzomo and S.R.S. Ferreira. 2019. Supercritical CO2 to recover extracts enriched in antioxidant compounds from beetroot aerial parts. *Biocatalysis and Agricultural Biotechnology.* 19: 101169.

Le, H.V. and V.V.M. Le. 2012. Comparison of enzyme assisted and ultrasound assisted extraction of vitamin C and phenolic compounds from acerola (*Malpighia emarginata* DC.) fruit. *International Journal of Food Science and Technology.* 47(6): 1206–1214.

Lin, L., and J. Zhang. 2017. Role of intestinal microbiota and metabolites on gut homeostasis and human diseases. *BMC Immunology.* 18(1): 1–25.

Liu, S., S. Zhang, X. Lv, J. Lu, C. Ren, Z. Zeng and Y. Chen. 2019. Limonin ameliorates ulcerative colitis by regulating STAT3/miR-214 signaling pathway. *International Immunopharmacology.* 75: 105768.

Lopez, N., E. Eduardo Puertolas, S. Santiago Condon, J. Raso and I. Alvarez. 2009. Enhancement of the solid-liquid extraction of sucrose from sugar beet (*Beta vulgaris*) by pulsed electric fields. *LWT-Food Science and Technology.* 42(10): 1674–1680.

Lopusiewicz, L., E. Drozlowska, P. Siedlecka, M. Mezynska, A. Bartkowiak, M. Sienkiewicz and P. Kwiatkowski. 2019. Development, characterization, and bioactivity of non-dairy kefir-like fermented beverage based on flaxseed oil cake. *Foods*. 8(11): 544.

Lujan, R.D., R.O. Morawicki, G.M. Montealvo and Y.J. Wang. 2014. Production of a high-protein meal and fermentable sugars from defatted soybean meal, a co-product of the soybean oil industry. *International Journal of Food Science and Technology*. 49(3): 904–910.

Mahato, N., K. Sharma, M. Sinha and M.H. Cho. 2018. Citrus waste derived nutra-pharmaceuticals for health benefits: Current trends and future perspectives. *Journal of Functional Foods*. 40: 307–316.

Maier, T., A. Goppert, D.R. Kammerer, A. Schieber and R. Carle. 2008. Optimization of a process for enzyme assisted pigment extraction from grape (*Vitis vinifera* L.) pomace. *European Food Research and Technology*. 227(1): 267–275.

Malhotra, S.K., V.N. Hubballi and M.G. Nayak. 2017. *Cashew: Production, Processing, and Utilization of by-Products. Processing and Utilization of by-Products Directorate of Cashewnut and Cocoa Development*. Cochin, Kerala, India.

Man, S., J. Ma, C. Wang, Y. Li, W. Gao and F. Lu. 2016. Chemical composition and hypoglycaemic effect of polyphenol extracts from Litchi chinensis seeds. *Journal of Functional Foods*. 22: 313–324.

Markhali, F.S., J.A. Teixeira and C.M. Rocha. 2020. Olive tree leaves-A source of valuable active compounds. *Processes*. 8(9): 1177.

Mart, M.L., D.O. Labuckas and A.L. Lamarque. 2010. Walnut (*Juglans regia* L.): Genetic resources, chemistry by-products. *Journal of the Science of Food and Agriculture*. 90(12): 1959–1967.

Martinez, E., R.G. Martinez, M. Alvarez, A. Rabadan, A. Pardo Gimenez and J. E. Pardo. 2021. Elaboration of gluten-free cookies with defatted seed flours: Effects on technological, nutritional, and consumer aspects. *Foods*. 10(6): 1213–1219.

Martins, S., S.I. Mussatto, G.M. Avila, J.M. Saenz, C.N. Aguilar and J.A. Teixeira. 2011. Bioactive phenolic compounds: Production and extraction by solid-state fermentation. A review. *Biotechnology Advances*. 29(3): 365–373.

Mateos, A.I., R. Armada, M.L.P. Cozar, P. Ruperez, A.R. Cuenca, and M.J.V. Suarez. 2020. Apple by-product dietary fibre exhibits potential prebiotic and hypolipidemic effects in high-fat fed Wistar rats. *Bioactive Carbohydrates and Dietary Fibre*. 23: 100219.

Mazzutti, S., R.C. Pedrosa and S.R.S. Ferreira. 2021. Green processes in foodomics. *Supercritical Fluid Extraction of Bioactives* Academic Press, London, UK (pp. 725–743).

Medeiros, I.U., J.D.S. Aquino, N. Cavalcanti, A.R.N. Campos, A.M. Cordeiro, K.S. Damasceno, and R.T. Hoskin. 2019. Characterization and functionality of fibre-rich pomaces from the tropical fruit pulp industry. *British Food Journal*. 122(3): 813–826.

Mehmood, T., F. Siddique, S.A. Malik, Q. Tabassam, A.S. Khan, A. Karim and M.A. Shaheen. 2016. Antioxidant attributes, pheno lics acids composition and biological activity of extracts from japanese persimmon (*Diospyros kaki*) as effected by organic solvents and drying technique. *Oxidation Communications*. 39(31): 2260–2279.

Mehta, M., S. Ahmed and G. Dryden. 2017. Immunopathophysiology of inflammatory bowel disease: How genetics link barrier dysfunction and innate immunity to inflammation. *Innate Immunity*. 23(6): 497–505.

Mgaya, J., G.B. Shombe, S.C. Masikane, S. Mlowe, E.B. Mubofu and N. Revaprasadu. 2019. Cashew nut shell: A potential bioresource for the production of bio-sourced chemicals, materials and fuels. *Green Chemistry*. 21(6): 1186–1201.

Michielan, A. and R. D'Inca. 2015. Intestinal permeability in inflammatory bowel disease: Pathogenesis, clinical evaluation, and therapy of leaky gut. *Mediators Inflamm*ation. 2015: 628157.

Miller, M.A. and J.F. Zachary. 2017. *Mechanisms and Morphology of Cellular Injury, Adaptation, and Death. Pathologic Basis of Veterinary Disease Expert Consult* (p. 2). Academic Press, London, UK

Mocciaro, G., L. Bresciani, M. Tsiountsioura, D. Martini, P. Mena, M. Charron, D. Del-Rio. 2019. Dietary absorption profile, bioavailability of (poly) phenolic compounds, and acute modulation of vascular/endothelial function by hazelnut skin drink. *Journal of Functional Foods*. 63: 103576.

Mohod, A., S. Jain and A. Powar. 2011. Cashew nut shell waste: Availability in small-scale cashew processing industries and its fuel properties for gasification. International Scholarly Research Notices, Article ID 346191 | https://doi.org/10.5402/2011/346191

Mosunova, O., J.C. Navarro-Munoz and J. Collemare. 2020. The bio synthesis of fungal secondary metabolites: From fundamentals to biotechnological applications. In *Encyclopedia of Mycology,* Academic Press, London, UK (pp. 458–476).

Multescu, M., I.C. Marinas, I.E. Susman, N. Belc. 2022. Byproducts (flour, meals, and groats) from the vegetable oil industry as a potential source of antioxidants. *Foods*. 11(3): 253.

Nantarat, N., M. Mueller, W.C. Lin, S.C. Lue, H. Viernstein, S. Chansakaow, J. Sirithunyalug and P. Leelapornpisid. 2020. Sesaminol diglucoside isolated from black sesame seed cake and its antioxidant, anti-collagenase and anti-hyaluronidase activities. *Food Bioscience*. 36:100628.

Naqash, F., F.A. Masoodi, S.A. Rather, S.M. Wani and A. Gani. 2017. Emerging concepts in the nutraceutical and functional properties of pectin-a review. *Carbohydrate Polymers*. 168: 227–239.

Nath, P., C. Kaur, S.G. Rudra and E. Varghese. 2016. Enzyme assisted extraction of carotenoid-rich extract from red capsicum (*Capsicum annuum*). *Agricultural Research*. 5(2): 193–204.

Nayak, A. and B. Bhushan. 2019. An overview of the recent trends on the waste valorization techniques for food wastes. *Journal of Environmental Management*. 233: 352 370.

Naziri, E., N. Nenadis, F.T. Mantzouridou and M.Z. Tsimidou. 2014. Valorization of the major agrifood industrial by-products and waste from Central Macedonia (Greece) for the recovery of compounds for food applications. *Food Research International*. 65: 350–358.

Nguyen, T.H.L., L.D. Ngoan, G. Bosch, M.W.A. Verstegen and W.H. Hendriks. 2012. Ileal and total tract apparent crude protein and amino acid digestibility of ensiled and dried cassava leaves and sweet potato vines in growing pigs. *Animal Feed Science and Technology*. 172(34): 171–179.

Nunes, M.A., F.B. Pimentel, A.S.G. Costa, R.C. Alves and M.B.P.P. Oliveira. 2016. Olive by-products for functional and food applications: Challenging opportunities to face environmental constraints. *Innovative Food Science and Emerging Technologies*. 35: 139–148.

Ogutu, F.O. and T.H. Mu. 2017. Ultrasonic degradation of sweet potato pectin and its antioxidant activity. *Ultrasonics Sonochemistry*. 38: 726–734.

Ok, S. and E. Yilmaz. 2019. The pretreatment of the seeds affects the quality and physicochemical characteristics of watermelon oil and its by-products. *JAOCS, Journal of the American Oil Chemists Society*. 96(4): 453–466.

Ordonez, T.A., C.T. Leon, A. Hernandez-Almanza, T. Flores-Guıa, D. Luque Contreras, C.N. Aguilar, and J. Ascacio-Valdes. 2020. Ultrasound-microwave-assisted extraction of poly-phenolic compounds from Mexican "Ataulfo" mango peels: Antioxidant potential and identification by HPLC/ESI/MS. *Phytochemical Analysis*. 32(4): 495–502.

Panja, P. 2018. Green extraction methods of food polyphenols from vegetable materials. *Current Opinion in Food Science*. 23: 173–182.

Pap, N., L. Hamberg, J.M. Pihlava, J. Hellstrom, P. Mattila, M. Eurolaandand A. Pihanto. 2020. Impact of enzymatic hydrolysis on the nutrients, phytochemicals and sensory properties of oil hemp seed cake (*Cannabis sativa* L. FINOLA Variety). *Food Chemistry*. 320: 126530.

Parker, E.A., T. Roy, C.R. D'Adamo and L.S. Wieland. 2018. Probiotics and gastrointestinal conditions: An overview of evidence from the cochrane collaboration. *Nutrition*. 45: 125–134.

Parniakov, O., F.J. Barba, N. Grimi, N. Lebovka and E. Vorobiev. 2016. Extraction assisted by pulsed electric energy as a potential tool for green and sustainable recovery of nutritionally valuable compounds from mango peels. *Food Chemistry*. 192: 842–848.

Paz, M., P. Gullon, M.F. Barroso, A.P. Carvalho, V.F. Domingues, A.M. Gomes and C. Delerue-Matos. 2015. Brazilian fruit pulps as functional foods and additives: Evaluation of bioactive compounds. *Food Chemistry*. 172: 462–468.

Pereira, J.A., I. Oliveira, A. Sousa, I.C.F.R. Ferreira, A. Bento and L. Estevinho. 2008. Bioactive properties and chemical composition of six walnut (*Juglans regia* L.) cultivars. *Food and Chemical Toxicology*. 46(6): 2103–2111.

Perez-Armada, L., S. Rivas, B. Gonzalez and A. Moure. 2019. Extraction of phenolic compounds from hazelnut shells by green processes. *Journal of Food Engineering*. 255: 1–8.

Perry, R.J., L. Peng, N.A. Barry, G.W. Cline, D. Zhang, R.L. Cardone, G.I. Shulman. 2016. Acetate mediates a microbiome-brain-β-cell axis to promote metabolic syndrome. *Nature*. 534(7606): 213–217.

Petkova, C. and Z. Antova. 2015. Proximate composition of seeds and seed oils from melon (*Cucumis melo* L.) cultivated in Bulgaria. *Cogent Food and Agriculture*. 1(1): 1018779.

Prakash, K., S.N. Naik, D. Vadivel, P. Hariprasad, D. Gandhi and S. Saravanadevi. 2018. Utilization of defatted sesame cake in enhancing the nutritional and functional characteristics of biscuits. *Journal of Food Processing and Preservation*. 42(9): 13751.

Prasad, K.N., B. Yang, J. Shi, C. Yu, M. Zhao, S. Xue and Y. Jiang. 2010. Enhanced antioxidant and antityrosinase activities of longan fruit pericarp by ultra-high pressure-assisted extraction. *Journal of Pharmaceutical and Biomedical Analysis*. 51(2): 471–477.

Prgomet, I., B. Goncalves, R. Domınguez-Perles, N. Pascual-Seva and A. Barros. 2017. Valorization challenges to almond residues: Phytochemical composition and functional application. *Molecules*. 22(10): 1774.

Rakariyatham, K., D. Zhou, N. Rakariyatham and F. Shahidi. 2020. Sapindaceae (Dimocarpus longan and Nephelium lappaceum) seed and peel byproducts: Potential sources for phenolic compounds and use as functional ingredients in food and health applications. *Journal of Functional Foods*. 67: 103846.

Rico, X., B. Gullon, J.L. Alonso and R. Yanez. 2020. Recovery of high value-added compounds from pineapple, melon, watermelon and pumpkin processing by-products: An overview. *Food Research International*. 132: 109086

Robles-Almazan, M., M. Pulido-Moran, J. Moreno-Fernandez, C. Ramirez-Tortosa, C. Rodriguez-Garcia and J.L. Quiles. 2018. Hydroxytyrosol: Bioavailability, toxicity and clinical applications. *Food Research International*. 105: 654–667.

Rodrigues, F., F.B. Pimentel and M.B.P.P. Oliveira. 2015. Olive by-products: Challenge application in cosmetic industry. *Industrial Crops and Products*. 70: 116–124.

Rodriguez De Luna, S.L., R.E. Ramirez-Garza and S.O. Serna Saldivar. 2020. Environmentally friendly methods for flavonoid extraction from plant material: Impact of their operating conditions on yield and antioxidant properties. *Scientific World Journal*. 2020: 6792069.

Rodriguez-Perez, C., B. Gilbert-Lopez, J.A. Mendiola, R. Quirantes-Pine, A. Segura-Carretero and E. Ibanez. 2016. Optimization of microwave-assisted extraction and pressurized liquid extraction of phenolic compounds from Moringa oleifera leaves by multiresponse surface methodology. *Electrophoresis*. 37(13): 1938–1946.

Romero-Garcia, J.M., L. Nino, C. Martinez-Patino, C. Alvarez, E. Castro and M.J. Negro. 2014. Biorefinery based on olive biomass. State of the art and future trends. *Bioresource Technology*. 159: 421–432.

Rosset, M., V.R. Acquaro and A.D. Beleia. 2014. Protein extraction from defatted soybean flour with viscozyme L pretreatment. *Journal of Food Processing and Preservation*. 38(3): 784–790.

Russell, D.A., R.P. Ross, G.F. Fitzgerald and C. Stanton. 2011. Metabolic activities and probiotic potential of bifidobacteria. *International Journal of Food Microbiology*. 149(1): 88–105.

Saatchi, A., H. Kiani and M. Labbafi. 2019. A new functional protein-polysaccharide conjugate based on protein concentrate from sesame processing by-products: functional and physico-chemical properties. *International Journal of Biological Macromolecules*. 122: 659–666.

Sagar, N.A., S. Pareek, S. Sharma, E.M. Yahia, M.G. Lobo. 2018. Fruit and vegetable waste: bioactive compounds, their extraction, and possible utilization. *Comprehensive Reviews in Food Science and Food Safety*. 17: 512–531.

Saha, D. and S. Bhattacharya. 2010. Hydrocolloids as thickening and gelling agents in food: A critical review. *Journal of Food Science and Technology*. 47: 587–597.

Salehi, B., M. Gultekinzguven, C.O. Kirkin, M.F.B. Morais-Braga and M. Carneiro. 2019. Anacardium plants: Chemical, nutritional composition and biotechnological applications. *Biomolecules*. 9(9): 465.

Sanchez-Mesa, N., J.U. Sepulveda-Valencia, H.J. Ciro-Velasquez, M.A. Meireles. 2020. Compuestos bioactivos de cá scara de mango (Mangifera indica L. var. Tommy Atkins) obtained by supercritical fluids and pressurized liquids extraction. *Revista Mexicana de Ingeniera Quimica*. 19(2): 755–766.

Santos, T.R.J., and L.C.L. de-Aquino Santana. 2019. Antimicrobial potential of exotic fruits residues. *South African Journal of Botany*. 124: 338–344.

Sarkhail, P., M. Salimi, P. Sarkheil and H.M. Kandelous. 2017. Anti- melanogenic activity and cytotoxicity of Pistacia vera hull on human melanoma skmel-3 cells. *Acta Medica Iranica*. 55(7): 422–428.

Sauceda, A.E.Q., R. Pacheco-Ordaz, J.F. Ayala-Zavala, A.H. Mendoza, A.F. Gonzalez-Cordova, B. Vallejo-Galland and G.A. Gonzalez-Aguilar. 2018. Impact of fruit dietary fibres and polyphenols on modulation of the human gut microbiota. In *Fruit and Vegetables Phytochemicals: Chemistry and Human Health*, Edited by: Elhadi M Yahia, John Wiley & Sons, NJ, USA(2nd Ed., pp. 405–422).

Sharma, P., V.K. Gaur, S. Gupta, S. Varjani, A. Pandey, E. Gnansounou, S. You, N.H. Ngo and J.W.C. Wong. 2022. Trends in mitigation of industrial waste: global health hazards, environmental implications and waste derived economy for environmental sustainability. *Science of Total Environment*. 811: 152357.

Shataer, D., R. Abdulla, Q.L. Ma, G.Y. Liu and H.A. Aisa. 2020. Chemical composition of extract of corylus avellana shells. *Chemistry of Natural Compounds*. 56: 338–340.

Shin, S.K., S.J. Cho, U.J. Jung, R. Ryu and M.S. Choi. 2016. Phlorizin supplementation attenuates obesity, inflammation, and hyperglycemia in diet-induced obese mice fed a high-fat diet. *Nutrients*. 8(2): 92.

Sigurdson, G.T., P. Tang and M.M. Giusti. 2017. Natural colorants: Food colorants from natural sources. *Annual Review of Food Science and Technology*. 8: 261–280.

Siro, I., E. Kapolna, B. Kapolna and A. Lugasi. 2008. Functional food. Product development, marketing and consumer acceptance-A review. *Appetite*. 51(3): 456–467.

Smeriglio, A., M. Denaro, D. Barreca, A. Calderaro, C. Bisignano and G. Ginestra. 2017. In vitro evaluation of the antioxidant, cytoprotective, and antimicrobial properties of essential oil from *Pistacia vera* L. variety bronte hull. *International Journal of Molecular Sciences*. 18(6): 1212.

Soeng, S. 2015. Antioxidant and hypoglycemic activities of extract and fractions of Rambutan seeds (*Nephelium lappaceum* L.). *Biomedical Engineering*. 1: 1.

Sojka, M., K. Kołodziejczyk, J. Milala, M. Abadias, I. Vinas, S. Guyot and A. Baron. 2015. Composition and properties of the polyphenolic extracts obtained from industrial plum pomaces. *Journal of Functional Foods*. 12: 168–178.

Stillitano, T., A.I. De-Luca, N. Iofrida, G. Falcone, E. Spada, and G. Gulisano. 2017. Economic analysis of olive oil production systems in Southern Italy. *Quality Access to Success*. 18(157): 107–112.

Struck, S., M. Plaza, C. Turner and H. Rohm. 2016. Berry pomace-A review of processing and chemical analysis of its polyphenols. *International Journal of Food Science and Technology*. 51(6): 1305–1318.

Sudha, M.L., S.M. Dharmesh, H. Pynam, S.V. Bhimangouder, S.W. Eipson, R. Somasundaram and S.M. Nanjarajurs. 2016. Antioxidant and cyto/DNA protective properties of apple pomace enriched bakery products. *Journal of Food Science and Technology*. 53(4): 1909–1918.

Surek, E. and A.O. Buyukkileci. 2017. Production of xylooligosaccharides by autohydrolysis of hazelnut (*Corylus avellana* L.) shell. *Carbohydrate Polymers*. 174: 565–571.

Surmen, Y. and A. Demirba. 2003. Cofiring of biomass and lignite blends: resource facilities; technological and environmental issues. *Energy Sources*. 25(3): 175–187.

Tabaraki, R. and F. Ghadiri. 2016. Comparative study of extraction methods for pistachio hull antioxidants by multiple assays. *Journal of Applied Chemistry*. 10: 19–29.

Talasila, U. and K.B. Shaik. 2015. Quality, spoilage and preservation of cashew apple juice: A review. *Journal of Food Science and Technology*. 52(1): 54–62.

Tas, N.G. and V. Gokmen. 2015. Bioactive compounds in different hazelnut varieties and their skins. *Journal of Food Composition and Analysis*. 43: 203–208.

Thomas, M., A. Badr, Y. Desjardins, A. Gosselin and P. Angers. 2018. Characterization of industrial broccoli discards (*Brassica oleracea* var. italica) for their glucosinolate, polyphenol and flavonoid contents using UPLC MS/MS and spectrophotometric methods. *Food Chemistry*. 245: 1204–1211.

Trandafir, I., S. Cosmulescu, M. Botu and V. Nour. 2016. Antioxidant activity, and phenolic and mineral contents of the walnut kernel (*Juglans Regia* L.) as a function of the pellicle color. Fruits 71(3):177–184.

Trigo, J.P., E.M. Alexandre, J.A. Saraiva and M.E. Pintado. 2022. High value-added compounds from fruit and vegetable byproducts characterization, bioactivities, and application in the development of novel food products. *Critical Reviews in Food Science and Nutrition*. 60(8): 1388–1416.

Troise, A.D., A. Fiore, A. Colantuono, S. Kokkinidou, D.G Peterson and V. Fogliano. 2014. Effect of olive mill wastewater phenol compounds on reactive carbonyl species and maillard reaction end-products in ultrahigh-temperature-treated milk. *Journal of Agricultural and Food Chemistry*. 62(41): 10092–10100.

Um, M., T.H. Han and J.W. Lee. 2018. Ultrasound-assisted extraction and antioxidant activity of phenolic and flavonoid compounds and ascorbic acid from rugosa rose (Rosa rugosa Thunb.) fruit. Food *Science and Biotechnology*. 27(2): 375–382.

USDA. 2021. Oilseeds: World Markets and Trade. Foreign Agricultural Service/USDA. https://apps.fas.usda.gov/psdonline/app/index.html#/app/home.%0Ahttps://apps.fas.usda.gov/psdonline/circulars/oilseeds.pdf.

Vallejo-Castillo, V., J. Munoz-Mera, M.F. Perez-Bustos and A. Rodriguez- Stouvenel. 2020. Recovery of antioxidants from papaya (Carica papaya L.) peel and pulp by microwave-assisted extraction. *Revista Mexicana de-Ingenieria Quimica*. 19(1): 85–99.

Varastegani, B., W. Zzaman and T.A. Yang. 2015. Investigation on physicochemical and sensory evaluation of cookies substituted with papaya pulp flour. *Journal of Food Quality*. 5: 175–183.

Vasquez-Villanueva, R., M. Plaza, M.C. Garcia and M.L. Marina. 2020. Recovery and determination of cholesterol-lowering compounds from Olea europaea seeds employing pressurized liquid extraction and gas chromatography-mass spectrometry. *Microchemical Journal*. 156: 104812.

Vazquez-Flores, A.A. 2018. By-products of the nut and peanut agro industry as sources of phytochemicals suitable for the nutraceutical and food industries. In: Plant food by-products: Industrial relevance for food additives and nutraceuticals, Edited by J. Fernando Ayala-Zavala, Gustavo González-Aguilar, Mohammed Wasim Siddiqui (p. 201220). Apple Academic Press, New York, USA.

Vieira da Silva, B., J.C.M. Barreira and M.B.P.P. Oliveira. 2016. Natural phytochemicals and probiotics as bioactive ingredients for functional foods: Extraction, biochemistry and protected-delivery technologies. *Trends in Food Science and Technology*. 50: 144–158.

Vieira, A.D.S., R. Bedani, M.A.C. Albuquerque, V. Biscola and S.M.I. Saad. 2017. The impact of fruit and soybean by-products and amaranth on the growth of probiotic and starter microorganisms. *Food Research International*. 97: 356–363.

Vieira, A.D.S., V. Biscola, M.A.C. de-Albuquerque, R. Bedani and S.M.I. Saad. 2019. Impact of acerola (*Malpighia emarginata* DC) by-product and probiotic strains on technological and sensory features of fermented soy beverages. *Journal of Food Science*. 84(12): 3726–3734.

Volpe, R., S. Messineo, M. Volpe and A. Messineo. 2015. Carbon footprint of tree nuts-based consumer products. *Sustainability*. 7(11): 14917–14934.

Vu, H.T., C.J. Scarlett and Q.V. Vuong. 2018. Phenolic compounds within banana peel and their potential uses: A review. *Journal of Functional Foods*. 40: 238–248.

Vulic, J.J., T.N. Cebovic, J.M. Canadanovic-Brunet, G.S. Cetkovic, V.M. Canadanovic, S.M. Djilas and V.T.T. Saponjac. 2014. In vivo and in vitro antioxidant effects of beetroot pomace extracts. *Journal of Functional Foods*. 6: 168–175.

Wang, B., M. Yao, L. Lv, Z. Ling and L. Li. 2017. The human micro-biota in health and disease. *Engineering*. 3(1): 71–82.

Wang, F., J. Gong, X. Zhang, Y. Ren and J. Zhang. 2018. Preparation of biocolorant and eco-dyeing derived from polyphenols based on laccase-catalyzed oxidative polymerization. *Polymers*. 10(2): 154

Wang, X., X. Sun, H. Liu, M. Li and Z. Ma. 2011. Barrier and mechanical properties of carrot puree films. *Food and Bioproducts Processing*. 89(2): 149–156.

Wang, Z., B. Xu, H. Luo, K. Meng, Y. Wang, M. Liu and T. Tu. 2020. Production pectin oligosaccharides using Humicola insolens Y1-derived unusual pectate lyase. *Journal of Bioscience and Bioengineering*. 129(1): 16–22.

World Health Organization, e-Library of Evidence for Nutrition Actions (eLENA). Increasing fruit and vegetable consumption to reduce the risk of noncommunicable diseases, (2019). Retrieved July 4, 2021, from https://www.who.int/elena/titles/fruit_vegetables_ncds/en/.

Wunder, S., K. McFarland, M. Hirschnitz-Garbers, J. Parfitt, K. Luyckx, D. Jarosz and E. van-Herpen. 2018. Food Waste Prevention and Valorisation, Relevant EU Policy Areas. *Review of EU Policy Areas with Relevant Impact on Food Waste Prevention and Valorisation*. ReFresh Deliverable, 3.

Xu, Y., T. Bao, W. Han, W. Chen, X. Zheng and J. Wang. 2016. Purification and identification of an angiotensin I-converting enzyme inhibitory peptide from cauliflower by-products protein hydrolysate. *Process Biochemistry*. 51(9): 1299–1305.

Yaemchuen, N., J. Wichaphon and W. Klangpetch. 2018. Antioxidant and antibacterial activities of natural red colorants from red dragon fruit peel and roselle. In *The International Conference on Food and Applied Bioscience*, Chiang Mai, Thailand, (pp. 163–167).

Yin, M., N. Jiang, L. Guo, Z. Ni, A.Y. Al-Brakati, M.S. Othman and R.B. Kassab. 2019. Oleuropein suppresses oxidative, inflammatory, and apoptotic responses following glycerol-induced acute kidney injury in rats. *Life Sciences*. 232: 116634.

Zari, Y.U.O.F., B. Bilesimi and A. Aktivite. 2018. Hazelnut testa as a by-product: Nutritional composition, antioxidant activity, phenolic compound profile and dietary fiber content. *Journal of Faculty of Pharmacy of Ankara University*. 42(3): 38–57.

Zaynab, M., M. Fatima, S. Abbas, Y. Sharif, M. Umair, M.H. Zafar and K. Bahadar. 2018. Role of secondary metabolites in plant defense against pathogens. *Microbial Pathogenesis*, 124: 198–202.

Zhang, N., Z. Ju and T. Zuo. 2018. Time for food: The impact of diet on gut microbiota and human health. *Nutrition*. 51: 80–85.

Zhang, Q., M.M. Zhou, P.L. Chen, Y.Y. Cao and X.L. Tan. 2011. Optimization of ultrasonic-assisted enzymatic hydrolysis for the extraction of luteolin and apigenin from celery. *Journal of Food Science*. 76(5): 680–685.

8 Emerging Non-thermal Food Preservation Techniques

Waseem Khalid
University Institute of Food Science and Technology, The University of Lahore

Izza Faiz ul Rasool, Hadia Hanif, Muhammad Sajid Arshad, and Fareed Afzal
Government College University Faisalabad

Hadiqa Faiz ul Rasul
University of Agriculture Faisalabad

8.1 INTRODUCTION

Humans have been using different methods to preserve food for storage since ancient times. High temperature, drying, pickling, smoking, preserving, cooling, and freezing are some of the methods used to prevent microbial spoilage, which can result in significant economic loss. Approximately 25% of the world's food supply is lost to spoilage. Food that has been spoiled no longer has the same flavor, texture, look, color, or odor. The environment is full of microbes, which can potentially contaminate food anywhere from the farm to the table. In order to ensure safety, prevent spoilage, extend shelf life, improve maintaining quality, prevent food poisoning, and reduce financial loss, food preservation is crucial (Pal, 2017). The increasing population has led to the need for wholesome and nourishing food, which poses a new problem. Critical health problems and new diseases are possible as a result of the microbiological spread in the intricate food supply system. As a result, sanitation and disinfection are of utmost significance in the food sector. Thermal treatment or chemical use is the most typical method of food disinfection, but these methods can have adverse effects such as changes in sensory and functional properties and reductions in nutritional content, making the food less appealing (Paniwnyk, 2017). Conventional techniques can leave chemical residue and have a major adverse effect on both sensory and nutritional qualities (Dai et al., 2020). The use of heat, known as thermal treatment, is a popular method to eliminate microorganisms that cause foodborne illnesses. However, it is not always effective. The application of heat can change the structural properties of proteins and alter sensory characteristics

 DOI: 10.1201/9781003412489-8

such as appearance, flavor, texture, and chemical properties, which can affect the tenderness and toughness of food. Boiling, pan-frying, roasting, deep-frying, and grilling are some of the most commonly used methods in thermal treatment (Khalid et al., 2023). Currently, there is a growing significance of food safety and nutrition. With the advancements in technology and the widespread access to information through the internet, consumers in the 21st century are more informed and have higher expectations for the safety and quality of the food they consume. They want safe, healthy, and high-quality foods that contain the necessary nutrients but without the usage of preservatives (Pollock et al., 2017). Historically, the food industry has relied on preservatives and thermal processing to prolong the shelf life of food. However, it is now well known that heat can have negative effects, and the industry is moving away from minimum and non-thermal technologies due to their detrimental effects on the nutritional value and sensory food quality (Mandal et al., 2020). Non-thermal food dispensation is the term for procedures that kill bacteria in food without using heat generated (Bhattacharjee et al., 2019). Traditional thermal food manufacturing techniques are progressively being replaced by these cutting-edge innovations, including pulse electromagnetic current, elevated processing, ozone treatment, non-thermal serum plasma, and ultrasonic technology. They are often combined with hurdle technology to achieve the best results. Physical processes, which also include pulsed electric fields, elevated processing, ultraviolet radiation, transmitted at high, ultrasound, and radiation, and biochemical mechanisms, which include ozone treatment or cold plasma, can be broadly divided into two categories (Chacha et al., 2021). Alternatives to conventional thermal food processing techniques are growing in popularity. Examples include zonation non-thermal blood serum plasma, pulsed electromagnetic current (peak expiratory), elevated processing, and ultrasonic technology. These techniques preserve food's nutritive, sensory, and physicochemical qualities while being more energy-efficient and speeding up processing time. Moreover, these methods can significantly reduce the food product's allergenicity (Pi et al., 2021).

8.2 DIFFERENT NON-THERMAL TECHNIQUES FOR FOOD PROCESSING

Non-thermal techniques for processing food are becoming an increasingly popular alternative to traditional thermal methods. These new technologies bring several benefits, such as enhancing food quality and safety, providing higher level of automation and precision in food processing. Conditioning food before cooking using non-thermal techniques can increase safety and quality of the cooked meal. These cutting-edge food heating technologies can take the place of conventional heating methods and thermal energy in preparation. Additionally, they can shorten the cooking process, improve quality elements, improve processing effectiveness, and ensure product safety (Chen et al., 2022) (Table 8.1).

TABLE 8.1
Different Non-thermal Food Preservation Techniques and Their Role in Food Preservation

Food Product	1. Sub-Food Product	2. Preservation Technique	3. Preservative Role	4. References
Meat product	Beef nugget	—	Increase shelf life and functional properties of beef nuggets	Arshad et al. (2022)
	Chicken and ostrich patties	Gamma irradiation	Increase shelf life	Khalid et al. (2021)
	White chicken meat	High hydrostatic pressure	Improve the overall texture and quality of cooked chicken meat	Ros-Polski et al. (2015)
	Cooked beef	Field of electric pulsations	In vitro simulations of the gastrointestinal protein metabolism of cooked beef are affected by pulsed voltage	Khalid et al. (2022)
Bakery products	Fresh sliced bread	—	Maintain baked goods fresh for longer than just a few days	Rahman et al., 2022
	Pastries	—	Increase shelf life	Wickramarachchi et al. (2015)
Dairy products	Yogurt	Sonication	Increasing food shelf life by improving product quality and increasing durability	Zamani Mazdeh et al. (2017)
	Cheese	Processing at high pressure	Increasing food shelf life by improving product quality and increasing durability	Zamani Mazdeh et al. (2017)
Beverage Products	Wine	Ultrasonication	Preventing microbiological deterioration of wine during production and storage	Van Wyk & Silva (2019)
	Apples, tomato, orange, sour cheerful nectar	Processing at high pressure	Handling good nutritional value and excellent sensory and textural qualities	Dasan & Boyaci (2018)

8.2.1 Processing at High Pressure

The increasing claim for high-quality, convenient, and natural-tasting food products has commanded to a growing interest in minimally processed foods. This has led to eliminate the usage of humectants and preservatives as additions in food production. To meet these requirements while ensuring food safety, newer preservation technologies are being implemented in the food industry. High-pressure processing, commonly referred to as cold pasteurized, is one such method. High pressure is employed in this non-thermal meat processing method to kill pathogens at room temperature, prolonging the product's shelf life without affecting its security (Khan et al., 2017).

Processing at high pressure is a non-thermal preservation method that utilizes high pressure to sterilize food. In this process, a steady pressure of 300–600 MPa is applied to a vessel containing food samples. The kind of food being consumed and the microorganisms that are being processed will determine the processing duration and pressure. The food processing sector frequently uses this technique to preserve a variety of food items like carrots, cauliflower, crab, meat, and more (Barbhuiya et al., 2021).

One improvement of this technology is that it does not depend on the food's packaging or volume, enabling the processing of various food volumes in the same batch. With no need for preservatives and while preserving the nutritional value and natural tastes of the natural food source, this method also guarantees the food's microbiological safety.

Processing at high pressure is described as minimum handling technology that protects food safety and flavor. It is completed around room temperature which decreases the amount of energy used throughout the heating and cooling processes. In addition, food is packed to guard against subsequent food contamination after pasteurization. Also, because the heat transfer medium may be recycled after processing, this method of food preservation is environmentally benign. High-pressure processing (HPP) technology is becoming a more widely utilized technology in the processing of foods because to its benefits of low energy usage and reduced contamination risk (Huang et al., 2017).

The manufacturing of several food and beverage items, including meat, dairy, marine, vegetable, and fruit products, makes extensive use of processing at high-pressure technology. Over 500,000 tons of goods processed at high pressure are sent around the world each year, with ready-to-eat beef goods like burger patties being the most popular use. The popularity of HPP drinks, particularly coconut water, has increased as a result of the clean label trend in industrialized nations. In the future, in the HPP food market, juice goods may have a greater share of the market than meat products (Peerun - Visiongain, 2015).

In the beverage industry, HPP is widely used to improve the quality as well as microbiological safety of juices. Many researches have been done to determine how HPP affects the enzymatic activity of various fruit juices, nectars, pulp purees, smoothies, and pastes, showing a substantial effect on the preservation of these products (Roobab et al., 2022).

8.2.2 Pulsed Electric Field

The pulsed electric field (PEF) technology uses electricity with a high voltage and short duration, in the 20–80 kV range, as a non-heat-based processing technique. Food sample is placed between electrodes or is exposed to an electrical current. Capacitor is a crucial component in PEF, as it stores the electrical energy and releases it in short bursts to prevent heating of the food. This results in a higher quality of the food sample without affecting its nutritional value (Asaithambi et al., 2021). Electroporation is a technology that uses electric fields to create temporary pores in the surface of a food sample. This process inactivates unfit or pathogenic microorganisms existing in the sample, making it a popular and effective method for food and beverage industries. Products like soups, juices, seeds, and dairy are commonly treated with electroporation (Jan et al., 2017).

Juices, dairy products, milk, soups, and even liquid eggs are just few of the rare food products that could be pasteurized by means of the PEF technology. There are several restrictions, though, namely that the air bubbles must not exist in food products, and they must have a relatively low electrical conductivity and have particles smaller than the treatment region's gap. These limitations are important to consider to ensure appropriate PEF treatment of the food product. PEF technology is often viewed as unsuitable for solid food items. However, research has shown that certain solid food products can still benefit from PEF treatment. Beyond pasteurization, PEF could also be utilized to increase the yield of bioactive constituents as well as sugars from plant cells (Siemer et al., 2015).

PEF knowledge is a promising approach for preserving and processing food products, as it has been shown to have minimal impact on their quality. Research into the usage of PEF as a non-thermal technique should go beyond just microbial inactivation and also consider its potential to reduce spoilage caused by chemical and enzymatic reactions, as well as maintain dietary ingredients that are useful after and during treatment. Information on the effectiveness of pulsed electric processing, including enzyme inactivation or enhancement of proteolytic cleavage in nutrient foods, and its influence on specific food quality attributes, can be found in the literature. However, more study is required to fully comprehend the mechanics behind these functions (Syed et al., 2017). According to research on PEF's use in food processing, PEF can extract useful ingredients, kill microorganisms, and inactivate them. It can also change bio macromolecules to speed up chemical reactions and reduce the shelf life of fermented foods. PEF is seen as a response to consumer expectations for greater sensory and qualitative characteristics, including enhanced flavor, increased nutrition, wholesomeness, and ecologically friendly processing techniques. PEF-based food processing generates fewer industrial waste products compared to traditional technologies. This technology has higher energy efficiency, less water consumption, and little hazardous gas emissions. Modeling electroporation phenomena will be necessary to progress PEF further and will involve multidisciplinary cooperation between researchers from different scientific and technical areas. Optimization of PEF protocols centers around economic factors, such as increased gas yield per unit of energy applied, reduced hydraulic retention time, and decreased bio-solids content after digestion. In conclusion, PEF is a beneficial as well as sustainable technology for food dispensation (Arshad et al., 2021).

8.2.3 Ultrasound

Ultrasound is a method of non-invasive processing that utilizes sound waves having frequencies higher than the human hearing threshold of 20 kHz. There are three distinct forms of ultrasound, differentiated by their frequency range. These include diagnostic ultrasound, medicinal ultrasound, with such a spectrum of frequencies between 100 kHz and 1 MHz, and with a frequency band of 1–500 MHz, and power ultrasound, which operates within the 20–100 kHz range (Zhang et al., 2019).

In the food industry, ultrasound is utilized for a range of processes, including the thawing, drying, brining, pickled, degassing, and defaming of many different foods (Bhargava et al., 2021). The longitudinal waves produced by ultrasound generate vibrations in the food matrix, leading to increased pressure and temperature as a result of compression and rarefaction cycles. The creation and collapse of cavitation bubbles result in changes to the food's structure, as weak intramolecular interactions are disrupted and mass transfer is improved (Barbhuiya et al., 2021).

Fruits and vegetables are prone to microbial infection during storage because of their high water content. The before- and after-harvest characteristics of vegetables and fruits are well preserved by ultrasound both during storage and processing. This is particularly crucial in fruit and vegetable juices, whose enhanced water activity makes them more likely to contain bacteria. It has been shown that using ultrasound to treat these issues is a successful solution (Khaire et al., 2022).

In addition to maintaining the properties of fruits and vegetables, ultrasound has been demonstrated to enhance the functional attributes or quality of several fruit juices and dairy products. The use of ultrasound is not limited to the food industry, as it has shown auspicious applications in pharmaceuticals, sewage treatment, medical care, and other things (Gajendragadkar & Gogate, 2016; Gogate & Pandit, 2015). Ultrasound is mostly used in the food sector to assure the quality of food ingredients through slicing, packaging, or the identification of foreign objects. Another important feature of ultrasound's use in the sector is its use for cleaning and microbial inactivation. The main effects of ultrasound on microbes include shear pressures, which cause cell membranes to rupture, hotspots to form, and the generation of free radicals. The DNA of the microorganisms is harmed by these processes (Majid et al., 2015).

When fruits and vegetables are subjected to low-frequency, high-power ultrasound, the tissue may soften as a result of the loss of turgor pressure brought on by the extreme temperatures and pressures produced by collapsing cavitation bubbles. However, the use of low-intensity ultrasound on several fruits and vegetables, such as carrots, potatoes, and pears, has been shown to enhance the structural and textural properties of these products. The meat industry also extensively utilizes ultrasound, including for cleaning equipment as well as evaluating live animals to determine fat depth. Another significant use of ultrasonic therapy is the tenderization of meat. Researchers have looked into the potential mechanisms behind the tenderizing effect of ultrasound on meat (Barbhuiya et al., 2021).

The influence of ultrasound technology is of great awareness in food industry owing to the changing trend of customers who are drawn to functional foods. Because of the many benefits it offers, ultrasound has been utilized to elevate the qualitative features of premium foods and confirm the safety of a broad range of food items while minimalizing any detrimental impact on their sensory appearances. Moreover, the non-invasive nature of this technology provides several possibilities for the investigation of food composition (Arvanitoyannis et al., 2017).

8.2.4 Pulsed Light

Pulsed light, often referred to as PL, has become a topic of great interest in recent years. This non-thermal technology has been explored by both the scientific community and the food processing business as a viable substitute for conventional chemical and thermal procedures for cleaning surfaces that come into touch with food. Unlike thermal methods, PL uses high-intensity, broad-spectrum light pulses to eliminate microbes, making it a cutting-edge and innovative solution in the field of food protection (Mahendran et al., 2019).

PL, which uses ultraviolet light, is a highly effective cleaning method in industrial food applications. With this technique, strong light pulses with such a wide variety of wavelengths, between 200 and 1100 nm, are used to food surfaces, packaging materials, and equipment. The light pulses used in PL are short in duration, typically lasting between 1 microsecond and 0.1 seconds, and they deliver a high level of energy, with a range of 0.01–50 J/cm^2. Peak emissions commonly take place between 400 and 500 nm: this wavelength range covers the ultraviolet, visible (VI), and near-infrared (IR) ranges (Mandal et al., 2020; Pinela & Ferreira, 2017).

On average, PL is comprised of ultraviolet light, visible light, or infrared radiation. It is commonly used in the fruit juice, vegetable, and meat processing industries. PL operates by using high voltage, stored in a capacitor and then released in the form of short pulses via a xenon light source. The food samples to be decontaminated are positioned between two electrodes (Kramer et al., 2017).

PL is considered one of the most effective non-thermal processing techniques for extending the shelf life of food products like apples, cheese, and many others (Barbhuiya et al., 2021).

Effectiveness of PL in inactivating microorganisms is primarily due to its wide wavelength range, which includes the UV-C region. UV-C wavelengths are absorbed by microbial DNA or RNA, causing damage to these essential cellular components. This is why PL is so effective in eliminating harmful microbes (Fan et al., 2017).

The technological potential of PL is vast and continues to be explored in various applications. One area of interest is using PL to recover the nutritional and sensory quality of fruits and vegetables. Researchers have found that low levels of radiation from PL can increase the phenolic content of these foods during processing and storage, a phenomenon known as hormesis. This means that low levels of PL radiation can actually stimulate beneficial responses in plants, making them healthier. In addition, low doses of UV-C radiation from PL have been shown to alter the susceptibility of plants to diseases and reduce harmful enzymatic activity, such as polygalacturonase (PG) and pectin methylesterase (PME). This can help extend the shelf life of fruits and vegetables, making PL a valuable tool in the food industry (Vargas-Ramella et al., 2021).

There is no denying the potential of PL technology as just a non-thermal decontamination method for food, food-contact surfaces, and packaging materials. Its use can quickly inactivate harmful microorganisms like *Listeria monocytogenes*, providing a safer food supply for consumers. While PL may not be suitable because of their opaqueness and uneven surfaces, sterilizing food can nevertheless lower their microbial burden. Technology PL also offers a greener alternative to traditional thermal processing, which often leads to the degradation of food quality. In addition, PL can be used successfully to disinfect the surfaces of food, packaging, equipment, and clear liquids. PL also has potential in the nutraceutical and natural health industries. Its ability to improve the nutritional attributes of foods makes it a potential tool for bio-fortification. The future potential of PL technology in the food industry is vast and holds much promise for enhancing food safety and quality (Mandal et al., 2020).

8.2.5 Ultraviolet Light

Ultraviolet light is a cutting-edge knowledge in the world of non-thermal food processing. It offers a variety of benefits for food and beverage products, including longer shelf life, increased safety, and improved nutritional value. The primary resolution of using UV light in food dispensation is to extend the shelf life of the products and eliminate any health hazards caused by pathogenic microorganisms. Fresh juices, drinks, beverages, and milk products are popular candidates for UV treatment due to its non-thermal nature, which minimizes the negative impact on quality and other functional attributes. Additionally, the application of UV light has the potential to enrich the nutritional value of food goods, as well as provide additional health benefits and reduce waste (Koutchma, 2019).

There has been a lot of interest in the use of UV irradiation in the food sector for a while now. Particularly, ultraviolet radiation (UVC) has shown to be an effective technique for deactivating a variety of microorganisms, counting bacteria, fungi, yeasts, molds, and viruses, thus effectively preventing foodborne illnesses. Using UVC technology has the benefit of extending food's shelf life without degrading its quality. The efficiency of UVC nuclear material in the food business depends on a number of variables, including the UVC equipment's operational settings, the traits of the organisms, and the makeup of the food. Recent advancements in technology have prompted the creation of more effective UVC-based disinfection systems that are more intense and more permeable, making it even more effective in inactivating harmful pathogens. Compared to traditional methods, it has been demonstrated that UVC technologies in the food business are more productive and economical (Singh et al., 2021).

The antimicrobial properties of ultraviolet (UV) light within the 200–280 nm range have been well documented. This type of UV light is frequently used in the food sector to preserve food items that are liquid food and drink. One of the main benefits of just using UV light instead of conventional heat pasteurization is that it allows for the preservation of food quality while still ensuring its safety. With UV light pasteurization, food manufacturers are able to offer consumers food that is both wholesome and high quality, something that can be difficult to achieve with other methods (Ramesh et al., 2016).

Use of ultraviolet (UV) light for the preservation of post-harvest fruits and vegetables has gained significant attention in recent years. Researchers and technologists have become progressively interested in the implementation of UV light in this area, leading to an exponential growth in research in this field (Bisht et al., 2021). UV light has been established to be an effective method for killing microorganisms while causing minimal depletion of nutrients, making it an ideal solution for consumers seeking sustainable, additive-free, clean, and nutritious foods (Islam et al., 2022).

8.2.6 Irradiation

Food preservation through irradiation is a developing method that involves the use of radioactive isotopes such as cobalt (27Co60) and cesium (55Cs137) to emit ionizing radiation that creates physical, chemical, and biological changes in substances (Barbhuiya et al., 2021). This process has been a topic of concern for consumers due to the potential presence of radioisotopes in food. To address these concerns, researchers have been exploring alternative forms of irradiation, such as electron beam and X-ray technologies. These methods utilize high-speed electrons or X-rays as the energy source for ionizing radiation, effectively sterilizing food deprived of the use of radioactive isotopes (Asaithambi et al., 2021; Pan et al., 2021).

Due to its specific impact on food quality, electron beam radiation (EBI) is receiving more attention from the food business and researchers than X-ray technology. EBI is therefore being used in a number of food industries, counting spices, cereals, meat, fish, fruits, and vegetables (Pillai and Pillai, 2021). Another approach to enhancing food safety is through the use of gamma ray irradiation, which is recognized for its powerful penetration strength and the ability to reduce harmful microorganisms. This technology is also considered safe for the environment as it does not use any chemicals (Zarbakhsh and Rastegar, 2019).

The use of gamma irradiation has been shown to have a significant influence on the quality and safety of indigenous chicken meat. Among the different treatments, the best results were seen at a dose of 2.0 kGy in terms of improved shelf life, microbiological examination, biochemical analysis, and sensory evaluation. It can be concluded that gamma irradiation provides a means of delivering a larger quantity of high-quality indigenous chicken meat with a longer shelf life (Islam et al., 2019).

Application of food irradiation as a new knowledge has garnered attention as the more efficient and effective method of preserving food when compared with other thermal and non-thermal techniques. This method is seen as a solution for reducing post-harvest losses in fresh produce, confirming food safety, and getting past trade and quarantine barriers. Several fresh horticulture items can be preserved in their physical, biochemical, sensory, and nutritional qualities using irradiation, which is regarded as one of the most dependable and secure methods.

Food irradiation has been shown to be highly effective in sterilization and deactivating harmful microbes, slowing down aging and maturity without considerably raising the temperature and coming in direct contact with the food material. The process removes all chemicals from food, lowers anti-nutritional constituents, and prevents food allergies. With little quality loss and faster processing times, this method is used on a variety of fruits and vegetables. However, despite its advanced capabilities, the technology faces challenges in terms of its high capital costs, consumer resistance to accepting irradiated food, potential local radiation dangers, and undesirable flavor changes caused by lipid oxidation. Therefore, more scientific evidence is desired to grow its commercialization (Bhatnagar et al., 2022).

8.2.7 Oscillating Magnetic Field

Another efficient technique for food preservation is the use of an oscillating magnetic field or OMF (oscillatory magnetic field). Unlike traditional thermal methods, the application of a pulsating magnetic field has minimal impact on the food's dietary value, flavor, fragrance, or practical qualities. The result is a high-quality final product that is maintained through the use of magnetic fields. The DNA molecule receives energy from the pulsed magnetic field (PMF) through paramagnetic molecules, which causes chemical bonds to be broken and the cell structure to suffer permanent harm. Research has confirmed the effectiveness of strong magnetic fields in inactivating vegetative forms of microbes. The pulse duration can vary from 10 microseconds through 1 millisecond, with such a maximum frequency of up to 5 GHz, to prevent overheating of the product. It is crucial to remember that using magnetic fields in conjunction with other unorthodox food preservation techniques will produce the best outcomes (Rudy et al., 2020).

The claim of PMF technology in food processing has been shown to have a significant effect on the inactivation of bacteria. Unlike traditional thermal sterilization methods, PMF sterilization does not have an undesirable impact on the nutritional content or flavor of the product, nor does it leave harmful chemical residue like other non-thermal sterilization methods (Li et al., 2021).

The magnetic fields in PMF technology have the ability to alter the growth, reproduction, and migration direction of microorganisms, resulting in a remarkable bactericidal effect. However, studies have shown that compared to asexual cells, PMF is less efficient against bacterial spores (Impe et al., 2018). It has been proved that the application of PMF technology is a very efficient approach for sterilizing juice, including both fruit and vegetable juices, while retaining their nutritional components. Compared to high-temperature sterilization, PMF has been shown to effectively inactivate bacteria in juice without compromising its nutritional value (Qian et al., 2020).

The PMF sterilization process can be used to preserve a variety of other foods, including ground beef, sufu, and raw beer, in addition to juice and milk. The adoption of PMF technology inside the food industry is essential because, when related to conventional thermal sterilization techniques, it has the capacity to decrease processing time as well as temperature, minimize quality loss, or reduce energy usage. Moreover, PMF has shown significant promise when paired with other technologies, improving the effectiveness of inactivating microbes during food processing. The food industry has yet to adopt PMF technology, despite its many benefits, and more study is required to completely comprehend its sterilizing mechanism, broaden its application areas, and get beyond its drawbacks (Guo et al., 2022).

8.3 NON-THERMAL TECHNIQUES FOR PROTECTION AND QUALITY ENHANCEMENT OF DIFFERENT FOOD PRODUCTS

When food is handled for preservation, food quality is the top priority. Conventional food protection methods subject food to extremely high heats, which undoubtedly decreases food contamination or microbial load but also leads to some unfavorable fluctuations in food, such as the damage of nutrient content that are temperature-sensitive, fluctuations in food texture as a result of heat, and alterations in the organoleptic qualities of food (Hernández-Hernández et al., 2019). Food is exposed to high temperature for a prolonged period of time during thermal processing, which results in obvious fluctuations in food properties and reduction in the quality of food (Iqbal et al., 2019). In non-thermal handling, food is only handled at room temperature for an extremely brief period of time, for example, for one minute or less, maintaining the meal's nutritional makeup, texture, and mouthfeel (Beyrer et al., 2020). In the past few decades, a number of non-thermal food processing methods have been discovered, including pulsed electromagnetic field, cold ionization, ozonation, microwaves, and supercritical technology. These non-thermal therapies subject food to various treatments for a brief period, lowering the microbial load while also prolonging the food's shelf life and enhancing the sensual and textural qualities (Thirumdas et al., 2015). The effect of non-thermal methods over thermal tools is increased because there is no possibility of the production of any undesirable products or byproducts in the food. After all, it is not visible at higher temperatures (Thirumdas et al., 2020). The action of enzymes can also be stopped by non-thermal treatment, causing fruits and vegetables to spoil; protein and carbohydrate in food are often enhanced physiologically using cold plasma technology so they can be employed in a variety of food processing applications; grain cooking and textural qualities have been improved via gas cold plasma processing (Thirumdaset al., 2017). Ultrasonication is a non-thermal energy-efficient treatment typically used to speed up operations like food synthesis, extraction, and preservation. Food benefits from the load current and exposure time of ultrasonication, using ultrasonication, it is possible to create food that is both safe and healthy by using the ideal ratio of load current and exposure time (Bhargava et al., 2021). Such non-thermal actions are used for a diversity of purposes, like, the extraction of bioactive components from plants and animals for use in nutraceutical food claims, dehydration, improving the chemical and physical characteristics of food parts in addition to decreasing the microbial load (Mazzutti et al., 2021).

8.3.1 Meat Products

High-energy gamma rays, X-rays, and fast electrons are permitted irradiations for use in the food processing sectors and are employed to produce photons with high energy. In food processing, the energy of X-rays is employed and the food industry uses the energy of high-speed electrons for a variety of purposes (Farkas, 2006). Effects of irradiation are achieved without raising food temperature; there is no risk of damaging the food's heat-sensitive components because the temperature of the food is not raised (Jadhav et al., 2021). The food processing industry uses irradiation primarily to preserve food goods, and it is efficient against harmful microorganisms like *Salmonella*, *Staphylococcus*, and *Escherichia coli* (Gaougaou et al., 2020). The ready-to-eat chicken showed great taste and textural structures even after 15 days of protection after being exposed to gamma radiation of varying intensities, which produced outstanding results for the activation of *Listeria monocytogenes*, *E. coli*, and *Salmonella typhimurium* (El-Din et al., 2010). A relatively affordable, non-thermal technique is ultraviolet technology; in general, according to Barton and d'Errico (2012), it is used to reduce the microbial load on the surfaces of food items that are passively subjected to radiation, specifically ultraviolet radiation. The newly published study by Bulut and Karatzas (2021) looked at high hydrostatic pressure's (HHP) effectiveness against *E. coli* in common foods. The destruction of germs in a wide range of foods, such as processed fruits, meat, and meat products, is accomplished by pulsed electric therapy.

8.3.2 Bakery Products

The primary ingredients in bread (*Triticum aestivum*) flour are carbohydrate, water, protein, and other minor ingredients like ash, lipids, and non-starch polysaccharides (Scheuer et al., 2011). The growth of the kernel bread doughs used in numerous crops depends on these ingredients; the principal elements that impact the adaptive system of wheat flour are, exactly, gluten proteins (Thavarajah et al., 2011), surveyed by carbohydrate and other minor components. Wheat flour's low ability for gas absorption, which is caused by the little amount and poor quality of protein present in it and results in a weak viscoelastic network, can limit its use for bread making (Melnyk et al., 2012). The non-thermal techniques that are primarily physical techniques that can be employed instead of or along with chemical additions can be used as developing "green" and environmentally approachable technologies because they often do not produce trash; low processing temperatures, lower energy costs, and a better skill to preserve flavor and nutrients are further benefits of non-thermal technologies (vo Kientza, 2021). Non-thermal technologies have now been successfully utilized in various latest studies to alter the constituents of foods, and as a result, their technological features are often used to inactivate microbes and enzymes (de Jesus et al., 2020). The technological structures of wheat flour can be affected by these technologies by changing the protein structures; for instance, its contact with water as well as the development of the viscoelastic networks can encourage the production of stronger doughs, which in turn results in better-quality bread (Misra et al., 2015).

8.3.3 Dairy and Beverage Products

Ultrasonication is an innovative non-thermal technique in food manufacturing (Dolatowski et al., 2007). Simply put, an ultrasonic is a comprehensive wave with a frequency that is higher than the range of human hearing (Clark & Macquarrie, 2008). The dairy and beverage sectors value the effectiveness of the ultrasonic-assisted filtration process; the membrane filtering method is employed in the production of cheese in order to completely separate protein from other dairy (Saxena et al., 2009). Additionally, ultrasonication helps with the chilling, drying, and thaw of food goods (Cheng et al., 2014). As a good alternative to pasteurization and sterilization in the deduction of bacterial count in food and food-related products, ultrasonication is especially advantageous for the technique of degassing in beverages (Bhargava et al., 2021). Besides solid, liquid, and gas, the fourth matter state is plasma; in the year 1925, Langmuir coined the term "plasma" (Jadhav et al., 2021). Juice retained its color well and its bioactive components due to the brief exposure duration, the preservation of tomatoes juice (Starek et al., 2019), hazy apple liquid (Illera et al., 2019), apples, tomato, orange, sour cheerful nectar (Dasan & Boyaci, 2018), and whey grape juice (Amaral et al., 2018) all yielded comparable results used by cold plasma. HHP applies pressure to the product being treated using a relatively common medium, water. Food that has undergone HHP handling has good nutritional value and excellent sensory and textural qualities because it is only briefly subjected to treatment circumstances (Huanget al., 2020). It has been shown that HHP therapy is successful in killing germs in a wide range of foods, such as fermented fruits, poultry, and dairy products, which are excellent microbe development media. The efficacy of HHP against *E. coli* in liquid meals was examined in a recent survey by Bulut and Karata (2021).

8.4 ROLE IN SENSORY ACCEPTANCE OF DIFFERENT PRODUCTS

Determining how people will respond to a novel food product is one of the most challenging aspects of its construction. Many factors influence consumer acceptability, including price, convenience, and packaging, but one critical issue that requires careful consideration is the sensory experience that buyers enjoy with the food (Singh-Ackbarali and Maharaj, 2013).

Sensory testing evaluates food products by employing human senses such as sight, smell, taste, touch, and hearing. While taste seems to be a major concern, it is not the only one. Foods that taste flavorful but have an unsatisfying appearance lose their attention. Consumer consideration is dependent on visual appeal. Food texture has an impact on how people perceive it. When eating by hand, the way the food feels to the touch is critical. Foods that leave your palm oily or sticky are less enjoyable to consume. Sensory testing is a process of examining and evaluating all components that contribute to a food product's success or failure (Singh-Ackbarali and Maharaj, 2014).

Sensory testing is frequently used to determine whether consumers will approve of a food product. The most prevalent methods for sensory analysis include difference tests, descriptive analysis, and customer acceptance testing. There are various tests, such as the triangle test, where a panelist must determine which of three samples is distinct from the remaining two, and the duo-trio test, where a panelist must determine which of two samples deviates from the accepted norm (Fiorentini et al., 2020). The nature of the differences is not defined in difference tests, which quantify the number of sensory differences between samples. It is common practice to utilize a combination of various tests and descriptive sensory analysis to solve difficulties. The descriptive sensory analysis uses a variety of techniques to distinguish between various product varieties based on their sensory quality and to evaluate a quantitative depiction of the detectable sensory differences rather than just the difficulties. Because this is not the goal of the evaluation, no "excellent" or "bad" judgments are given, as in traditional quality judging methods. Consumer acceptance is measured using acceptability, preference, and hedonic (degree of liking) tests (Mihafu et al., 2020).

The vegetable harvesting season and growth conditions are all aspects that influence the interior sensory qualities of vegetables. There is consideration given to handling factors, various stowage, heating, and flavoring options, as well as the integration of vegetables into meals (Hoppu et al., 2021).

All five senses are used in food perception, and vegetables can be perceived in all five flavor categories (bitter, sour, sweet, salty, and umami). Van Stokkom et al. (2016) observed strongly flavored ten various vegetables (raw, cooked, mashed, and cold-pressed juice). They found that cooking techniques affected taste intensity, but the impact varied depending on the variety of vegetables and that most vegetables comprised low amounts of the five flavor modalities (Van Stokkom et al., 2016). Adults' differences in flavor perception (Chamoun et al., 2018) could be additionally connected to how vegetables' taste modalities are perceived (Puputti et al., 2018).

The flavor and other sensory aspects of vegetables are difficult for many consumers. Vegetables are considered more bitter and tougher than other basic dietary groups (Poelman et al., 2017). Several vegetables taste highly bitter to people who are susceptible to certain harsh compounds related to phenotype or genotype (Sandell et al., 2014). Shen et al. (2016), on the other hand, found that since demographic variables were important, pungent flavor genotype or phenotype on its own could not foresee vegetable predilection and intake. Unexpectedly, variations in odor perception due to genetics may be associated with vegetable intake. According to Jaeger et al. (2012), participants who were susceptible to the smell of cis-3-hexen-1-ol (green/grassy) chose to consume more salad greens and cucumber (Duffy et al., 2010).

For instance, visual cues are the main determinant of how food is received generally, and the colors of vegetables are linked to their ripeness and freshness (Arce-Lopera et al., 2015). Customers' perceptions of carrots of different hues' sensory and functional qualities were found to be affected by color hue and intensity (Schifferstein et al., 2019). According to Spence et al. (2010), people have different color–flavor associations, and specific client categories can be found based on their color preferences, such as those who favor blue potatoes (Paakki et al., 2016). As a result, in future studies integrating sensory evaluation of vegetables, consumer background characteristics and distinct consumer types should be established.

REFERENCES

Amaral, G. V., Silva, E. K., Costa, A. L. R., et al. (2018). Whey-grape juice drink processed by su carbon dioxide technology: Physical properties and sensory acceptance. *Lwt*, *92*, 80–86.

Arce-Lopera, C., Masuda, T., Kimura, A., Wada, Y., and Okajima, K. 2015. Model of vegetable freshness perception using luminance cues. *Food Quality and Preference*, *40*, 279–86. doi: 10.1016/j.foodqual.2014.06.010

Arshad, M. S., Hina, G., Anjum, F. M., & Suleria, H. A. R. (2022). Effect of milk-derived bioactive peptides on the lipid stability and functional properties of beef nuggets. *Scientific Reports*. *12*(1):1–12

Arshad, R. N., Abdul-Malek, Z., Roobab, U., et al. (2021). Pulsed electric field: A potential alternative towards a sustainable food processing. *Trends in Food Science & Technology*, *111*, 43–54.

Arvanitoyannis, I. S., Kotsanopoulos, K. V., & Savva, A. G. (2017). Use of ultrasounds in the food industry-Methods and effects on quality, safety, and organoleptic characteristics of foods: A review. *Critical Reviews in Food Science and Nutrition*, *57*, 109–128

Asaithambi, N., Singh, S. K., & Singha, P. (2021). Current status of non-thermal processing of probiotic foods: A review. *Journal of Food Engineering*, *303*, Article 110567.

Barbhuiya, R. I., Singha, P., & Singh, S. K. (2021). A comprehensive review on impact of non-thermal processing on the structural changes of food components. *Food Research International*, *149*, Article 110647

Barton, N., & d'Errico, F. (2012). North African origins of symbolically mediated behaviour and the Aterian. In *Developments in Quaternary Sciences* (Vol. *16*, pp. 23–34). Elsevier.

Beyrer, M., Pina-Perez, M. C., Martinet, D., & Andlauer, W. (2020). Cold plasma processing of powdered Spirulina algae for spore inactivation and preservation of bioactive compounds. *Food Control*, *118*, 107378.

Bhargava, N., Mor, R. S., Kumar, K., & Sharanagat, V. S. (2021) Advances in application of ultrasound in food processing: a review. *Ultrasonics Sonochemisry*, *70*, 105293

Bhatnagar, P., Gururani, P., Bisht, B., Kumar, V., Kumar, N., Joshi, R., & Vlaskin, M. S. (2022). Impact of irradiation on physico-chemical and nutritional properties of fruits and vegetables: A mini review. Heliyon, 8, e10918.

Bhattacharjee, C., Saxena, V., & Dutta, S. (2019) Novel thermal and non-thermal processing of watermelon juice. *Trends in Food Science and Technology*, *93*, 234–243

Bisht, B., Bhatnagar, P., Gururani, P., et al. (2021). Food irradiation: Effect of ionizing and non-ionizing radiations on preservation of fruits and vegetables-a review. *Trends in Food Science & Technology*, *114*, 372–385.

Bulut, S., & Karatzas, K. A. (2021). Inactivation of *Escherichia coli* K12 in phosphate buffer saline and orange juice by high hydrostatic pressure processing combined with freezing. *LWT*, *136*, 110313.

Chacha, J. S., Zhang, L., Ofoedu, C. E., et al. (2021). Revisiting non-thermal food processing and preservation methods-action mechanisms, pros and cons: A technological update (2016-2021). Foods, 10(6), 1430.

Chamoun, E., Mutch, D. M., Allen-Vercoe, E., Buchholz, A. C., Duncan, A. M., Spriet, L. L., Haines, J., & Ma, D. W. L. 2018. Guelph Family Health Study. (2018). A review of the associations between single nucleotide polymorphisms in taste receptors, eating behaviors, and health. *Critical Reviews in Food Science and Nutrition* *58*(2), 194–207. doi: 10.1080/10408398.2016.1152229.

Chen, F., Zhang, M., Fan, K., & Mujumdar, A. S. (2022). Non-thermal technology and heating technology for fresh food cooking in the central kitchen processing: a review. *Food Reviews International*, *38*, 608–627. doi: 10.1080/87559129.2020.1740246.

Cheng, X. F., Zhang, M., & Adhikari, B. (2014). Effect of ultrasonically induced nucleation on the drying kinetics and physical properties of freeze-dried strawberry. *Drying Technology*, *32*(15), 1857–1864.

Clark, J. H., & Macquarrie, D. J. (Eds.). (2008). Handbook of Green Chemistry and Technology. John Wiley & Sons.

Dai, J., Bai, M., Li, C., Cui, H., & Lin, L. (2020). Advances in the mechanism of different antibacterial strategies based on ultrasound technique for controlling bacterial contamination in food industry. *Trends in Food Science & Technology*, *105*, 211–222.

Dasan, B. G., & Boyaci, I. H. (2018). Effect of cold atmospheric plasma on inactivation of *Escherichia coli* and physicochemical properties of apple, orange, tomato juices, and sour cherry nectar. *Food and Bioprocess Technology*, *11*, 334–343.

De Jesus, A. L. T., Cristianini, M., Dos Santos, N. M., & Júnior, M. R. M. (2020). Effects of high hydrostatic pressure on the microbial inactivation and extraction of bioactive compounds from açaí (Euterpe oleracea Martius) pulp. *Food Research International*, *130*, 108856.

Dolatowski, Z. J., Stadnik, J., & Stasiak, D. (2007). Applications of ultrasound in food technology. *Acta Scientiarum Polonorum Technologia Alimentaria*, *6*(3), 88–99.

Duffy, V. B., Hayes, J. E., Davidson, A. C., Kidd, J. R., Kidd, K. K., & Bartoshuk, L. M. (2010). Vegetable intake in college-aged adults is explained by oral sensory phenotypes and TAS2R38 genotype. *Chemosensory Perception*, *3* (3-4), 137–148. doi: 10.1007/s12078-010-9079-8.

El-Din, H. M. N., Abd Alla, S. G., & El-Naggar, A. W. M. (2010). Swelling and drug release properties of acrylamide/carboxymethyl cellulose networks formed by gamma irradiation. Radiation Physics and Chemistry, *79*(6), 725–730.

Fan, X., Huang, R., & Chen, H. (2017). Application of ultraviolet C technology for surface decontamination of fresh produce. *Trends in Food Science & Technology*, *70*, 9–19.

Farkas, J. (2006). Irradiation for better foods. *Trends in Food Science & Technology*, *17*(4), 148–152.

Fiorentini, M., Kinchla, A. J., & Nolden, A. A. (2020). Role of sensory evaluation in consumer acceptance of plant-based meat analogs and meat extenders: A scoping review. *Foods*, *9*(9), 1334.

Gajendragadkar, C. N., & Gogate, P. R. (2016). Ultrasonics sonochemistry intensified recovery of valuable products from whey by use of ultrasound in processing steps-a review. *Ultrasonics Sonochemistry*, *32*, 102–118.

Gaougaou, G., Shankar, S., Liot, Q., Constant, P., Déziel, E., & Lacroix, M. (2020). Gamma irradiation triggers a global stress response in *Escherichia coli* O157: H7 including base and nucleotides excision repair pathways. Microbial Pathogenesis, *149*, 104342.

Gogate, P. R., & Pandit, A. B. (2015). Design and scale-up of sonochemical reactors for food processing and other applications. In J. A. GallegoJuarez & K. F. Graff (Eds.), *Power Ultrasonics* (pp. 725–756). Elsevier Ltd.

Guo, L., Azam, S. R., Guo, Y., Liu, D., & Ma, H. (2022). Germicidal efficacy of the pulsed magnetic field against pathogens and spoilage microorganisms in food processing: An overview. *Food Control*, *136*, 108496.

Hernández-Hernández, H. M., Moreno-Vilet, L., & Villanueva-Rodríguez, S. J. (2019). Current status of emerging food processing technologies in Latin America: Novel non-thermal processing. *Innovative Food Science & Emerging Technologies*, *58*, 102233.

Hoppu, U., Puputti, S., & Sandell, M. (2021). Factors related to sensory properties and consumer acceptance of vegetables. *Critical Reviews in Food Science and Nutrition*, *61*(10), 1751–1761.

Huang, H. W., Hsu, C. P., & Wang, C. Y. (2020). Healthy expectations of high hydrostatic pressure treatment in food processing industry. *Journal of Food and Drug Analysis*, *28*(1), 1–13.

Huang, H. W., Wu, S. J., Lu, J. K., Shyu, Y. T., & Wang, C. Y. (2017). Current status and future trends of high-pressure processing in food industry. Food Control, *72*, 1–8.

Illera, A. E., Chaple, S., Sanz, M. T., Ng, S., Lu, P., Jones, J., & Bourke, P. (2019). Effect of cold plasma on polyphenol oxidase inactivation in cloudy apple juice and on the quality parameters of the juice during storage. *Food Chemistry: X*, *3*, 100049.

Impe, J. V., Smet, C., Tiwari, B., et al. (2018). State of the art of nonthermal and thermal processing for inactivation of micro-organisms. *Journal of Applied Microbiology, 125*, 16–35.

Iqbal, A., Murtaza, A., Hu, W., Ahmad, I., Ahmed, A., & Xu, X. (2019). Activation and inactivation mechanisms of polyphenol oxidase during thermal and non-thermal methods of food processing. *Food and Bioproducts Processing, 117*, 170–182.

Islam, A., Sadakuzzaman, M., Hossain, M. A., Hossain, M. M., & Hashem, M. A. (2019). Effect of gamma irradiation on shelf life and quality of indigenous chicken meat: Irradiation on chicken meat. *Journal of the Bangladesh Agricultural University, 17*(4), 560–566.

Islam, F., Saeed, F., Afzaal, M., Ahmad, A., Hussain, M., Khalid, M. A., & Khashroum, A. O. (2022). Applications of green technologies-based approaches for food safety enhancement: A comprehensive review. *Food Science & Nutrition.*

Jadhav, H. B., Annapure, U. S., & Deshmukh, R. R. (2021). Non-thermal technologies for food processing. *Frontiers in Nutrition, 8*, 657090.

Jaeger, S. R., Pineau, Bava, B., C. M., et al. 2012. Investigation of the impact of sensitivity to cis-3-hexen-1-ol (green/grassy) on food acceptability and selection. *Food Quality and Preference, 24*(2), 230–42. doi: 10.1016/j.foodqual.2011.11.005.

Jan, A., Sood, M., Sof, S. A., & Norzom, T. (2017). Non-thermal processing in food applications: a review. *International Journal of Food Science and Nutrition 2*(6), 171–180.

Khaire, R. A., Thorat, B. N., & Gogate, P. R. (2022). Applications of ultrasound for food preservation and disinfection: A critical review. *Journal of Food Processing and Preservation, 46*(10), e16091.

Khalid, W., Arshad, M. S., Yasin, M., Imran, A., & Ahmad, M. H. (2021). Quality characteristics of gamma irradiation and kale leaf powder treated ostrich and chicken meat during storage. *International Journal of Food Properties, 24*(1), 1335–1348.

Khalid, W., Maggiolino, A., Kour, J., et al. (2023). Dynamic alterations in protein, sensory, chemical and oxida-tive properties occurring in meat during thermal and non-thermal processing techniques: a comprehensive review. Frontiers in Nutrition, 9, 3310.

Khalid, W., Maqbool, Z., & Arshad, M. S. (2022). Recent Advances and Innovation in Meat with Reference to Processing Technologies.

Khan, S., Keshavalu, S. G., Amaresh, R. P. M., Badhautiya, S., & Bhat, S. A. (2017). High pressure processing in food industry. *Bulletin of Environment, Pharmacological and Life Science, 6*, 28–31.

Koutchma, T. (2019). Ultraviolet Light in Food Technology: Principles and Applications (2nd Ed.)

Kramer, B., Wunderlich, J., & Muranyi, P. (2017). Recent findings in pulsed light disinfection. *Journal of Applied Microbiology, 122*(4), 830–856.

Li, W., Ma, H., He, R., Ren, X., & Zhou, C. (2021). Prospects and application of ultrasound and magnetic fields in the fermentation of rare edible fungi. *Ultrasonics Sonochemistry, 76*(1), 105613

Mahendran, R., Ramanan, K. R., Barba, F. J., Lorenzo, J. M., LópezFernández, O., Munekata, P. E. S., Roohinejad, S., Sant'Ana, A. S., & Tiwari, B. K. (2019). Recent advances in the application of pulsed light processing for improving food safety and increasing shelf life. *Trends in Food Science & Technology, 88*, 67–79

Majid, I., Nayik, G. A., & Nanda, V. (2015). Ultrasonication and food technology: A review. *Cogent Food & Agriculture, 1*(1), 1071022. doi: 10.1080/23311932.2015.1071022.

Mandal, R., Mohammadi, X., Wiktor, A., Singh, A., & Pratap Singh, A. (2020). Applications of pulsed light decontamination technology in food processing: An overview. *Applied Sciences, 10*(10), 3606.

Mazzutti, S., Pedrosa, R. C., & Ferreira, S. R. S. (2021). Green processes in foodomics. Supercritical fluid extraction of bioactives.

Melnyk, J. P., Dreisoerner, J., Marcone, M. F., & Seetharaman, K. (2012). Using the Gluten Peak Tester as a tool to measure physical properties of gluten. *Journal of Cereal Science*, *56*(3), 561–567.

Mihafu, F. D., Issa, J. Y., & Kamiyango, M. W. (2020). Implication of sensory evaluation and quality assessment in food product development: A review. *Current Research in Nutrition and Food Science Journal*, *8*(3), 690–702.

Misra, N. N., Kaur, S., Tiwari, B. K., Kaur, A., Singh, N., & Cullen, P. J. (2015). Atmospheric pressure cold plasma (ACP) treatment of wheat flour. *Food Hydrocolloids*, *44*, 115–121.

Paakki, M., Sandell, M., & Hopia. A. 2016. Consumers's reactions to natural atypically colored foods: An investigation using blue potatoes. *Journal of Sensory Studies 31* (1):78–89. doi: 10.1111/joss.12193.

Pal, M. (2017). Pulsed electric field processing: an emerging technology for food preservation. *Journal of Experimental Food Chemistry*, *3*(02), 10–12.

Pan, M. F., Yang, J. Y., Liu, K. X., et al. (2021). Irradiation technology: An effective and promising strategy for eliminating food allergens. *Food Research International*, *148*(12), 110578. doi: 10.1016/j.foodres.2021.110578

Paniwnyk, L. (2017). Applications of ultrasound in processing of liquid foods: A review. *Ultrasonics Sonochemistry*, *38*, 794–806

Peerun, S. (2015). The Food High Pressure Processing (HPP) Technologies Market Forecast 2015-2025: Pascalization & Bridgmanization. Visiongain Ltd.: London, UK.

Pi, X., Yang, Y., Sun, Y., et al. (2021). Food irradiation: a promising technology to produce hypoallergenic food with high quality. *Critical Reviews in Food Science and Nutrition*, 1–16.

Pillai, S. D., & Pillai, E. T. (2021). Agriculture: Electron beam irradiation technology applications in the food industry. *Encyclopedia of Nuclear Energy*, 313–329. doi: 10.1016/B978-0-12-819725-7.00141-0.

Pinela, J., & Ferreira, I. C. (2017). Nonthermal physical technologies to decontaminate and extend the shelf-life of fruits and vegetables: Trends aiming at quality and safety. *Critical Reviews in Food Science and Nutrition*, *57*(10), 2095–2111.

Poelman, A. A. M., C. M. Delahunty, and de Graaf, C. (2017). Vegetables and other core food groups: A comparison of key flavor and texture properties. *Food Quality and Preference*, *56*, 1–7. doi: 10.1016/j.foodqual.2016.09.004.

Pollock, A. M., Pratap Singh, A., Ramaswamy, H. S., & Ngadi, M. O. (2017). Pulsed light destruction kinetics of L. monocytogenes. *LWT Food Science and Technology*, *84*, 114–121

Puputti, S., Aisala, H., Hoppu, U., & Sandell, M. 2018. Multidimensional measurement of individual differences in taste perception. *Food Quality and Preference 65*, 10–7. doi: 10.1016/j.foodqual.2017.12.006.

Qian, J.,Zhang, M., Dai, C., Huo, S., & Ma, H. (2020). Transcriptomic analysis of Listeria monocytogenes under pulsed magnetic field treatment. *Food Research International*, *133*, 109195.

Rahman, M., Islam, R., Hasan, S., Zzaman, W., Rana, M. R., Ahmed, S., & Sunny, A. R. (2022). A Comprehensive review on bio-preservation of bread: An approach to adopt wholesome strategies. *Foods*, *11*(3), 319.

Ramesh, T., Nayak, B., Amirbahman, A., Tripp, C. P., & Mukhopadhyay, S. (2016). Application of ultraviolet light assisted titanium dioxide photocatalysis for food safety: A review. *Innovative Food Science & Emerging Technologies*, *38*, 105–115.

Roobab, U., Abida, A., Afzal, R., Madni, G. M., Zeng, X. A., Rahaman, A., & Aadil, R. M. (2022). Impact of high-pressure treatments on enzyme activity of fruit-based beverages: an overview. *International Journal of Food Science & Technology*, *57*(2), 801–815.

Ros-Polski, V., Koutchma, T., Xue, J., Defelice, C., & Balamurugan, S. (2015). Effects of high hydrostatic pressure processing parameters and NaCl concentration on the physical properties, texture and quality of white chicken meat. *Innovative Food Science & Emerging Technologies*, *30*, 31–42.

Rudy, M., Kucharyk, S., Duma-Kocan, P., Stanisławczyk, R., & Gil, M. (2020). Unconventional methods of preserving meat products and their impact on health and the environment. *Sustainability*, *12*(15), 5948.

Sandell, M., Hoppu, U., Mikkilä, V., Mononen, N., Kähönen, M., Männistö, S., Rönnemaa, T., Viikari, J., Lehtimäki, T., & Raitakari, O. T. (2014). Genetic variation in the hTAS2R38 taste receptor and food consumption among Finnish adults. *Genes & Nutrition*, *9*(6), 433. doi: 10.1007/s12263-014-0433-3.

Saxena, A., Tripathi, B. P., Kumar, M., & Shahi, V. K. (2009). *Adv.*(44) Bergers, JJ; Vingerhoeds, MH; van Bloois, L. Colloid Interface Sci, *145*, 1–22.

Scheuer, P. M., FRANCISCO, A. D., MIRANDA, M. D., & Limberger, V. M. (2011). Trigo: características e utilização na panificação. *Revista Brasileira de Produtos Agroindustriais*, *13*(2), 211–222.

Schifferstein, H. N. J., Wehrle, T., and Carbon, C. 2019. Consumer expectations for vegetables with typical and atypical colors: The case of carrots. *Food Quality and Preference*, *72*, 98–108. doi: 10.1016/j.foodqual.2018.10.002.

Shen, Y., Kennedy, O. B., & Methven, L. (2016). Exploring the effects of genotypical and phenotypical variations in bitter taste sensitivity on perception, liking and intake of brassica vegetables in the UK. *Food Quality and Preference*, *50*, 71–81. doi: 10.1016/j.foodqual.2016.01.005.

Siemer C, Aganovic K, Toepfl S, et al. (2015). Application of pulsed electric fields in food. *Conventional and Advanced Food Processing Technologies*, 645–672

Singh, H., Bhardwaj, S. K., Khatri, M., Kim, K. H., & Bhardwaj, N. (2021). UVC radiation for food safety: An emerging technology for the microbial disinfection of food products. *Chemical Engineering Journal*, *417*, 128084.

Singh-Ackbarali, D., & Maharaj, R. (2013). Sensory Evaluation As A Tool In Determining Acceptability of Processed Local Agricultural Products (No. 537-2016-38575).

Singh-Ackbarali, D., & Maharaj, R. (2014). Sensory evaluation as a tool in determining acceptability of innovative products developed by undergraduate students in food science and technology at the University of Trinidad and Tobago. *Journal of Curriculum and Teaching*, *3*(1), 10–27.

Spence, C., Levitan, C. A., Shankar, M. U., & Zampini, M. (2010). Does food color influence taste and flavor perception in humans? *Chemosensory Perception*, *3*(1), 68–84. doi: 10.1007/s12078-010-9067-z.

Starek, A., Pawłat, J., Chudzik, B., Kwiatkowski, M., Terebun, P., Sagan, A., & Andrejko, D. (2019). Evaluation of selected microbial and physicochemical parameters of fresh tomato juice after cold atmospheric pressure plasma treatment during refrigerated storage. Scientific Reports, *9*(1), 1–11.

Syed, Q. A., Ishaq, A., Rahman, U. U., Aslam, S., & Shukat, R. (2017). Pulsed electric field technology in food preservation: a review. *Journal of Nutritional Health & Food Engineering*, *6*(6), 168–172.

Thavarajah, P., Sarker, A., Materne, M., Vandemark, G., Shrestha, R., Idrissi, O., & Vandenberg, A. (2011). A global survey of effects of genotype and environment on selenium concentration in lentils (*Lens culinaris* L.): Implications for nutritional fortification strategies. *Food Chemistry*, *125*(1), 72–76.

Thirumdas, R., Sarangapani, C., & Annapure, U. S. (2015). Cold plasma: a novel non-thermal technology for food processing. *Food Biophysics*, *10*, 1–11.

Thirumdas, R., Sarangapani, C., & Barba, F. J. (2020). Pulsed electric field applications for the extraction of compounds and fractions (fruit juices, winery, oils, by-products, etc.). In *Pulsed Electric Fields to Obtain Healthier and Sustainable Food for Tomorrow* (pp. 227–246). Academic Press.

Thirumdas, R., Trimukhe, A., Deshmukh, R. R., & Annapure, U. S. (2017). Functional and rheological properties of cold plasma treated rice starch. Carbohydrate Polymers, *157*, 1723–1731.

Van Stokkom, V. L., P. S. Teo, M. Mars, C. de Graaf, O. van Kooten, and M. Stieger. 2016. Taste intensities of ten vegetables commonly consumed in the Netherlands. *Food Research International 87*, 34–41. doi: 10.1016/j.foodres.2016.06.016.

Van Wyk, S., & Silva, F. V. (2019). Nonthermal preservation of wine. *Preservatives and Preservation Approaches in Beverages*, 203–235.

Vargas-Ramella, M., Pateiro, M., Gavahian, M., Franco, D., Zhang, W., Khaneghah, A. M., & Lorenzo, J. M. (2021). Impact of pulsed light processing technology on phenolic compounds of fruits and vegetables. *Trends in Food Science & Technology*, *115*, 1–11.

vo Kientza, H. T. (2021). Cooking. In *Handbook of Molecular Gastronomy* (pp. 157–161). CRC Press.

Wickramarachchi, K. S., Sissons, M. J., & Cauvain, S. P. (2015). Puff pastry and trends in fat reduction: an update. *International Journal of Food Science & Technology*, *50*(5), 1065–1075.

Zamani Mazdeh, F., Sasanfar, S., Chalipour, A., Pirhadi, E., Yahyapour, G., Mohammadi, A., & Hajimahmoodi, M. (2017). Simultaneous determination of preservatives in dairy products by HPLC and chemometric analysis. International Journal of Analytical Chemistry, 2017.

Zarbakhsh, S., & Rastegar, S. (2019). Influence of postharvest gamma irradiation on the antioxidant system, microbial and shelf life quality of three cultivars of date fruits (Phoenix dactylifera L.). *Scientia Horticulturae*, *247*, 275–286.

Zhang, Z. H., Wang, L. H., Zeng, X. A., Han, Z., & Brennan, C. S. (2019) Non-thermal technologies and its current and future application in the food industry: a review. *International Journal of Food Science*, *54*(1):1–13.

9 Circular Models and Strategies to Reduce and Reuse Food Waste

Sharath Kumar, N.
ICAR-Central Institute of Temperate Horticulture

Thottiam Vasudevan Ranganathan
Karunya Institute of Technology and Sciences

9.1 INTRODUCTION

With the human population expected to grow at an increasing rate, there is an equal demand or rather an exponential need for an increase in food production. This might lead to a cascading effect of an increase in waste generation and greenhouse gas emission. This has become a global concern and matter of interest in recent years to research groups, academicians, public and private players in the economy, as it puts a tremendous pressure on managing the available but limited natural resources and may negatively impact the social, environmental and economic health of the society. The fact is that the food waste utilisation and recovery has come into limelight as it is not only concerned with the policies and regulations but has more to do with the consumer behaviour, mind-set and cultural patterns. Of late, we should not underscore the rise of malnutrition complications leading to food security challenges in the days to come. The plausible solution lies in adopting a sustainable approach towards zero wastage by an effective recycling and re-usage of food waste in a closed loop pattern, employing the principles of circular economy. This would go a long way in mitigating the climate change impacts (Tamasiga et al., 2022).

Presently the world is facing few of the complex challenges like loss of biodiversity, climate change, raise in greenhouse gases (GHGs), resource depletion, and fragmentation of cultivable land coupled with the increase in population. Thus, it needs a sustainable approach that can benefit the growing population with the limited available resources. This approach focuses on developing a long-term strategy of utilising the available resources, wherein the needs of both the current and future generations are effectively met (Assessment, 2005). Maximum utilisation of the minimal resources as well as enhancing the renewable flow of the resources in the natural ecosystem is the solution.

DOI: 10.1201/9781003412489-9

Circular economy (CE) can be viewed as a system that follows the concept of an effective regeneration and restoration of the invaluable available resources by a process of recycle, reuse and remanufacture of existing products. Thus, the concept of end of life for products is made redundant, ensuring minimal waste generation and emissions (MacArthur et al., 2015). It can be seen as an engine of sustainability and as an alternative to the currently followed 'take-do-disposal' model (EU Action Plan for Circular Economy, 2015). Circular system works on the recovery and reuse of material and energy balances, thus reducing the amount of waste and pollution. It ideally revolves around the efficient and effective utilisation of resources through minimisation of waste and also adding value to the waste generated. This minimises the usage of primary resources, avoids generation of closed loop products, product parts and materials, thus protecting the environment while being economically beneficial to the populace. In a nutshell, practicing the CE is expected to provide human population a wide range of advantages like improvement in human health, nutrition and wellbeing, increased food security, better job opportunity, reduced emission and wastage, reducing the impact of climate change and improvement in biodiversity (Lisca et al., 2021; Olson et al., 2021). Adding to this, the popular CE concepts and strategies includes recover, recycle, repurpose, remanufacture, refurbish, repair, reuse, reduce, rethink and refuse can be used for mitigating the waste and loss (Morseletto, 2020).

9.2 FOOD WASTE

According to FAO reports, the world population is set to reach around 2.3 billion people by 2050. Around one-third of the food being produced is either lost or wasted, which might lead to a possible problem of world hunger taking into account the increasing population (FAO, 2014) which may have the possibility to keep around 800 million people undernourished (Gustavsson et al., 2011) at various stages of distribution and consumption. Waste generated in the world, along the supply chain, accounted for 1.3 billion tonnes (FAO, 2019) with a cost incurring of more than 1000 billion dollars per year (FAO, 2019) mainly caused by mismanagement of resources and supply chain inefficiency along with the unsustainable consumption practices by consumers (Taghikhah et al., 2019). Pressurising on limited resources to produce food for the skyrocketing population is not a sustainable approach as this may lead to larger costs incurred in the production of food produce which gives double burden on farmers by reducing their profit margins, and on the other hand, reducing the productivity of fertile land on account of excessive fertiliser usage. It also generates food wastage sumptuously.

Food loss and waste refer to those parts of plants and animals which are left unconsumed by humans, though they are produced and harvested primarily for human consumption (Parfitt et al., 2010; Lipinski et al., 2013). Although food loss and waste seem to be used interchangeably, but there exists a subtle difference in their definitions. While food loss refers to post-harvest loss occurring during the supply chain (handling, storage, transportation, processing, distribution), food waste is more concentrated towards the retail operations and consumption behaviour towards the food (food being discarded in spite of its edible nature and good quality). Food wastage accounts for the loss factors from both food loss and food waste.

In industrialised and developed nations, food wastage happens predominantly at the retail level and by the consumers, while in low-income countries, production and post-harvest losses due to insufficient or improper primary processing appear to be the major reasons for food wastage (FAO 2011). Corrado and Sala (2018) found that current estimates of food loss and waste generation varied between 194 and 389 kg per person per year on a global scale and between 158 and 298 kg per person per year on a European scale. In addition to food loss, there is substantial loss of resources and energy that have contributed in producing, transporting and distributing these food commodities. The problem of food security and undernutrition arises because the produced food is unable to reach the less fortunate members of society. Effective prevention of even 25% of the food that is wasted in this entire chain (production, processing and transportation) would be sufficient enough to feed the entire undernourished global population (Basher et al., 2013).

The United Nations in the year 2015 adopted 17 SDGs as a part of the 2030 Agenda for Sustainable Development. SDG 12 stands for ensuring sustainable consumption and production patterns and its third target SDG 12.3 calls for cutting to half the per capita global food waste at the retail and consumer level and reducing the food loss along the production and supply chain which even includes the post-harvest losses by the year 2030.

9.2.1 Food Waste Quantification

Food waste quantification has been an area of interest and top priority for research as the sources of food waste need to be identified to have control measures. This will greatly assist in streamlining food distribution and also for taking the preventive, reduction, management decisions and valorisation approach accordingly. It is also required for the policy decisions and interventions. In this line, the project FUSIONS, which stands for Food Use for Social Innovation by Optimising Waste Prevention Strategies which was funded under the European Union Seventh Framework Programme for research, technological development and demonstration, had developed an extensive list of available methodologies at different stages of the food chain to reduce the food wastage (Møller et al., 2014). It had the privilege to collaborate with 21 partners from 13 countries with the primary aim of increasing the resource efficiency in Europe by harmonising the monitoring patterns of food waste thereby reducing the food wastage.

Food wastes are generally quantified by territorial and consumption based approaches. The territorial-based approach allocates the food waste generation to the geographical area where the food waste is produced, and the consumption-based approach allocates to the area where the food is consumed. The formal approach is generally followed to quantify the actual amount of food waste which need to be valorised or managed (Corrada et al., 2019).

Food wastes can be categorised as avoidable food waste and unavoidable food waste (Papargyropoulou et al., 2014). This helps in formulating strategies for the effective management decisions. Avoidable food wastes are those that are thrown away because it is no longer wanted or has been allowed to go past its best. This includes food and food parts which are considered edible by vast majority of people.

Unavoidable food refers to those foods which has lost its edibility and are considered waste by majority people. It includes peel and skin of fruits, seeds, etc.

9.2.2 Stages of Food Waste

Food gets wasted right after the harvest, and it continues until the final consumption. EU wastes an estimated quantity of 88 million tonnes every year which includes edible and non-edible components. Out of these, 12% comes from food service, 53% is derived from households, 5% comes from wholesale and retail sections, and the remaining 30% comes from the production and processing sectors (Stenmarck et al., 2016).

Food supply chain starts with the harvesting of main products which is of commercial interest (grains, fruits, vegetables, fish, chicken) and thereby producing the by-products (stalks, manures). The food gets lost or wasted starting from the agriculture and continues up to consumption stage (Smil, 2004; Parfitt et al., 2010). The problem of food loss and food waste starts with the faulty storage, transport and distribution channels. Various activities undertaken during the movement of the produce lead to waste and loss generation. An average of 2600 kcal per capita per day gets wasted (including the animal feed and waste) in terms of caloric approach derived for food waste quantification (Smil, 2000).

9.2.3 Characterisation of Food Waste

Food waste thus collected from the different sources (households, agriculture, food processing industries, commercial, hospital, institutions, etc.) needs to be characterised so as to fit the right treatment for the particular type of waste. Various parameters that need to be assessed for the food waste are primarily physical property, chemical property, biodegradability nature and methods of handling the waste (Cheng & Lo, 2016).

1. Physical parameters - moisture content, bulk density, solids content, pH, particle size
2. Chemical parameters - nutritional contents (vitamins, fats, minerals), C/N ratio

9.2.4 Impact of Food Waste

Wastage of food leads to significant impact on the economic, social and environmental aspects. Brief description of the impacts caused is provided below.

9.2.4.1 Economic Impact

Food waste leads to a substantial loss to the farmer/producer who has used all the resources to raise the crops. Avoidable food losses will have a direct and negative impact on the income of farmers and consumers. It has an impact on farmers' livelihoods since waste prevents them from making the profit margins they should, and it deprives consumers, particularly those living in poverty, and deprive them from wholesome and reasonably priced food (Gustavsson et al., 2011; Lundqvist et al., 2008). US Environmental Protection Agency (EPA) highlights the economic impacts of food waste and encourages the food producers, retailers and the food service sector to reduce food waste in order to achieve their cost savings and profit margins. These expenses will add upon to the lower purchasing cost and to the ultimate cost of disposing the waste (EPA 2003, 2012).

9.2.4.2 Environmental Impact

Food waste upon being landfilled has been shown to emit methane and carbon dioxide, the prominent players in GHGs which are currently the major agents in climate change due to emissions. It is seen that methane has the potential to trap 21 times more heat as compared with carbon dioxide (Adhikari et al., 2006). The waste bioconversion in the soil ecosystem can have a negative impact by disturbing the existing biogenic cycles of nitrogen and phosphorous (Rockström et al., 2009; Smil, 2002) and it can harm the growth of succeeding crops.

9.2.4.3 Social Impact

The social impact is mostly determined by an individual moral and ethical opinions of wasting food. In particular, it sheds light on the amount of wasted food on one end and the food poverty on the other end. The emergence of new concepts like gleaning and freeganism in Australia has more to deal on the social impact of food wastage as the groups following the movement shall consume the disposed foods to reduce the environmental impact and to reduce the socio-economic disparity related to food availability (Edwards & Mercer, 2007).

9.2.5 Indicators for Food Waste

To choose the best practices for food waste management, it often becomes important to identify the nature and composition of food that is going to be considered as waste, after which the management practices can be taken up. It also helps in the design and development of optimal and feasible waste management solutions by considering the technical, regulatory and environmental considerations. Few of the indicators which can be used for describing the food waste are as follows (García-García et al., 2016) (Figure 9.1).

9.3 FOOD WASTE DISPOSAL METHODS

Generally, management of the food waste is quite complex due to the heterogeneity of associated food and its sources. Reduction and management of food waste require the collaborative efforts between the different stakeholders of the society. It also requires the change in consumer and consumption behaviour. Waste generated from the households, restaurants, cafeterias and food service operators need to be collected, graded and then sent for suitable treatment/management procedures. Currently followed methods for waste management is being discussed in detail.

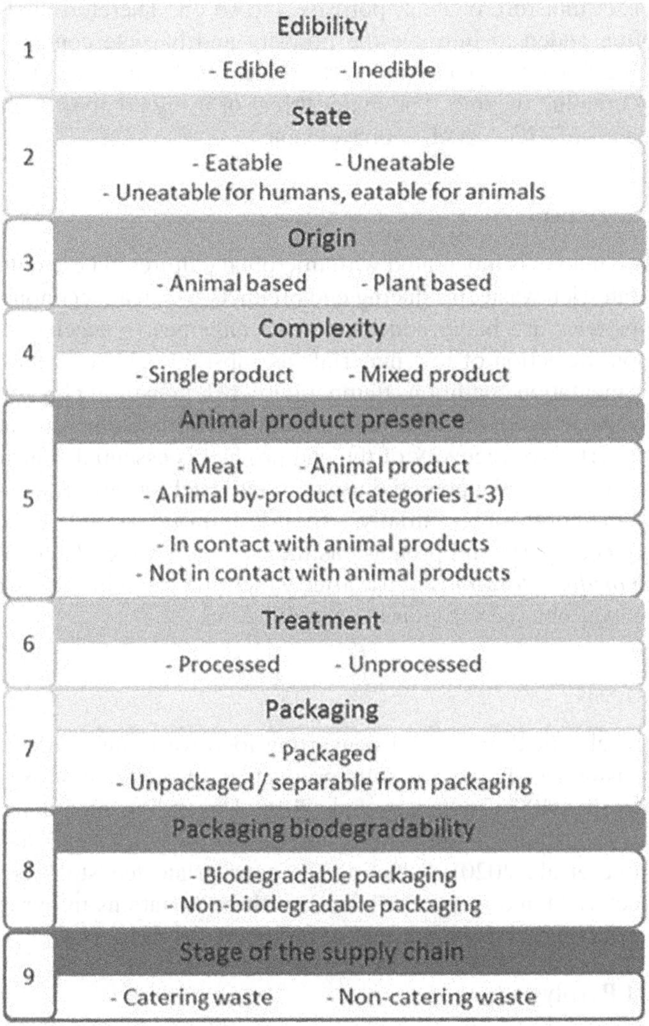

FIGURE 9.1 Indicators of food waste. Adopted from García-García, G., Woolley, E., Rahimifard, S., Colwill, J., White, R., & Needham, L. (2016). A methodology for sustainable management of food waste. Waste and Biomass Valorization, 8(6), 2209–2227.

9.3.1 Composting

It is viewed as an alternative method to recycle the food waste as the food waste creates conducive environment for the growth and development of microbes and insects. Microbial growth and proliferation promotes aerobic breakdown and degradation of organic matter, resulting in the production of compost, which improves soil health and reduces greenhouse gas emissions (Kumar et al., 2010). It is an eco-friendly approach for converting the organic waste into compost. The end product is characterised by the particle size, moisture content, porosity and so on. Therefore, various bulking agents are often added to improve the porosity and particle content. Few among them are zeolite, lime, biochar, wood bark and leaves. It can also act to increase the livelihood by creating income. The major threat of using compost technology is its potential release of GHGs which is unavoidable.

9.3.2 Fermentation

The food waste matrix is inoculated with microbial cultures. These cultures ferment the carbohydrate-rich waste, producing end products which are economically viable. Often the substrates are being added with the enzymes to accelerate the process of fermentation. Selection of raw materials as substrate and selection of microbial strains and fermentation conditions (temperature, pH, moisture) play important roles in the process. Apart from these, selection of an appropriate downstream processing method for the effective recovery of the end product is essential. Carmona-Cabello et al. (2021) investigated the potential use of potato peel for bio-diesel production by following two-step procedure. Initially, *Aspergillus awamori* fungi fed on the potato peels using glucoamylase and protease as the enzymes, which later was fermented with *Rhodosporidium toruloids* yeast after which the microbial oil was extracted, and bio-diesel was obtained from trans-esterification.

9.3.3 Pyrolysis

It is the thermochemical process of converting food waste into carbonaceous compounds comprising gas, liquid and solid products in the absence of oxygen at a temperature higher than 400°C (Oladejo et al., 2018). Depending upon the feedstock, the end product obtained can be gases, biochar, bio-oil and bio-chemicals (Kim et al., 2018, 2019; Lee et al., 2020). Selecting the appropriate feedstock is essential for the end product yield and recovery. There are three variations in pyrolysis reaction depending on the time and temperature of treatment to feedstocks.

9.3.3.1 Fast Pyrolysis

This is characterised by the exposure of waste to a shorter heating time and a higher temperature (often more than 500°C), yielding oils as a major fraction (Balat et al., 2009).

9.3.3.2 Slow Pyrolysis

Here the feedstock are subjected to a longer heat time and lower heating conditions which yields biochar (up to 35%) as the main product (Ahmad et al., 2014).

9.3.3.4 Catalytic Pyrolysis

This method seems promising as it yields high-quality bio-oils (Lee et al., 2020) and hydrogen-rich fuel as the end product (Chen et al., 2003). Key advantages include reduced pyrolysis temperatures for shorter time and greater bio-oil production compared to conventional pyrolysis. Catalysts commonly used are KCl, Ni-Mo/γ alumina, Cu/Al_2O_3, Al_2O_3/Na_2CO_3, chicken manure biochar, and Al-SBA-15.

9.3.4 Landfills

Open dumps or landfills are a common method followed for disposal of the food waste where the waste dumped in an open space and allowed for degradation by itself. The major limitation lies in the release of GHGs approx. 8% (Adhikari et al., 2009) and the possibility of insect associated vector diseases arising from the dumpyards.

9.3.5 Incineration

It is the process where solid food waste is burnt at high temperatures for conversion to ash. High amounts of external heat are supplied to ensure complete combustion of the waste biomass. As a result, there will be emission of toxic gases with the presence of heavy metals and dioxins (Katami et al., 2004).

9.3.6 Hydrothermal Carbonisation

This is a thermochemical process used to convert the food waste into value-added products. It converts biomass to hydrogen-rich biochar (hydrochar), oils at the temperature range of 180°C–260°C under 1–8 MPa (Tekin et al., 2014). The major limitation lies in the initial high cost for the setup of the infrastructure.

9.4 CIRCULAR MODELS IN THE CONTEXT OF FOOD WASTE MANAGEMENT

Circular model in waste management leads to the reduction in residual and household waste generation coupled with re-entry of these waste in a useful and beneficial format back into the production system (González-Sánchez et al., 2023). The main goal is to convert the linear flow to circular model, and thus, the material and energy flow is being recirculated and is not being lost. Adopting CE models and tools in the agri-food system is essential for mitigating the problems of food waste and food loss, thereby achieving the SDGs.

Major issues in the agri-food sector revolve around the food safety, traceability, product quality, and respect for the environmental and human resources. In addition to these, (i) ways and means to reduce the food waste, (ii) assessment of the waste generated, (iii) streamlining of the process used to recover and valorise the waste and (iv) conversion into yet another value-added product is the need of the hour. Further the impetus should be on developing a process that focuses both on material and energy conservation and recovery, thus reducing its environmental cost. Few of the circular models used in food waste management are discussed.

9.4.1 Bioactive Compounds Recovery from Food Waste

Wasted food accounts for not only mere food, but it also contains numerous bioactive compounds which has tremendous nutritional and health benefits. Bioactive compounds are the functional components present in the food matrix which provides immense health benefits. It includes amino acids, vitamins, minerals, fatty acids, long-chain fatty acids and polyphenols. Extracting bioactive components from food waste and delivering as animal feed seems to be a promising way for food waste recovery. Food waste collected will have a higher moisture content; therefore, appropriate heat treatments are required to reduce the moisture content necessary for further applications. Processing technologies like cooking, extrusion, pelletising, dehydration, ensiling (Westendorf et al., 1999) and probiotic treatments (Saray et al., 2014) are currently used for waste reduction by removing the moisture. Various procedures have been developed and demonstrated by research groups which report that the transformation of different kinds of food waste into bioactive-rich formulations can be utilised in the animal diet. In this way, low-cost feed formulations can be derived from the food waste which can be a way of lowering the input cost factors to the livestock farmers. Hossein and Dahlan (2015) studied and optimised the proportion of dehydrated restaurant waste which can be incorporated in the poultry diet and found that incorporating upto 20 % of processed restaurant waste in regular poultry feed had no major impact but more than 20 % had a negative impact in the polyunsaturated fatty acids (PUFA) content of the meat. This implies that homogeneity of the food waste flow stream (composition and properties) is important for the valorisation approach (Girotto et al., 2015). Polyphenols, vitamins and minerals are some of the bioactive components present in the food wastes (particularly from fruit and vegetable waste fractions) which can be extracted and converted to value-added products like nutraceuticals, single-cell protein, bio-surfactants and bioplastics. Food waste was utilised in a two-stage fermentation approach by *Yarrowia lipolytica* which produced a single-cell protein output with the protein content output of 38.8% ± 0.2% (w/w) of biomass dry weight (Yang et al., 2022). Most of the developments are in the lab scale which further needs refinement for the pilot scale and commercial level adoption.

9.4.2 Anaerobic Digestion

Anaerobic digestion seems to be a promising strategy for the food waste containment due to the fact that the end product derived is mainly a renewable energy in the form of biogas (Polo et al., 2018). The major end products of the process are heat and the digestate. The microbial digestion of organic components in the anaerobic condition lead to the generation of biogas. The digestate can be used as a fertiliser as soil amendments. Adoption of anaerobic digestion in the food waste seems to cover the principles of recycling, reduction of landfills and recovery of the by-product. But the emission of toxic sulphur-containing gases (Chen et al., 2008) generated along with the biogas is a matter of concern and the initial process set-up is associated with the high capital investments (De Baere, 2006).

9.4.3 Biomaterials Production

Food wastes are good and potential source of carbohydrates. Microbes can be cultured on such substrates for the biosynthesis of polymers. These polymers are biodegradable as it is obtained from the microbial source. Examples include polylactate a plastic constituent obtained from lactate and lactic acid, polyhydroxyalkanoates

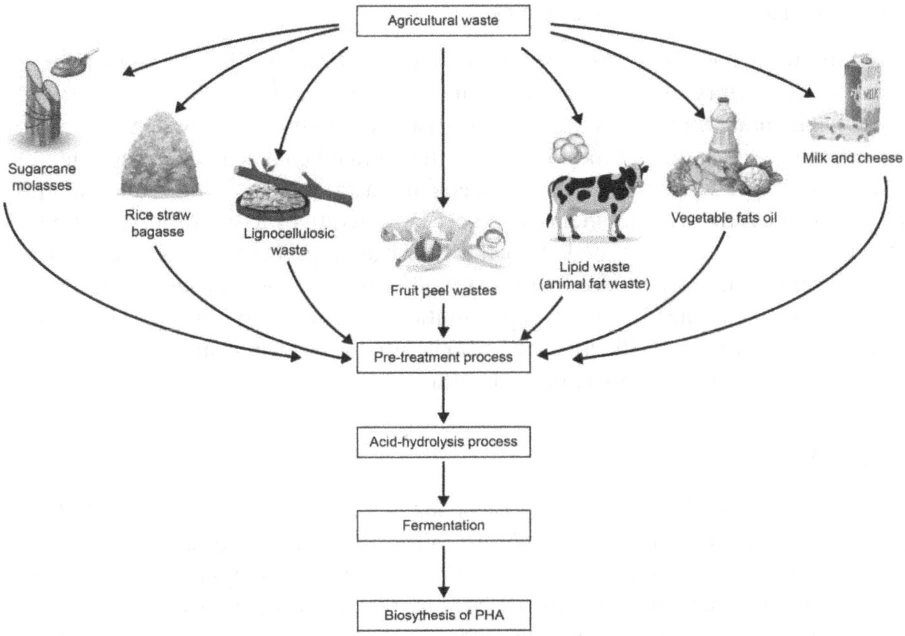

FIGURE 9.2 Flowchart of the PHA production by fermentation. Adopted from Vigneswari, S., Noor, M. S. M., Amelia, T. S. M., Balakrishnan, K., Adnan, A., Bhubalan, K., Amirul, A. A., & Ramakrishna, S. (2021). Recent advances in the biosynthesis of polyhydroxyalkanoates from lignocellulosic feedstocks. Life, 11(8), 807.

and polyhydroxybutyrate. In the studies of Hassan et al. (2013) for the sustainable production of PHAs from oil palm biomass, they observed that oil palm empty fruit bunch (OPFB) which is a waste remnant left over of oil extraction is a good source of cellulose, hemicelluloses, lignins which can be used as feedstock for fermentation and biopolymer synthesis. A generalized method for PHA production is depicted in (Figure 9.2).

9.4.4 Bio-Refinery Production

Food processing by-products can be used as a potential candidate for the production of bio-refineries (Kamm & Kamm, 2004; Ghatak, 2011). A bio-refinery can be viewed as integrated bio-based industry encompassing a variety of technologies to produce saleable products in the form of chemicals, bio-fuels, food and feed ingredients, biomaterials and energy from the biomass raw materials. A bio-refinery takes into account the environment, social and economy factors for the value addition and conversion. The main aim is to maximise the biomass value addition and resource use efficiency. The steps for bio-refinery are waste collection, pre-treatment and fractionation steps, extraction and separation technologies, as well as physical, chemical and/or biochemical conversion, modification technologies, downstream processing and distribution.

9.4.5 Agricultural Co-operatives

Agricultural co-operatives are usually created within a specific production sector (cereals, wine, fruits, etc.). The production facility creates the main product of economic value along with the generation of waste and by-product flows. Autonomous and voluntary association of individuals for the mutual benefits and common interest towards social, economic and cultural needs and aspirations are the foundation pillars for co-operative development. The agro co-operatives assist in the creation of infrastructure development needed for the valorisation process. The business model is not generally diversified and focuses on the most accessible product valorisation pathway with guaranteed market opportunities (Donner et al., 2020). This model seems to be sustainable in the long run, but needs commitment from the stakeholders and a strong investment support in the beginning.

9.4.6 Agro-Parks

Agro-parks are the cluster of closely associated economic activities and agro-functions. It actually relies on the principles of industrial symbiosis (that brings together companies/industries from different sectors in an effort to promote the valorisation of waste, improvement of resource efficiency and reduction of environmental impact) and circular bio-economy. Essentially, these clusters (consisting of mutually dependant business) operate synergistically leading to the development among the different units. In this case, the waste of one unit can become a feedstock to the nearest related unit. This efficiently maximises the recycling of resources, without the much needed external resource requirements like fossil fuel to transport. In this way, creation of agro-parks is also a sustainable model which can mutually benefit the stakeholders with minimum impact on the environment and also leads to support and drive the local economy (Montero et al., 2012). However, there exists an uncertainty with regard to technological, institutional and market development due to the fact that if one component breaks down due to any of the technological reasons, the rest of the operations will have a cascading negative impact.

9.4.7 Creation of Support Structures

Creation of support structures helps close the material loops through cooperation between different players (Mentink, 2014). It is relied upon the coordination, networking, bringing together of normally disconnected players, technological and logistic intelligence and promotion. In this way, providing these structures helps companies develop their own business-oriented circular models. Three types of support structures have been identified and proposed (Donner et al., 2020).

a. Geographical approach: This creates substitution and mutualisation between players within a well-defined territory. Raising awareness among the stakeholders and supporting local economy are the major activities.
b. Valorisation approach: Here, valorisation techniques are aimed to create innovative solutions and technologies for the value addition. This approach

in turn creates new market opportunities for the potential business model and economic sustainability.

c. Waste flow approach: This approach relies on the nutrient rejuvenation of the soil and increasing the quality and productivity of the soil, and the valorisation techniques for the by-products generated in the field.

9.4.8 Insect Farming

An emerging and radical method of food waste recycling where saprophagous insects are being fed on the organic food waste, thus facilitating the nutrient recycle. The most commonly farmed insects include mealworms, crickets and black soldier flies (BSFs) (Elleby et al., 2021). In particular, black soldier fly has been shown to be one of the most promising candidate in terms of its ability to convert food waste into proteins and fat rich biomass (Oonincx et al., 2015). In this way, insects can be seen as a future way of mitigating the food waste in terms of converting the waste to high protein fractions and the oil meal from the insect which is high valued for its benefits. The full-grown insect can be used as a feed to animals, and the frass of the insects can be used as a soil amendment as it replenishes the nutrient composition in the soil. The major advantage of biomass conversion strategy by insect farming lies in its short life span, lower space requirement for its cultivation, feeding and recovery, higher feed conversion rate of insects, lower gas emissions. These advantages can significantly help develop a long-term solution for managing food waste.

9.4.9 Waste Management Hierarchy

The waste management hierarchy has become a gold standard approach for the waste reduction and management and has been adopted worldwide. The major goals revolves in choosing the best possible option for reducing the waste accumulation, and hence, prevention is at the base of the framework with maximum importance and the most favourable option. Prevention of food wastage can be achieved by creating necessary infrastructure for storage, logistics and distribution so that waste generation can be reduced in the supply chain. Other methods like proper food labelling, packaging and consumer planning can prevent the food wastage at the retail and consumer levels. If all the options for prevention strategy are exhausted, then reuse of the wasted food needs to be practiced, where the food surplus gets distributed to the needy sections of the society through establishment of food banks. After the reuse, the next best strategy is to recycle the food waste by following technologies like composting and converting to animal feeds. Then, in the recovery step, the maximum recovery of biomass is being made by following anaerobic digestion which leads to generation of renewable biogas. At the end, the last option should be of disposal by means of incineration and landfills. The major lacuna associated with the waste hierarchy was it focusses more importantly on environmental impact. Economic factors were not taken for formulation to be viewed as a business strategy (Figure 9.3).

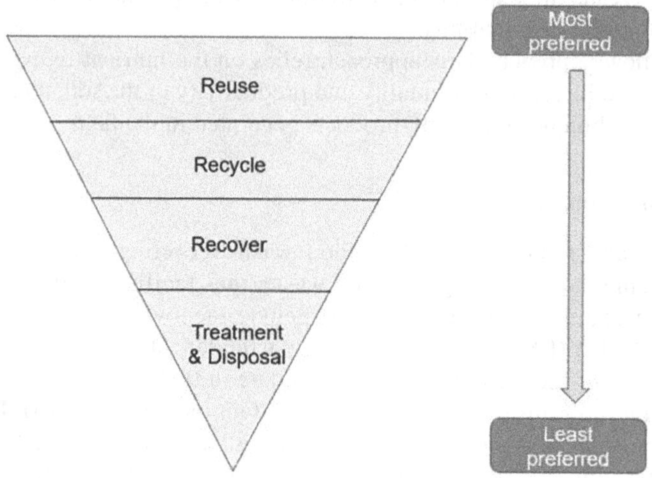

FIGURE 9.3 Waste management hierarchy.

9.4.10 Sustainable Consumption and Production Approach

It can be seen as a strategy to achieve the sustainable goals using the technological and social innovation taking into account the social, economic and environmental health. United Nations Environmental Program, defines Sustainable Consumption and Production (SCP) as the production and utilisation of goods and services which are the basic needs and can improve the quality of life with the minimum utilisation of natural resources and reduced generation of toxic materials and emissions of waste and pollutants over the life cycle so as to protect the needs of future generation.

9.5 CONCLUSION

It is estimated that worldwide, around 33% of all produced food gets unused, and it can be in the form of either food wasted or food lost (Buzby & Hyman, 2012; Kojima & Ishikawa, 2013). The presently followed sustainable elements in CE like economic, social and environmental dimensions should also include circular agriculture which at present is seen as a subsidised sector (Hamam et al., 2021) but due to the complexity of the existing agri-food supply chain, it almost becomes a difficult task to define a specific CE model for the whole sector (Esposito et al., 2020). To accelerate the shift to circular models, collaborative efforts are needed both horizontally (which is among the supply chain stakeholders) and vertically (among the different stages of stakeholders in the supply chain). Following eco-innovation principles which is comprised of cradle-to-cradle concept, CE will definitely lead us to the zero-waste economy (Mirabella et al., 2014). Innovative technologies for food waste reduction and management are the need of the hour along with routing of these technologies into an economic perspective, and thereby it becomes a self-revolving business model that gives substantial revenue and profits to the stakeholders with a minimum impact on the resources. Reducing food loss and waste can boost the productivity of the food suppliers involved in the food supply chain and also helps build stronger customer relationships and enhance their reputation for environmental stewardships. Various intelligent approaches can be used for the successful management of food waste like artificial intelligence (AI), machine learning (ML) and Internet of Things (IoT), which facilitate real-time decision making and process optimisation that need to be incorporated in the existing framework.

But the major limitation in implementing the CE is that it requires change at a system level and an involvement of all the stakeholders within the food supply value chains (suppliers, manufacturers, retailers, consumers). We need to extend this circular model into more of CE-derived activity, wherein the aim is not only to cycle the energy and material flow but also to maintain a sound financial health and long-term viability of the firm.

REFERENCES

Adhikari, B., Barrington, S., & Martinez, J. (2006). Predicted growth of world urban food waste and methane production. *Waste Management & Research*, *24*(5), 421–433.

Adhikari, B., Barrington, S., Martinez, J., & King, S. (2009). Effectiveness of three bulking agents for food waste composting. *Waste Management*, *29*(1), 197–203.

Ahmad, M., Rajapaksha, A. U., Lim, J., Zhang, M., Bolan, N., Mohan, D., Vithanage, M., Lee, S. S., & Ok, Y. S. (2014). Biochar as a sorbent for contaminant management in soil and water: A review. *Chemosphere*, *99*, 19–33.

Assessment, M. E. (2005). *Ecosystems and Human Well-Being* (Vol. 5, p. 563). Island Press, Washington, DC.

Balat, M., Balat, M., Kırtay, E., & Balat, H. (2009). Main routes for the thermo-conversion of biomass into fuels and chemicals. Part 1: Pyrolysis systems. *Energy Conversion and Management*, *50*(12), 3147–3157.

Abul Basher, S., Raboy, D., Kaitibie, S., & Hossain, I. (2013). Understanding challenges to food security in dry Arab micro-states: Evidence from Qatari micro-data. *Journal of Agricultural & Food Industrial Organization*, *11*(1), 31-49.

Buzby, J. C., & Hyman, J. (2012). Total and per capita value of food loss in the United States. *Food policy*, *37*(5), 561-570.

Carmona-Cabello, M., García, I. L., Papadaki, A., Tsouko, E., Koutinas, A., & Dorado, M. (2021). Biodiesel production using microbial lipids derived from food waste discarded by catering services. *Bioresource Technology*, 323, 124597.

Chen, G., Andries, J., & Spliethoff, H. (2003). Catalytic pyrolysis of biomass for hydrogen rich fuel gas production. *Energy Conversion and Management*, *44*(14), 2289–2296.

Chen, Y., Ji, C., & Creamer, K. (2008). Inhibition of anaerobic digestion process: A review. *Bioresource Technology*, *99*(10), 4044–4064.

Cheng, J. Y., & Lo, I. M. (2016). Investigation of the available technologies and their feasibility for the conversion of food waste into fish feed in Hong Kong. *Environmental Science and Pollution Research*, 23, 7169–7177.

Corrado, S., & Sala, S. (2018). Food waste accounting along global and European food supply chains: State of the art and outlook. *Waste Management*, *79*, 120–131.

Corrado, S., Caldeira, C., Eriksson, M., Hanssen, O. J., Hauser, H. E., Van Holsteijn, F., Liu, G., Östergren, K., Parry, A. D., Secondi, L., Stenmarck, Å., & Sala, S. (2019). Food waste accounting methodologies: Challenges, opportunities, and further advancements. *Global Food Security*, 20, 93–100.

De Baere, L. (2006). Will anaerobic digestion of solid waste survive in the future? *Water Science and Technology*, *53*(8), 187–194.

Donner, M., Gohier, R., & De Vries, H. (2020). A new circular business model typology for creating value from agro-waste. *Science of the Total Environment*, 716, 137065.

Edwards, F., & Mercer, D. (2007). Gleaning from Gluttony: An Australian youth subculture confronts the ethics of waste. *Australian Geographer*, *38*(3), 279–296.

Elleby, C., Jensen, H. G., Domínguez, I. P., Chatzopoulos, T., & Charlebois, P. (2021). Insects reared on food waste: A game changer for global agricultural feed markets? *EuroChoices*, *20*(3), 56–62.

EPA, (2003). *Beyond RCRA Waste and Materials Management in the Year 2020*. Washington, DC.

EPA, (2012). *Putting Surplus Food to Good Use*. Washington, DC.

Esposito, B., Sessa, M. R., Sica, D., & Malandrino, O. (2020). Towards circular economy in the agri-food sector: A systematic literature review. *Sustainability*, *12*(18), 7401.

European Commission (EC). 2015. Closing the Loop - An EU Action Plan for the Circular Economy, COM (2015) 0614.

FAO. (2011). *Global Food Losses and Food Waste—Extent, Causes and Prevention*. Rome, Italy.

Food and Agriculture Organization of the United Nations. (FAO). (2019). The State of Food and Agriculture, Moving Forward on Food Loss and Waste Reduction. Food and Agriculture Organization of the United Nations: Rome, Italy.

Food and Agriculture Organization of the United Nations. (FAO). (2014). *Food: Global Initiative on Food Loss and Waste Reduction. Definitional Framework of Food Loss.* Food and Agriculture Organization of the United Nations: Rome, Italy.

García-García, G., Woolley, E., Rahimifard, S., Colwill, J., White, R., & Needham, L. (2016). A methodology for sustainable management of food waste. *Waste and Biomass Valorization*, *8*(6), 2209–2227.

Ghatak, H. R. (2011). Biorefineries from the perspective of sustainability: Feedstocks, products, and processes. *Renewable & Sustainable Energy Reviews*, *15*(8), 4042–4052.

Girotto, F., Alibardi, L., & Cossu, R. (2015). Food waste generation and industrial uses: A review. *Waste Management*, *45*, 32–41.

González-Sánchez, R., Alonso-Muñoz, S., & Medina-Salgado, M. S. (2023). Circularity in waste management: A research proposal to achieve the 2030 Agenda. *Operations Management Research*, 16(3), 1–21.

Gustavsson, J., Cederberg, C., & Sonesso, U. (2011). Global Food Losses and Waste- Extent, Causes and Prevention; Food and Agriculture Organization of The United Nations: Rome, Italy.

Hamam, M., Chinnici, G., Di Vita, G., Pappalardo, G., Pecorino, B., Maesano, G., & D'Amico, M. (2021). Circular economy models in agro-food systems: A review. *Sustainability*, *13*(6), 3453.

Hassan, M. A., Yee, L., Yee, P. L., Ariffin, H., Raha, A. R., Shirai, Y., & Sudesh, K. (2013). Sustainable production of polyhydroxyalkanoates from renewable oil-palm biomass. *Biomass and Bioenergy*, *50*, 1–9.

Hossein, S., & Dahlan, I. (2015). Growth performance of free-range village chickens fed dehydrated processed food waste. *Malaysian Journal of Animal Science*, *18*(1), 77–86.

Kamm, B., & Kamm, M. (2004). Biorefinery-systems. *Chemical and Biochemical Engineering Quarterly*, *18*(1), 1–7.

Katami, T., Yasuhara, A., & Shibamoto, T. (2004). Formation of dioxins from incineration of fallen leaf. *Bulletin of Environmental Contamination and Toxicology*, *72*(1), 114–118.

Kim, S., Kwon, E. E., Kim, Y. T., Jung, S., Kim, H. J., Huber, G. W., & Lee, J. (2019). Recent advances in hydrodeoxygenation of biomass-derived oxygenates over heterogeneous catalysts. *Green Chemistry*, *21*(14), 3715–3743.

Kim, S., Tsang, Y. F., Kwon, E. E., Lin, K. A., & Lee, J. (2018). Recently developed methods to enhance stability of heterogeneous catalysts for conversion of biomass-derived feedstocks. *Korean Journal of Chemical Engineering*, *36*(1), 1–11.

Kojima, R., & Ishikawa, M. (2013) *Prevention and Recycling of Food Wastes in Japan: Policies and Achievements*; Kobe University: Kobe, Japan.

Kumar, M., Ou, Y. L., & Lin, J. (2010). Co-composting of green waste and food waste at low C/N ratio. *Waste Management*, *30*(4), 602–609.

Lee, J., Kwon, E. E., & Park, Y. (2020). Recent advances in the catalytic pyrolysis of microalgae. *Catalysis Today*, *355*, 263–271.

Lipinski, B., Hanson, C., Waite, R., Searchinger, T., & Lomax, J. (2013). Reducing food loss and waste.

Lisca, A., Feeley, J., Lozano, A. O., & Wang, K. (2021). *Circular Economy Action Agenda: Textiles*. Platform for Accelerating the Circular Economy: Hague, The Netherlands

Lundqvist, J., De Fraiture, C., & Molden, D. (2008). Saving water: From field to fork: Curbing losses and wastage in the food chain.

MacArthur, E., Zumwinkel, K., & Stuchtey, M. R. (2015). Growth within: a circular economy vision for a competitive Europe. *Ellen MacArthur Foundation*, *100*.

Mentink, B., (2014). Circular business model innovation: A process framework and a tool for business model innovation in a circular economy. Masters thesis. University of Leiden

Mirabella, N., Castellani, V., & Sala, S. (2014). Current options for the valorization of food manufacturing waste: A review. *Journal of Cleaner Production*, *65*, 28–41.

Møller, H., Hanssen, O. J., Svanes, E., Hartikainen, H., Silvennoinen, K., Gustavsson, J., Östergren, K., Schneider, F., Soethoudt, H., Canali, M., Politano, A., Gaiani, S., Redlingshöfer, B., Moates, G. K., Waldron, K. W., & Stenmarck, Å. (2014). *Standard approach on quantitative techniques to be used to estimate food waste levels* (pp. 1–108)

Montero, J. I., Claret, A., Antón, A., & Torrellas, M. (2012). Advantages of clustering.

Morseletto, P. (2020). Targets for a circular economy. *Resources, Conservation and Recycling*, 153, 104553.

Oladejo, J., Shi, K., Luo, X., Yang, G., & Wu, T. (2018). A review of sludge-to-energy recovery methods. *Energies*, *12*(1), 60.

Olson, S.; Lozano, A.O.; Wang, K. (2021) *Circular Economy Action Agenda: Food*; Platform for Accelerating the Circular Economy: Hague, The Netherlands.

Oonincx, D., Van Broekhoven, S., Van Huis, A., & Van Loon, J. J. A. (2015). Feed conversion, survival and development, and composition of four insect species on diets composed of food By-Products. *PLoS ONE*, 10(12), e0144601.

Papargyropoulou, E., Lozano, R., Steinberger, J., Wright, N., & Ujang, Z. (2014). The food waste hierarchy as a framework for the management of food surplus and food waste. *Journal of Cleaner Production*, *76*, 106–115.

Parfitt, J., Barthel, M., & Macnaughton, S. J. (2010). Food waste within food supply chains: Quantification and potential for change to 2050. *Philosophical Transactions of the Royal Society B*, 365(1554), 3065–3081.

Polo, C. M., Del Mar Cledera Castro, M., & Soria, B. Y. M. (2018). Reviewing the anaerobic digestion of food waste: From waste generation and anaerobic process to its perspectives. *Applied Sciences*, 8(10), 1804.

Rockström, J., Steffen, W., Noone, K. J., Persson, Å., Chapin, F. S., Lambin, É. F., Lenton, T. M., Scheffer, M., Folke, C., Schellnhuber, H. J., Nykvist, B., De Wit, C. A., Hughes, T. P., Van Der Leeuw, S., Rodhe, H., Sörlin, S., Snyder, P. K., Costanza, R., Svedin, U., ... Foley, J. A. (2009). A safe operating space for humanity. *Nature*, *461*(7263), 472–475.

Cheraghi Saray, S., Hosseinkhani, A., Janmohammadi, H., Zare, P., & Daghighkia, H. (2014). Thermal and probiotic treatment effects on restaurant waste for incorporation into poultry diet. *International Journal of Recycling of Organic Waste in Agriculture*, *3*, 1-7.

Smil, V. (2002). Nitrogen and food production: Proteins for human diets. AMBIO: *A Journal of the Human Environment*, *31*(2), 126–131.

Smil, V. (2004). Improving efficiency and reducing waste in our food system. *Environmental Sciences*, *1*(1), 17–26.

Smil, V., (2000). *Feeding the World: A Challenge for the Twenty-First Century*. The MIT Press, Cambridge, MA.

Stenmarck, Â., Jensen, C., Quested, T., Moates, G., Buksti, M., Cseh, B., ... & Östergren, K. (2016). *Estimates of European Food Waste Levels*. IVL Swedish Environmental Research Institute.

Taghikhah, F., Voinov, A., & Shukla, N. (2019). Extending the supply chain to address sustainability. *Journal of Cleaner Production*, *229*, 652–666.

Tamasiga, P., Miri, T., Onyeaka, H., & Hart, A. (2022). Food waste and circular Economy: Challenges and opportunities. *Sustainability*, *14*(16), 9896.

Tekin, K., Karagöz, S., & Bektaş, S. (2014). A review of hydrothermal biomass processing. *Renewable & Sustainable Energy Reviews*, *40*, 673–687.

Vigneswari, S., Noor, M. S. M., Amelia, T. S. M., Balakrishnan, K., Adnan, A., Bhubalan, K., Amirul, A. A., & Ramakrishna, S. (2021). Recent advances in the biosynthesis of polyhydroxyalkanoates from lignocellulosic feedstocks. *Life*, *11*(8), 807.

Westendorf, M. L., Schuler, T., Zirkle, E. W., Hays, V. W., & Wilson, L. L. (1999). Nutritional quality of recycled food plate waste in diets fed to Swine12. *The Professional Animal Scientist*, 15(2), 106–111.

Yang, R. Q., Chen, Z., Hu, P., Zhang, S., & Luo, G. (2022). Two-stage fermentation enhanced single-cell protein production by *Yarrowia lipolytica* from food waste. *Bioresource Technology*, 361, 127677.

10 Environmentally Friendly Food Processes, Biorefineries and 3D Food Printing

Afzal Rahman
Bangladesh Agricultural University

Juwel Rana
Daffodil International University

Md. Shamsur Rahman
Daffodil International University

Zannatul Ferdoush
Daffodil International University

Md. Anisur Rahman Mazumder
Bangladesh Agricultural University

Saroat Rawdkuen
Mae Fah Luang University

Thottiam Vasudevan Ranganathan
Karunya Institute of Technology and Sciences

10.1 INTRODUCTION

Food is not only a fundamental need but also an essential component of both our culture and our social life. On the other hand, as the number of people living in the world increases exponentially, so does the need for food, which in turn creates severe environmental problems (Tipu et al., 2022). The conventional processes of food production can result in considerable waste and pollution, making the food sector one of the significant contributors to the destruction of the environment. Furthermore, the

DOI: 10.1201/9781003412489-10

linear production model used in the food business, which entails extracting resources, making food, consuming food and then disposing of food, generates substantial waste (Sarker et al., 2022). Because the world is still dealing with the repercussions of climate change and environmental degradation, the hunt for food production methods that are both environmentally sound and respectful of the planet's ecosystems is becoming an increasingly important requirement. According to More, Jambrak and Arya (2022), high temperatures, high pressures and the inclusion of chemical compounds are frequently used in conventional methods of food preparation. This results in a substantial amount of energy consumption and pollution. Several novel technologies have surfaced in response to these challenges, including developing ecologically friendly food processing methods. Each of these technological advancements comes with its own set of benefits, particularly concerning reducing waste and enhancing production efficiencies in the food industry (Chemat et al., 2020). Methods of food processing that are environmentally friendly strive to produce the least amount of waste possible, use the least amount of energy and make the least amount of use of potentially hazardous chemicals.

The fruit and vegetable processing sector alone accounts for 16.5–20.5 million metric tons of the approximately 55 million metric tons of food waste produced worldwide (Saini et al., 2019). Fruits' and vegetables' physiologically active components (phenolic compounds, terpenes, terpenoids, alkaloids and so on) are responsible for their numerous health benefits (Azmir et al., 2013). Because of their antioxidant, antibacterial and anti-inflammatory activities, these bioactives are known to protect against oxidative stress and related chronic disorders (Saini et al., 2019). Peel, seed, pomace, stems and so on account for at least 25–30% of the waste produced by most fruits and vegetables. Bioactive substances, such as polyphenols, flavonoids, tannins, essential oils, natural colors, a wide range of organic acids, vitamins and fibers, are abundant in all these waste product fractions. There is now a significant issue with managing waste due to the tremendous volume of garbage being produced. Despite its wealth of bioactive components, it is often utilized as animal feed or left in landfills, releasing damaging greenhouse gases (GHG) and posing severe economic and ecological environmental hazards. In addition, the proliferation of harmful bacteria is primarily responsible for unpleasant odors and infectious illnesses because of the waste materials' high nutrition and moisture content (Sagar et al., 2018). By 2025, annual global biowaste production was projected to rise to 2.2 billion tons (Nizami et al., 2017). Biorefinery emerged as a sustainable approach and was recognized as a promising method that utilizes renewable sources to extract valuable components from agricultural waste streams. Bio-refining food waste might help achieve sustainable zero-waste growth. Agro-waste biorefinery techniques urgently need a circular economy and contribute to a low-carbon, cleaner environment that is ecologically beneficial, economically and technically feasible, and widely accepted. Therefore, it is essential to use this bioactive-rich waste (trash to treasure approach) by extracting those utilizing safer and more environmentally friendly methods, including ultrasound-assisted extraction (UAE), microwave-assisted extraction, pulsed electric field, high-pressure processing, etc. These green-derived bioactive materials may be employed as food additives and preservatives and in the formulation of functional and nutraceutical foods. They are generally accepted as safe (GRAS). Hence, they are utilized as a food additive, such as a flavor or coloring ingredient (Alexandre et al., 2018).

Three-dimensional food printing is a cutting-edge technique that efficiently and accurately creates individually tailored food items with little food waste. Digital design files may be used in 3D food printing to make individual meals, allowing for more individualized approaches to nutrition and culinary creation (Mantihal et al., 2020). The ability to alter the food shapes, textures and nutritional compositions is a significant benefit of 3D printing for this purpose (Derossi et al., 2018). Printing food also has a number of other advantages: (i) food may be tailored in terms of ingredients, composition and tastes to meet the demands, health conditions, lifestyles and appetites of various individuals; (ii) nutrient-dense items, including exotic insects, can be included; and (iii) food can be prepared quickly. Using many of the same techniques used to print organs and tissues in the biomedical industry, researchers have utilized this technology to print cell grown meat and meat analogs (Godoi, Prakash, and Bhandari, 2016). Additionally, a number of different solid or semi-solid foods, such as cookies, cakes, burgers, apples and chocolates, have been produced using 3D printing (Lipton et al., 2015; Fanli Yang et al., 2018; Feng, Zhang, and Bhandari, 2018; Wang et al., 2018; Pérez et al., 2019). A variety of 3D printer designs are generally accessible, including selective laser sintering, powder bed printing, ink-jet printing and extrusion printing; however, extrusion printing is now the most popular method of 3D printing food (Wegrzyn et al., 2012; Sun et al., 2015). Typically, 3D food printers include one or more extruders that draw shapes layer by layer from point to point to produce food items. Typically, food inks—which may include one or more ingredients—are introduced to an extruder, forced out by using external force and then printed into pre-designed shapes using spatial data recorded in the instrument software (Feng et al., 2019). Usually, attaching the printer to an air pressure device produces the force needed to push food ink through a nozzle. One of the most important elements in any successful 3D food printing application is the development of food inks with the necessary physicochemical and functional characteristics (Pérez et al., 2019). Important material characteristics, including rheology, surface tension and the phase behavior of the food ink, must be taken into account in order to properly print food. Additionally, because they affect the structure and stability of the printed foods, the 3D printer's operating parameters—such as nozzle height, nozzle diameter, flow rate and printing speed—are also very important. Hardness, springiness, plasticity and viscoelasticity are a few crucial factors that affect the success of 3D printing food (Wegrzyn et al., 2012; Sun et al., 2015; Dankar et al., 2018; Feng et al., 2019). Nevertheless, an integrated document on recent information on different aspects of 3D food printing is frequently overlooked. These topics include principles, properties of materials and food inks, design and operational parameters, printing technology, applications, quality assessment of printing process and printed object, safety, consumer attitude and future perspectives. This chapter will thus concentrate on the brief integration of information on those aspects, which are essential for evaluating the quality, customer acceptability and safety of 3D food but are sometimes overlooked in other publications. The chapter will go over environmentally friendly food processes and biorefineries, which may increase food quality while lowering energy usage and waste. Each of these breakthroughs has the potential to reduce our global environmental impact while also enhancing food quality, availability and nutrition. Finally, the chapter will conclude with a short integration of information on those areas that are critical for assessing the quality, consumer acceptance and safety of 3D food. Readers will have a greater knowledge of each technology by the end of this chapter, as well as how it is presently being used to build a more sustainable and equitable food system.

10.2 DEFINITION OF ENVIRONMENTALLY FRIENDLY FOOD PROCESSING

According to the summary of the other literature, environmentally friendly food processing refers to the methods and techniques used to manufacture and produce food products in a way that minimizes negative impacts on the environment involving a range of practices, such as reducing energy consumption, minimizing waste and pollution, sourcing sustainable ingredients and using environmentally friendly packaging. Several environmentally friendly techniques are high-pressure processing (HPP), ohmic heating, ultrasound and pulsed electric fields, etc. An environmentally friendly food processing technique uses renewable energy sources, optimizing production processes to reduce water usage, utilizing recycling and waste reduction programs, and using natural preservatives instead of synthetic ones.

10.3 TECHNIQUES FOR ENVIRONMENTALLY FRIENDLY FOOD PROCESSING

10.3.1 Ohmic Heating

Ohmic heating is an advanced thermal heating process that is quick and uniform, as electric current is transferred through food materials generating heat according to Joule's law (Gavahian, Chu, and Farahnaky, 2019). Ohmic heating is utilized during the extraction process to decrease the time required for treatment, resulting in minimal thermal harm to the beneficial components found in plants, such as polyphenols (Figure 10.1). This method's core premise entails inserting an electrode into a food ingredient and transmitting an alternating electrical current (AC) at a frequency ranging from 50 Hz to 100 kHz through it (Ebrahimi and Lante, 2022). According to the ohmic heating concept, an electrical conductor disperses electrical energy in heat; thus, the food converts the electrical energy it receives into heat energy. This process is known as the Joule effect. Within a matter of seconds to a few minutes, the food is quickly heated. The heat level is determined by the food's electrical conductivity, concentration, electrodes and the voltage difference applied (Wu et al., 2020). This heating system is very efficient (>90%). High-speed heating has been used on a variety of food products, including fruits, vegetables, dairy and meat . It has shown to be an effective technology in the food sector for a variety of operations such as sterilization, pasteurization, blanching, starch gelatinization, thawing, evaporation, heating, fermentation and extracting bioactive components from byproducts (Jan et al., 2021).

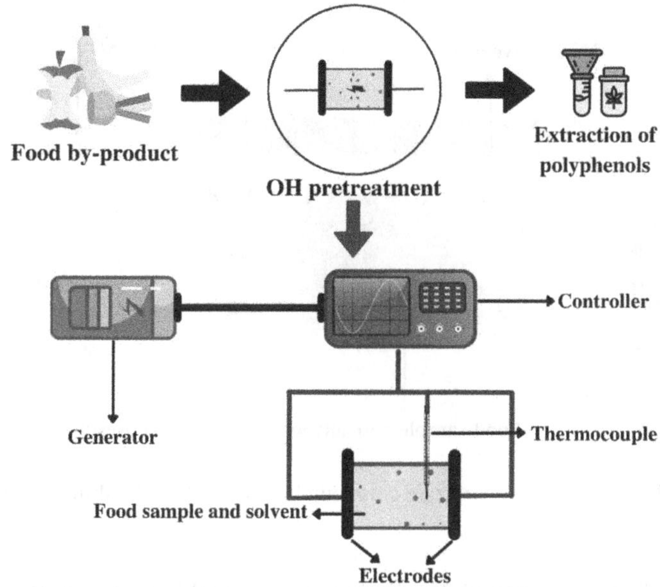

FIGURE 10.1 Extraction of polyphenols using ohmic heating (Ebrahimi and Lante, 2022).

In addition, ohmic heating was used for heterogeneous heating, which increased the extraction of pectin and minimized the changes in pectin properties, thus, production costs. Pectin isolated by ohmic heating at a different voltage of 15 V/cm at a temperature of 90°C for 30 minutes enhanced pectin extraction relative to conventional extraction (Saberian et al., 2018). According to Coelho et al. (2019), applying OH as a pretreatment in the extraction of polyphenols from tomato byproducts, using 70% ethanol as a solvent at 70°C for 15 minutes, resulted in a 77% higher recovery of rutin compared to traditional extraction methods.

10.3.2 Microwave-Assisted Extraction (MAE)

MAE is a modern and effective technique that combines traditional solvent extraction with microwave technology to extract polyphenols from food waste (Figure 10.2). Microwave technology refers to non-ionizing electromagnetic waves with a frequency range of 300 MHz to 300 GHz and a wavelength range of 1–1000 mm (Chuyen et al., 2018). Microwave energy is converted into heat energy without physical contact with the food through electromagnetic wave energy transfer. Microwave extraction is based on how microwave radiation interacts with polar molecules (such as water) in the solvent around the sample to create heat. Due to ionic conduction and dielectric polarization, the temperature rises, increasing cell wall permeability and facilitating the extraction of polyphenols (Routray and Orsat, 2014). Higher temperatures reduce solvent viscosity, enhance intermolecular contact and intracellular pressure, and increase molecular mobility, making polyphenols more soluble (Da Rosa et al., 2019). Since water is the primary absorber of MW energy in bioactive molecules and

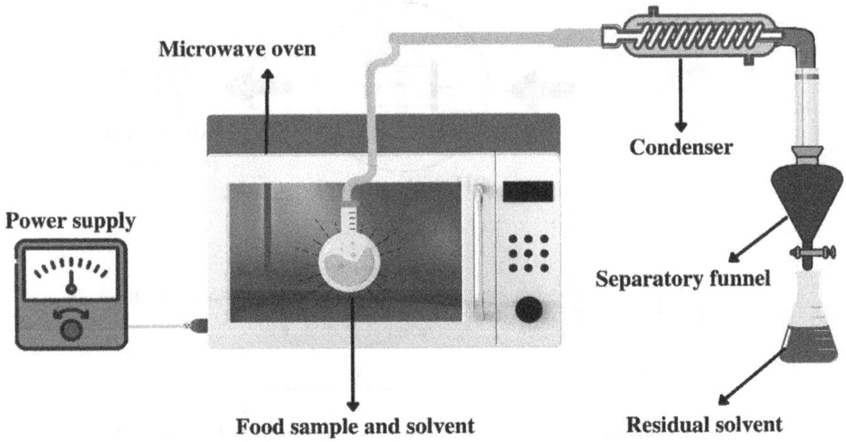

FIGURE 10.2 Extraction of bioactive compounds using MAE (Ebrahimi and Lante, 2022).

food, biomaterials with a high moisture content respond faster to MW heating. The extraction yield of MAE is significantly influenced by the solvent selection. Organic solvents like ethanol, acetone and methanol have proven effective. Da Rosa et al. (2021) found that MAE extracts polyphenols from Brazilian olive leaves better than UAE. MAE has been utilized to improve PE and has been proven to produce more pectin in less time than conventional extraction (CE) (Carbonell Capella et al., 2017). This approach has various limitations, including high technical complexity, limited energy input control, heat-sensitive bioactive component degradation, expensive preliminary cost and low extraction yield (Nüchter et al., 2004).

10.3.3 Ultrasound-Assisted Extraction (UAE)

Ultrasonic waves are electromechanical waves that occur above the human hearing range of 20 kHz (Dash, Pathak, and Pradhan, 2021). In UAE, ultrasonic waves with frequencies ranging from 20 to 100 kHz are often employed. The principle behind UAE is based on acoustic cavitation, which is the formation, growth and implosion of bubbles in a liquid medium under the influence of ultrasound waves. During UAE, high-frequency ultrasound waves are applied to the plant material and solvent mixture. The ultrasound waves create pressure fluctuations in the liquid medium, which cause tiny bubbles to form and grow in the solvent. These bubbles grow until they reach a critical size, at which point they become unstable and collapse violently, creating shock waves that propagate through the liquid. The collapse of these bubbles creates localized high-intensity pressure and temperature changes, which can disrupt the plant cell walls and release bioactive compounds into the solvent, thus facilitating extraction (Mazumder et al., 2023) (Figure 10.3). As bubbles expand and contract, they release energy at the molecular level, which may lead to very high temperatures (up to 5000 K) and pressures (up to 1000 atm). In other words, acoustic cavitation may hydrate and expand food matrices, rupturing cells

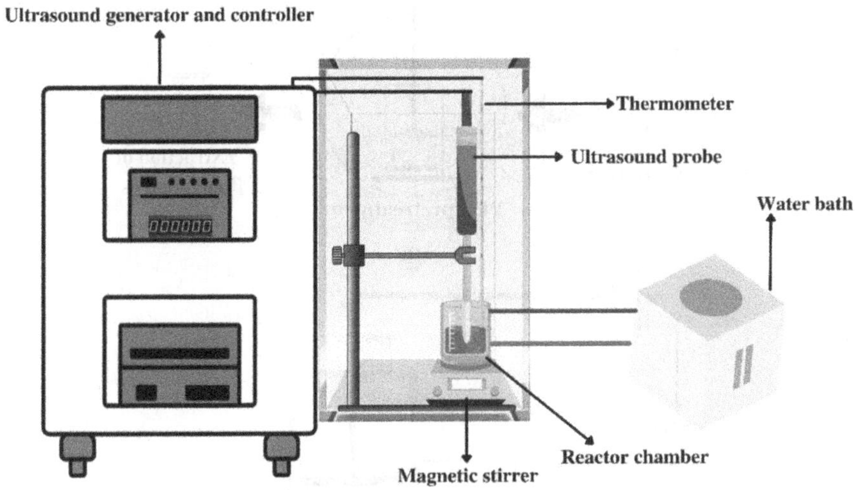

FIGURE 10.3 UAE of bioactive compounds (Ebrahimi and Lante, 2022).

and enabling solvent penetration to increase bioactive component extraction (Da Silva et al., 2018). The efficacy of the UAE is determined by the strength of the sonication, the duration of the extraction, the frequency, the kind of solvent, the temperature and the amplitude of the solvent in the extraction container (Lante and Friso, 2013). UAE resulted in a more remarkable recovery of pectins from tangerine (13.46%) and grapefruit (26.05%) in only 15 and 30 minutes, respectively. The pectin extracted has good rheological and functional qualities, making it suitable for use in the food sector (Polanco-Lugo et al., 2019).

10.3.4 Pulsed Electric Field (PEF) Extraction

PEF extraction is a technique that includes exposing food matrices to an electrical field (Figure 10.4). PEF refers to the process of treating the food matrix in order to extract phytochemicals. Electric pulses are produced by a transformer, which raises the voltage from 140 to 1000 V or even more than 25000 V. This results in electroporation in cell membranes, which may be reversible or irreversible. This high voltage is transformed into a brief pulse by a capacitor, which operates in an enclosed chamber of metallic electrodes (Rodríguez et al., 2020). When a few to several hundred pulses of electricity with a time ranging from nanoseconds to milliseconds are used, a process known as electroporation may cause cell change or rupture (Haberl et al., 2013). The formation of holes in the cell membrane may be linked to the cells being exposed to an external electric field. Because pore formation is a dynamic process that fluctuates with the strength of the PEF, electroporation can be reversible or irreversible.

Several variables influence the effectiveness of PEF pretreatment, including extraction duration, pretreatment temperature, frequency, pulse shape, specific electrical input, electric field strength, pH level, pulse depth, food density and size matrix, and chemical properties of the extracting end-product (Rifna et al., 2023).

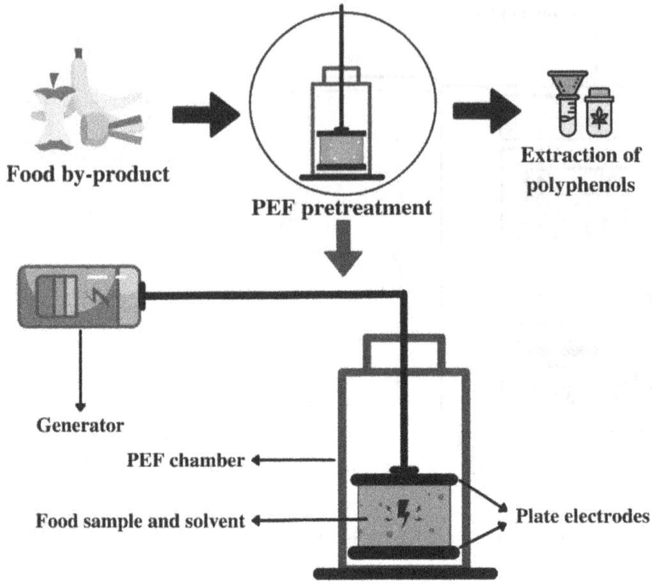

FIGURE 10.4 PEF extraction of bioactive compounds (Ebrahimi and Lante, 2022).

10.3.5 Pressurized Liquid Extraction (PLE)

Extracting bioactive chemicals from solid or semi-solid matrices using solvents under high-pressure and high-temperature conditions is the goal of the sample preparation method known as pressurized liquid extraction (PLE). Bioactive chemicals, flavor compounds and other components of interest have all been extracted using PLE, making it a staple in the food industry (Figure 10.5) (Mihaylova et al., 2014). PLE is the process of applying heat and pressure to a solid or semi-solid matrix in order to improve the solubility of analytes in the solvent and facilitate their extraction into the liquid phase. Sample processing, extraction and solvent recovery are the three stages that make up a standard PLE procedure. The sample is loaded into an extraction cell made of stainless steel, which is subsequently sealed and pressurized with an inert gas (nitrogen or argon). Creating a sealed system in this way helps to reduce the wasteful evaporation of volatile components and guarantees a more thorough extraction. A solvent is introduced to the cell and heated past its boiling point during the extraction process. Keeping the pressure at this level keeps the solvent from boiling and keeps it in a liquid form. The analytes in the matrix may be more easily dissolved by the solvent at higher temperatures and pressures. Once the analytes have been dissolved, they are moved to the liquid phase. The solvent is recovered after the liquid extract has been collected and concentrated by evaporation at low pressure. The concentrated extract is produced after the remaining solvent has been collected using a solvent recovery device like a rotating evaporator or a nitrogen-based evaporator (Ebrahimi and Lante, 2022).

Temperatures in PLE may go as high as 200°C. However, anything beyond 140 or 150°C may cause the breakdown of thermo-sensitive chemicals or the creation of

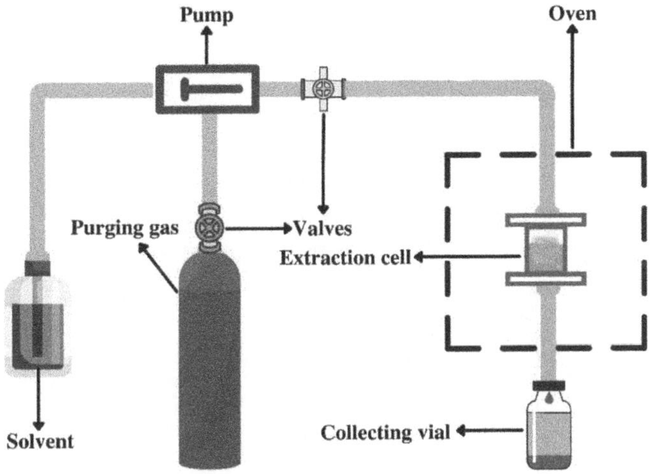

FIGURE 10.5 PLE of bioactive compounds (Ebrahimi and Lante, 2022).

undesired products owing to the Maillard reaction. Due to the application of high pressure, the process temperature may exceed the boiling point of the solvent. But the high pressure (about 8–15 MPa) preserves the solvent's liquid state (Pagano et al., 2021). The success of PLE relies on several variables, including the solvent used, the extraction's temperature and pressure, and the time it is allowed to run. To maximize extraction yield, it is necessary to pick a solvent with the proper polarity and selectivity for the desired analytes and to operate at the ideal temperature and pressure. The extraction time should be fine-tuned to avoid degrading the target analytes and maintain a high yield rate (Alvarez-Rivera et al., 2020).

10.3.6 Enzyme-Assisted Extraction (EAE)

Enzyme-assisted extraction (EAE) is a green technology that boosts extraction efficiency by incorporating enzymes into a mixture. To free bioactive components, the basic technique involves hydrolyzing food material using an enzyme as a catalyst under ideal extraction conditions (Nadar et al., 2018). When an enzyme is added to a sample, the cell wall becomes more permeable, allowing more solvent medium to penetrate the sample. This approach facilitates the release of bioactive compounds, polyphenols and other phytochemicals that dwell within cells but are difficult to remove otherwise. Cellulose, protease and pectinase are powerful enzymes used in extraction.

The food sector is looking for long-term extraction solutions, and it is now more vital than ever to discover alternatives that use less solvent. Enzymes are catalysts that speed up hydrolysis while maintaining high selectivity; as a result, they reduce the quantity of solvent needed while maintaining a high output. Because it uses water as its primary solvent medium, the technique is thought to be environmentally friendly. Furthermore, the extraction procedure is carried out at low temperatures and uses little energy, which helps keep the polyphenols from degrading (Figure 10.6) (Sridhar et al., 2021).

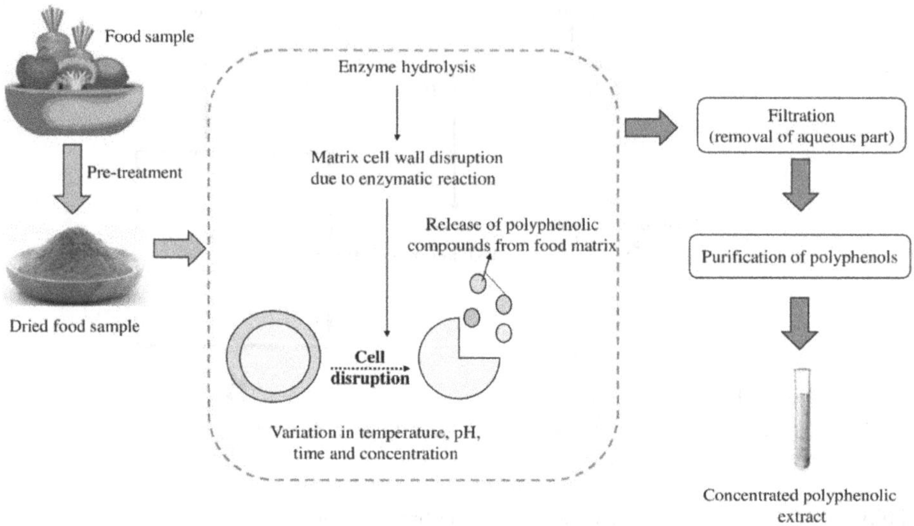

FIGURE 10.6 Enzyme-assisted extraction (EAE) of phenolic compounds (Sridhar et al., 2021).

10.3.7 Ultrasound-Microwave-Assisted Extraction (UMAE)

Extracting various bioactive components from plant materials has been made possible with the use of the ultrasound-microwave-assisted extraction (UMAE) technique, which is a combination of ultrasound and microwave energy. In the process of ultrasonic-microwave extraction, microwaves increase heat transmission, while ultrasound helps with the transfer of mass. By dissolving cell walls and enhancing mass transfer, ultrasound has the ability to increase the contact surface area, improve the solvent's diffusion into the reaction media, and aid in the solubilization of substances. Mass transfer is made easier by ultrasound. The sample is rapidly warmed in a microwave during the second phase, increasing the solubility of the solute and the mass transfer rate. This expedites desorption of the compounds of interest from the sample matrix and boosts the extraction's effectiveness (Vinatoru, Mason, and Calinescu, 2017).

10.4 BENEFITS OF ENVIRONMENTALLY FRIENDLY FOOD PROCESSING

10.4.1 Increased Food Safety and Quality

The food business is snowballing because customers want safe, nutritious and longer shelf life, yet, worries about the sector's impact on the environment have developed in recent years. Concerns about the environment, nutrition and the inactivation of microorganisms have led to the developing of environmentally friendly technologies that can replace traditional heat processes (Chemat et al., 2020; Chakka et al., 2021).

Moreover, applying a more incredible amount of heat during the treatment might result in the denaturation of proteins and the loss of volatile components, leading to a decline in food quality. Non-thermal methods' effectiveness in product quality and the safeguards they provide have been shown. Environmentally friendly food processing can increase food safety by reducing harmful chemicals and minimizing the environmental impact of food production.

10.4.2 Reduced Energy Consumption

Historically, fossil fuels have provided the heat necessary for food processing and preservation through thermal technologies, including pasteurization, sterilization, cooking and drying. As an alternative to these thermal technologies, the non-thermal ones speed up the cooking process, keep the food's quality intact and make it safer for consumers to consume (Fengying et al., 2020). Sustainable extraction practices prioritize conserving renewable energy sources, using non-conventional solvents and manufacturing extracts and products of the highest possible quality and safety standards (Chemat et al., 2020). Environmentally friendly food processing reduces energy consumption using renewable energy sources, energy-efficient equipment, process optimization, recycling and reuse of materials and energy management systems. By adopting these practices, food manufacturers may improve efficiency, save costs and lessen their negative influence on the environment.

10.4.3 Reduced Waste Generation

One-third of all food produced is byproducts, and the disposal of these goods is a significant source of greenhouse gases contributing to climate change (Roda and Lambri, 2019). Because of their high bioactive component content and distinct health benefits, waste products from fruit and vegetable processing have the potential to be employed in the food industry (Azmir et al., 2013). As a means of reducing food waste, bioactive packaging combined with nanotechnology is superior to traditional plastic food packaging (Sarker et al., 2023). Food scraps are turned into bioenergy through integrated biorefinery and used to manufacture valuable byproducts. Bioactive extraction from wasted fruits and vegetables is an eco-friendly method that maximizes the use of agricultural goods while minimizing waste and contamination. Food manufacturers may lessen their harmful environmental effects and save money on waste disposal and treatment by producing less garbage.

10.5 BIOREFINERIES

Biorefinery is mainly a facility that utilizes various biomass conversion processes and equipment to produce fuels, power, heat, chemicals and materials through integrated, multi-step processing operations (Liu et al., 2013, Drapcho et al., 2013, Clark and Deswarte, 2015). To put it another way, a biorefinery is a manufacturing facility that uses biomass to create a variety of products through a number of processes. These products include biofuels, biochemicals and biomaterials, which can be used as substitutes for fossil fuels and other non-renewable resources in various applications. The importance of biorefineries is increasing due to the need for a more sustainable and circular economy. Biorefineries may convert biomass feedstocks into sustainable energy, fuels and chemicals, reducing reliance on fossil fuels, cutting greenhouse gas emissions and increasing resource efficiency. In rural locations, they can also create jobs and aid in the development of local businesses. Biorefineries that effectively utilize sustainable biomass resources can help to meet the growing need for energy, food and materials brought on by the world's expanding population. Biorefineries convert biomass into valuable products, reducing waste and promoting sustainability, leading to a more resilient and sustainable future.

Biorefineries at present state are the results of continuous innovation and development aimed at utilizing renewable biomass resources to produce a wide range of products in an environmentally sustainable and economically viable way. When sugar and starch-based raw materials were first utilized to create ethanol, a biofuel, in the late 19th century, the idea of biorefineries was born (Fajgelj et al., 2007). However, the term "biorefinery" was first coined in the early 21st century to describe a facility that produces multiple products from biomass using various conversion processes. In order to create biofuels, chemicals and other value-added products from biomass, the US Department of Energy (DOE) started a research program to develop biorefinery technology in the 1990s (Perlack et al., 2005). This program led to the first generation of biorefineries, which focused primarily on producing ethanol from corn and sugarcane. Second-generation biorefineries began to appear in the early 2000s, and their raw materials included lignocellulosic biomass including wood, agricultural byproducts and municipal solid waste (Perlack et al., 2005). These biorefineries utilized various conversion processes such as hydrolysis, gasification and fermentation to produce biofuels and chemicals. Algae and other microorganisms are used as raw materials in third-generation biorefineries, which have recently been constructed. These biorefineries have the capacity to manufacture a variety of high-value goods, including biofuels, foods, feed and nutraceuticals.

10.6 PROCESSES INVOLVED IN BIOREFINERIES

Biorefineries use a variety of processes that may be divided into several phases, such as biomass pretreatment, hydrolysis, fermentation, and product separation and purification.

10.6.1 Biomass Pretreatment

By simplifying the separation of cellulose, hemicellulose and lignin, pretreatment enables enzymes to hydrolyze the complex carbohydrate units present in cellulose and hemicellulose into simpler sugars (Chakraborty et al., 2012). Breaking down the biomass's complex structure to reveal the cellulose and hemicellulose components for further processing can be accomplished by pretreatment using mechanical, chemical, physicochemical and biological methods (Agbor et al., 2011, Akhtar et al., 2016).

10.6.2 Hydrolysis

In order to convert biomass into fuels and chemicals, complex polysaccharides and lignin are broken down during hydrolysis (Zakzeski et al., 2010). Numerous hydrolysis techniques, such as acid hydrolysis and enzymatic hydrolysis, have been developed. Acid hydrolysis, which converts biomass into simple sugars by subjecting it to high temperatures and pressures of diluted acid, is a widely used technique for biomass hydrolysis (Mood et al., 2013). This process has received much research due to its efficiency in converting lignocellulosic biomass into biofuels and other compounds. The generation of inhibitory chemicals, which can decrease the effectiveness of downstream processes, and the acid's corrosive character, which can harm equipment and raise running costs, are some of the drawbacks of acid hydrolysis (Kumar et al., 2009).

Enzymatic hydrolysis is another widely used method for biomass hydrolysis. Enzymes break down complex carbohydrates into simple sugars at lower temperatures and pressures than acid hydrolysis (Kumar et al., 2009). This method has several advantages over acid hydrolysis, including higher selectivity, lower production of inhibitory compounds and milder operating conditions (Mosier et al., 2005). However, the high cost of enzymes and their susceptibility to deactivation by inhibitory compounds remain significant challenges for the commercialization of this method (Kumar et al., 2016).

10.6.3 Fermentation

It is possible to create a variety of products by fermentation, including organic acids, enzymes and biofuels, from sugars or other organic molecules. Ethanol generation is one of the fermentation methods that biorefineries utilize the most frequently. Microorganisms like *Saccharomyces cerevisiae* and *Zymomonas mobilis* are used in the fermentation process to manufacture ethanol from sugars obtained from lignocellulosic biomass (Li et al., 2010). The efficiency and output of ethanol generation from lignocellulosic biomass have greatly increased because to the use of genetically modified microbes and optimized fermentation conditions (Jönsson and Martn, 2016). Other biofuels including butanol, hydrogen and methane can also be produced by fermentation techniques in addition to ethanol. Using bacteria like *Clostridium acetobutylicum* and *Clostridium beijerinckii*, butanol is generated by digesting carbohydrates (Yang et al., 2013). Using microorganisms like *Clostridium thermocellum* and *Methanococcus maripaludis* to ferment organic materials, hydrogen and methane are created (Li et al., 2018).

In addition to biofuels, fermentation processes are also used to produce organic acids such as lactic acid, acetic acid and succinic acid. Lactic acid is produced by fermenting sugars using lactic acid bacteria such as *Lactobacillus* spp. (Lynd et al., 2002). Acetic acid is produced by fermenting ethanol using *Acetobacter* spp. (Hua et al., 2018). Succinic acid is produced by fermenting sugars using *Actinobacillus succinogenes* (Ahn et al., 2016). The fermentation procedures used in biorefineries to produce enzymes yield yet another useful product. Microorganisms that ferment food, such as *Trichoderma reesei* and *Aspergillus niger*, create enzymes like cellulases, amylases and proteases (Tuohy, 2016). These enzymes are used in a variety of industrial processes, including those that produce biofuels, process foods and make medications.

10.6.4 Product Separation

According to Mosier et al. (2005), product separation methods require extracting and purifying different products from a complicated combination of byproducts and contaminants. Examples of these products include biofuels, organic acids and enzymes. Distillation is one of the procedures that biorefineries most frequently utilize to separate products. In the process of producing ethanol, ethanol is separated by distillation from the fermentation broth (Basso et al., 2008). While distillation can separate some compounds, including organic acids and sugars, it is energy-intensive and has limits in other areas.

In biorefineries, numerous different product separation techniques are utilized in addition to distillation. Products are separated depending on their molecular size and charge using membrane separation techniques such as ultrafiltration, nanofiltration and reverse osmosis (Fröhlich et al., 2012). Low energy usage, excellent separation efficiency and simple scaling up are benefits of membrane separation techniques (Mulder and Mulder, 1996). Adsorption is another product separation process used in biorefineries. Adsorption involves separating products based on their affinity for a particular adsorbent material (Dąbrowski, 2001). Activated carbon, ion exchange resins and molecular sieves are commonly used adsorbent materials in biorefineries (Cherubini and Strømman, 2011). Crystallization is another product separation process used in biorefineries. Crystallization involves precipitation products from the solution by cooling or adding an anti-solvent (Myerson, 2002).

10.7 PRODUCTS FROM BIOREFINERIES

Several types of products can be produced from biorefineries, including biofuels, biochemicals and biopharmaceuticals. Products from biorefineries offer a promising solution to the challenges of reducing greenhouse gas emissions and increasing energy security. However, their production and use must be carefully managed to ensure economic viability and scalability.

10.7.1 Biofuels

According to Buhrke et al. (2013), biofuels are fuels that are either liquid or gaseous and are made from biomass and may be used for heating or transportation. Growing worries about climate change and energy security have brought attention to the development and use of biofuels. Ethanol, biodiesel and biogas are just a few of the several kinds of biofuels. According to Scragg et al. (2009), ethanol is primarily made by fermenting crops high in sugar and starch, such corn, sugarcane and wheat. Instead, animal or vegetable lipids are trans-esterified to create biodiesel (Knothe, 2009). Organic wastes including food scraps, animal manure and wastewater are digested anaerobically to generate biogas, which is predominantly constituted of the gas methane (Khalid et al., 2011). Biofuels are produced using various production methods, including biochemical and thermochemical processes (Dahiya, 2014). Biochemical processes like fermentation produce biofuels such as ethanol and biogas (Zhang, 2015). Biofuels like bio-oil and syngas are created by thermochemical processes such as pyrolysis and gasification (Bridgwater, 2012).

10.7.2 Bioplastics

Bioplastics are plastics derived from renewable biomass sources, such as plant-based materials, instead of traditional petroleum-based sources (Ibrahim et al., 2021). Bioplastics have emerged as a promising alternative to traditional plastics made from non-renewable fossil fuels and contribute to environmental pollution. Bioplastics have a range of properties, depending on the materials and production methods used (Ukaga et al., 2010). Physical characteristics like flexibility, toughness and transparency that are common to conventional plastics can also be found in bioplastics. Biodegradability, compostability and improved heat and UV radiation resistance are just a few of the distinctive qualities that bioplastics may possess.

Biodegradable and non-biodegradable bioplastics are the two primary categories. Non-biodegradable bioplastics cannot decay naturally in the environment and must be disposed of using particular procedures (Arikan et al., 2015). Biodegradable bioplastics can degrade naturally in the environment. Biodegradable bioplastics can be categorized into compostable and oxo-degradable bioplastics (Mohanty et al., 2005). Compostable bioplastics can be broken down in industrial composting facilities, while oxo-degradable bioplastics break down through exposure to oxygen and sunlight. The production of bioplastics involves various methods, including fermentation, polymerization and extrusion. Fermentation produces bioplastics from natural sugars and starches, while polymerization produces bioplastics from natural oils and fats (Flieger et al., 2003).

10.7.3 Biochemicals

Chemicals derived from living things including plants, animals and microbes are referred to as biochemicals (Berdy, 2005). They are used for various purposes, including pharmaceuticals, food additives, cosmetics and industrial applications. Biochemicals can be classified into several categories: carbohydrates, lipids, amino acids, nucleic acids and vitamins (Nelson et al., 2008). They can also be classified based on their origin, such as plant-based, animal-based and microbial-based biochemicals. Biochemicals can be produced through various methods, including extraction from natural sources, fermentation and biocatalysis (Angenent et al., 2004). Extraction involves using solvents to extract the desired biochemical from the natural source, while fermentation involves using microorganisms to produce the desired biochemical (Bommarius and Riebel-Bommarius, 2004). Biocatalysis involves using enzymes to catalyze biochemical reactions and produce the desired product. Pharmaceuticals, food and beverage, cosmetics, and industrial chemicals are just a few of the areas where biochemicals are used in a variety of ways. As an illustration, the pharmaceutical business uses amino acids and nucleic acids to make medicines. However, in the food sector, lipids and carbohydrates are employed to create flavorings and additives (Katoh et al., 2015). Biochemicals are also used in the cosmetics industry to produce skincare products and in the industrial sector to produce bioplastics and other sustainable materials.

10.7.4 Biopharmaceuticals

Biopharmaceuticals, also known as biological drugs or biologics, are pharmaceutical products produced using living organisms or their components (Sekhon et al., 2010). Biopharmaceuticals have revolutionized medicine by offering treatments for diseases that were previously untreatable or had limited treatment options. There are several types of biopharmaceuticals, including monoclonal antibodies, recombinant proteins, vaccines and gene therapies (Moo-Young, 2019). According to Köhler et al. (1975), monoclonal antibodies can cure a variety of illnesses, such as cancer, autoimmune conditions and infectious diseases. Recombinant proteins like insulin and growth hormone treat metabolic disorders and other diseases (Pavlou et al., 2004). Vaccines are used to prevent infectious diseases, while gene therapies are used to treat genetic disorders. The production of biopharmaceuticals involves various methods, including cell culture, gene expression and purification (Gomord et al., 2005). Biopharmaceuticals are typically produced using living cells, such as bacteria, yeast, or mammalian cells (Chon et al., 2011). These cells are genetically engineered to produce the desired protein or antibody, which is then extracted and purified.

10.8 ADVANTAGES OF BIOREFINERIES

10.8.1 Reduced Greenhouse Gas Emissions

Greenhouse gas emissions are a significant contributor to climate change, and reducing these emissions is critical for mitigating the impacts of climate change. By providing alternatives to fossil fuels and encouraging the use of renewable resources, biorefineries, which turn biomass into value-added products, offer a possible approach to reduce greenhouse gas emissions (Bajpai 2013). Multiple techniques exist for biorefineries to significantly lower greenhouse gas emissions. In the beginning, utilizing biomass as a feedstock for biorefineries might lessen reliance on fossil fuels and encourage the use of renewable resources (Ng et al., 2014). In addition, as biomass is a carbon-neutral resource, creating biofuels from it can dramatically lower greenhouse gas emissions when compared to fossil fuels (Lam et al., 2010). Third, using biomass to produce chemicals and materials can also reduce greenhouse gas emissions, as these products are often produced from fossil fuels (Klass, 1998).

10.8.2 Reduced Dependence on Fossil Fuels

Fossil fuels have been the primary source of energy for several decades. However, their use has led to environmental problems, such as greenhouse gas emissions, climate change and air pollution. In the place of fossil fuels, biorefineries use biomass to create a variety of products with added value, such as biofuels, biochemicals and biomaterials. In contrast to fossil fuels, biorefineries provide a number of advantages, such as less greenhouse gas emissions, greater energy security and economic growth (Fernando et al., 2006). Using renewable biomass in the place of fossil fuels allows biorefineries to cut greenhouse gas emissions. Due to their decreased reliance on foreign oil and use of locally accessible biomass resources, biorefineries also provide greater energy security (Fernando et al., 2006). In addition, biorefineries can create new economic opportunities by creating jobs and supporting local industries.

10.8.3 Creation of New Products and Markets

Biorefineries are crucial in creating new products and markets by transforming renewable biomass resources into value-added products. As a new source of income for farmers and rural communities, biorefineries have the potential to provide new employment and boost economic growth (Kumar et al., 2009). Small and medium-sized enterprises may benefit from the creation of new goods using biomass, and the biotech and chemical sectors may be encouraged to innovate as a result (Perlack, 2005). Developing biorefineries can also reduce dependence on imported oil and promote energy independence, which can benefit countries that rely on oil imports.

10.8.4 Reduced Waste Generation

By transforming waste biomass into useful goods, biorefineries have the potential to minimize waste output. By transforming waste biomass into beneficial products like biofuels, biochemicals and biomaterials, biorefineries can help minimize the production of waste (Ng et al., 2014). Biorefineries can assist in preventing organic waste from ending up in landfills and lowering greenhouse gas emissions by using waste biomass as a feedstock. Biorefineries have emerged as a promising solution to meet the global demand for sustainable energy and chemicals. Using renewable biomass sources, biorefineries can produce a wide range of value-added products such as biofuels, bioplastics, biochemicals and biopharmaceuticals. However, the development of biorefineries is still facing several challenges, such as the availability and cost of feedstock, technological barriers, regulatory issues and competition with existing fossil-based industries. Despite these challenges, the potential benefits of biorefineries in terms of environmental sustainability, energy security and rural development make them a critical component of the transition toward a more sustainable future.

10.9 INTRODUCTION TO 3D FOOD PRINTING

By applying successive thin layers of material, three-dimensional (3D) printing, also known as additive manufacturing (AM), uses digital data to produce actual three-dimensional objects. By employing a special technology called 3D printing, users may make extremely complex materials that would be challenging to make using conventional mechanical manufacturing methods. Personalized food items are made possible by 3D food printing, which combines 3D printing and digital gastronomy techniques. This is in contrast to robotics-based food manufacturing technologies, which aim to automate manual processes for mass production. The design parameters and the usage of various printing materials allow for a great deal of customization of 3D printed products. The sharp increase in publications in this field since 2010 demonstrates how these benefits have spurred quick advancements in 3D printing applications in recent years. 3D food printing has lately experienced tremendous development. Numerous articles about the use of 3D-printed food have been published just in the last 5 years. Using many of the same techniques used to print organs and tissues in the biomedical industry, researchers have utilized this technology to print cell grown meat and meat analogs (Godoi, Prakash, and Bhandari, 2016; Godoi, Prakash, and Bhandari, 2016). Additionally, a number of different solid or semi-solid foods, such as cookies, cakes, burgers, apples and chocolates, have been produced using 3D printing (Lipton et al., 2015; Fanli Yang et al., 2018; Wang et al., 2018; Pérez et al., 2019; Feng, Zhang, and Bhandari, 2018).

10.10 PRINCIPLES OF 3D PRINTING TECHNIQUES

The current 3D printing methods used to produce food items are categorized based on the printing processes used: inkjet, extrusion and heat source (powder layer binding). Extrusion and inkjet methods may both be used to deposit liquid-based materials. Powder-based structures are printed via deposition followed by a heating mode (laser or hot air) or binder jetting (inkjet technology). Figure 10.7 depicts a schematic categorization of 3D food printing technologies based on the kind of driving mechanism.

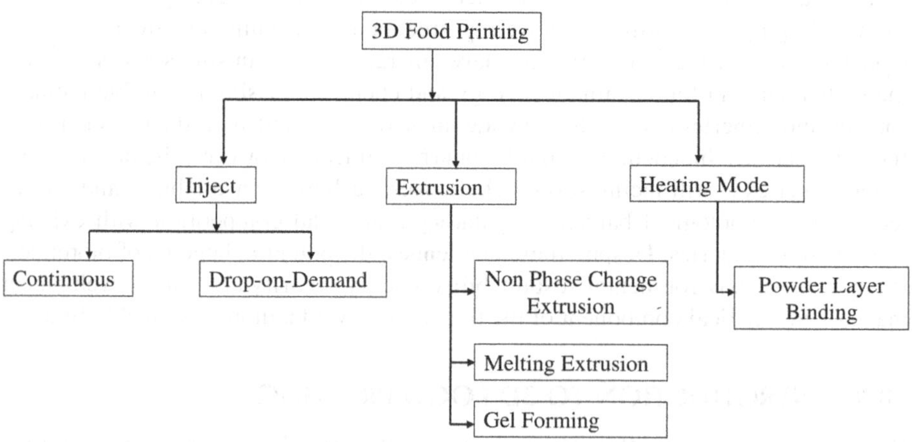

FIGURE 10.7 Schematic diagram of various 3D food printing techniques.

10.10.1 Inkjet-Based Printing

Inkjet printing (IJP) may be used to create 2D and 3D designs by depositing liquid droplets onto a substrate using computer-aided design technologies. Ink is generally jetted in channels of 20–50 mm by an IJP print head. A 1-pL (picoliter) ink droplet is usually 13 mm in diameter (Xaar 2012). Inkjet printing (IJP) may be done either continuously (C-IJP) or on demand (DoD-IJP). Both methods involve forcing a fluid through an orifice, which causes a stream of droplets with the same volume but less surface area instability (Rayleigh–Plateau instability). A high-pressure pump propels liquid ink via an orifice between 50 and 80 mm in diameter in C-IJP, resulting in a continuous ink flow. C-IJP generates droplets at a quicker pace than DoD-IJP; nevertheless, it needs the use of an electrically conducting fluid, limiting its applicability in food customization. The DoD-IJP has numerous heads (100–1000) and often uses thermal or piezoelectric heads. The print head of a thermal inkjet printer is electrically heated to create pressure pulses that force droplets out the nozzle. A piezoelectric crystal within the print head of a piezoelectric inkjet printer generates an acoustic pulse that separates the liquid into droplets at uniform intervals. When a voltage is applied to a piezoelectric material, it immediately changes form, resulting in the pressure required to expel droplets from the nozzle. Piezoelectric inkjet printers use a piezoelectric crystal within the printhead to generate an acoustic pulse that separates the liquid into droplets at regular intervals. When a voltage is applied to a piezoelectric material, it immediately changes form, producing the pressure required to expel droplets from the nozzle. An approximate solution to the Navier-Stokes equation was previously used to determine the fluid dynamics behavior of a droplet expelled from a tiny aperture. Three primary factors are used to predict droplet jetting behavior (Seerden et al., 2001): Re/We: the ratio of the Reynolds number (Re) to the Weber number (We), which represents trends in the viscous, inertial and surface tension forces on fluid flow. In Eq. 10.1, a, ρ, γ and η denote the fluid's characteristic length, density, surface tension and viscosity, respectively. Droplet spreading in the absence of solidification (ε): Bhola and Chandra's Eq. 10.2, where r_{max} and r are the maximum splat radius and starting drop radius, respectively, and θ is the droplet's equilibrium contact angle with the substrate. Splashing of liquid droplets (K): Droplet splashing during jetting causes dimensional instability and lack of homogeneity. When the parameter K surpasses a critical value, K_c (Eq. 10.3), splashing occurs.

$$\frac{\text{Re}}{\text{We}} = \frac{(\gamma \rho a)^{1/2}}{\eta} \tag{10.1}$$

$$\varepsilon = \frac{r_{\max}}{r} = \left[\frac{\text{We}^2 + 12}{3(1-\cos\theta) + 4\dfrac{\text{We}^2}{\text{Re}^{\frac{1}{2}}}} \right]^{1/2} \tag{10.2}$$

$$K = \text{WeRe}^{1/4} \tag{10.3}$$

10.10.2 Applicability of Inkjet Printing

Since low-viscosity materials are typically handled by inkjet printers, IJP does not find use in the creation of intricate food structures. The technique is often used to bioprinting, graphic ornamentation, fillings, microencapsulation, and, to a much lesser extent, nanoprinted 3D constructions. DoD deposition may be used to describe the graphic embellishment and fills method developed by Haenggi and Hanspeter (2011) for applying a liquid to layers. By using a variety of pneumatic membrane nozzle jets to deposit small droplets onto a moving item, this technology—also known as FoodJet printing—was commercialized. Microencapsulation procedures like the one developed by the Netherlands Organization for Applied Scientific Research (TNO) use inkjet technology. They claim that their printer can generate very mono-disperse droplets that, when dried, transform into highly mono-disperse powders. With a capacity of 100 L/h, the print head contains 500 nozzles. In 3D food printing applications, a sliced 2D profile created by a computer-generated 3D model is employed to direct the ejection of the binder solution onto a thin layer of powder. Binder jetting is the term used to describe the method. In order to create a 3D build, the binder plays a crucial function in connecting nearby particles. This could happen as a result of cross-linking or dissolution fusion of the particle surfaces (Peltola et al., 2008).

10.10.3 Extrusion-Based 3D Printing

Extrusion-based techniques, unlike DoD-IJP approaches, depend on the flow of a continuous ink in a layer-by-layer basis, often highly concentrated colloidal inks. Depending on the material's capacity to form a gel or achieve paste consistency, the concentration in total solids may vary from 5% to 50%. When the length (l) and a pressure gradient (ΔP) are applied, the ink flows through the nozzle. As a result, according to Eq. 10.4, where r denotes the radial location inside the nozzle, a radially variable shear stress (ζ_r) occurs. There is no velocity at the nozzle wall center (r=R), and there is a maximum velocity at the center (r=0) (Lewis, 2002).

$$\zeta_r = \frac{r\Delta P}{2l} \tag{10.4}$$

The extruding substance should ideally respond quickly and solid-like throughout the deposition process. This property may be assessed in relation to the yield stress (YS), which is divided into two categories for certain fluids: static YS, which is the stress needed to start a fluid flowing, and dynamic YS, which is the stress needed for a fluid already moving to continue flowing. Due to the need that the material flow through the nozzle and overcome the static YS, both YSs are crucial for ensuring printability. But to withstand high shear rates and still form objects with little deformation after printing, dynamic YS is needed (Lewis, 2002; Smay, Cesarano, and Lewis, 2002).

10.10.3.1 Melting Extrusion

The three appearance types of materials appropriate for melting extrusion are paste-like, powder (or solid bits) and filament (rare in culinary applications). For pastes high in fat or sugar that is typically amorphous, extrusion-based 3D printing requires temperature control. Fatty acids, for instance, that has more carbon atoms have melting points that are higher. On the other hand, a melting point that is lower is caused by a higher proportion of double bonds. To prevent the printed structure liquefying while baked, Lipton et al. (2010) optimized the butter fat concentration in conventional dough recipes. Due to its melting characteristics, chocolate is the most popular edible ink. It may be melted into a paste-like consistency or fed into the printer reservoir as a powder. The operating temperature used by Mantihal et al. (2017) was within the range shown in other research that used temperatures between 28°C and 40°C (Hao et al., 2010). The following criteria were identified as impacting the geometrical accuracy of the chocolate deposition by Hao et al. (2010): (i) nozzle aperture diameter, (ii) ideal nozzle height from the forming bed and (iii) extrusion axis movement.

10.10.3.2 Gel-Forming Extrusion

Based on the fraction and/or colloidal forces, gel rheology may be customized. They display shear-thinning flow behavior when strained beyond their gel yield point (τ), as predicted by Eq. 10.5, because particle–particle links within the gel begin to break down.

$$\tau = \tau_y + K\gamma^n \tag{10.5}$$

The elastic properties of 3DP should be taken into consideration together with the temporal management of the gelation processes in order to prevent premature gelation of the material inside the nozzle. Thermal-gelation, chemical cross-linking, ionotropic cross-linking, complex coacervate synthesis and enzymatic cross-linking are the five categories into which the gelation processes may be separated.

10.10.3.3 Thermal-gelation

Gelatin is a traditional printing medium that gels upon cooling owing to the creation of junctions by short segments of polypeptide chains returning to the collagen triple helix-like structure (typically within 15–20°C, depending on the raw material and extraction conditions). The gel setting time and temperature of gelation are important elements in adjusting printing speed and temperature. As previously proven by Lipton et al. (2010), this property may be leveraged to improve the printability of food systems.

10.10.3.4 Chemical Cross-Linking

This is a standard approach for increasing heat stability in gels like gelatin. Methacrylation of gelatin for bioprinting applications is one example, which is further cross-linked by UV radiation (Wang et al., 2017). Chemical cross-linking, on the other hand, is unlikely to be used for food printing since many cross-linking chemicals are toxic and must be entirely removed from the intended structure before consumption.

10.10.3.5 Ionotropic Cross-Linking

The food sector has frequently used ionotropic cross-linking, particularly in micro-encapsulation procedures (Bokkhim et al., 2016; Ching et al., 2015). The gel-like layers were cross-linked using two methods: (i) postimmersion of the 3D construct in $CaCl_2$ solution (concentration 300 mM) (Vancauwenberghe et al., 2017), and (ii) concurrent crosslink by coaxial extrusion ($CaCl_2$) of the outer flow (Vancauwenberghe et al., 2018).

10.10.4 Powder Layer Binding

The California-based Evil Mad Scientist Laboratories' CandyFab machines employ the selective hot air sintering and melting (SHASAM) process to print anything from sugar. With the use of a small, focused, low-velocity hot air beam, sugar powder is carefully fused together using SHASAM technology, creating a 2D image. A 3D object is progressively constructed by repeating this procedure. In order to disassemble the made model and save the leftover powder for the production of the next object, the bed is moved back to its initial position when the 3D object is finished (CandyFab, 2017). The selective laser sintering (SLS) procedure involves the layer-by-layer fusing of powder particles till the final fusing of the powder particles into a three-dimensional structure. The laser scans cross sections on the surface of each layer, fusing the powder in specific areas. Each cross-section is scanned, and the powder bed is discarded before a new layer is added. This process is carried out repeatedly till the finished product. Finally, the unfused powder is removed and recycled for use in the printing that follows (Diaz et al., 2016). SLS has been extensively used in the production of metal and ceramics, but there are a number of obstacles preventing its use in the food industry. These obstacles include (i) the requirement for suitable powdered materials that can fuse together without decomposing during the fabrication process and (ii) the requirement to construct various edible objects using a variety of food materials (Diaz et al., 2015). Generally speaking, SLS permits the production of complex, high-resolution, freestanding 3D structures; however, the type of material that can be employed is only powder, such as granules of sugar, fat, or starch. To broaden the use of this technology to conventional cuisine, more food options must be easily accessible. SLS has been applied to the construction of complex structures using sugar or sugar-rich powders. TNO researchers have used sugars and NesQuik powders to build intricate and complicated 3D structures (Gray, 2013). CandyFab Project's use of SLS has allowed them to successfully make a number of beautiful intricate structures out of sugar granules that were otherwise impossible to make (CandyFab, 2017).

TABLE 10.1
Comparison of Different 3D Food Technologies

		Extrusion-Based Printing	Selective Laser Sintering	Binder Jetting	Inkjet Printing
	Available material	Chocolate, soft materials such as dough, cheese, meat puree	Powdered materials such as sugar, chocolate, fat	Liquid binder and powdered materials such as starch, sugar, protein	Low-viscosity material such as pizza sauce
Factors affecting printing precision	Material properties	Rheological properties, mechanical strength, Tg	Melting temperature, flowability, particle size, wettability, Tg	Flowability, particle size, wettability and binder's viscosity and surface tension	Compatibility, ink rheological properties, surface properties
	Processing factors	Printing height, nozzle diameter, printing rate, nozzle movement rate	Laser types, laser power, laser energy density, scanning speed, laser spot diameter, laser thickness	Head types, printing rate, nozzle diameter, layer thickness	Temperature, printing rate, nozzle diameter, printing height
	Post-processing	Additive, recipe control	Removal of excess parts	Heating, baking, surface coating, removal of excess parts	No
	Advantages	More material choices, simple device	Complex 3D food fabrication, varying textures	Complex 3D food fabrication, full color potential, varying flavors and textures	More material choices, better printing quality, fast fabrication

(*Continued*)

TABLE 10.1 (*Continued*)
Comparison of Different 3D Food Technologies

	Extrusion-Based Printing	Selective Laser Sintering	Binder Jetting	Inkjet Printing
Limitations	Incapable of fabricating of complex food designs, difficult to hold 3D structures in post-processing	Limited materials, less nutritious products	Limited material, less nutritious products	Simple food design, only for surface filling or image decoration
Products	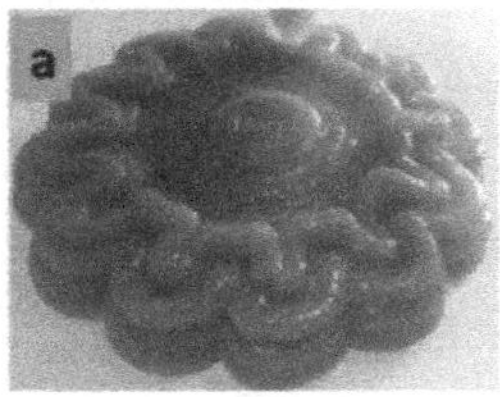			
	(a) Natural Machines Co., available at https://www.naturalmachines.com/;	(b) CandyFab Project, available at https://candyfab.org/TNO	(c) 3D Systems Co., available at https://www.3dsystems.com/culinary/gallery	(d) FoodJet Printing Systems, available at http://www.foodjet.com/

10.11 ASSESSMENT OF PRINTING QUALITY

Two crucial factors namely, shape fidelity in relation to the 3D design's dimensions and mechanical properties, may be used to evaluate the printing quality of the final 3D build. Shape fidelity is often evaluated.

10.11.1 Shape Fidelity

Lattice structures are frequently used to assess the printing quality of printed constructions that have a soft appearance. The characteristics of the material, such as gel setting time and viscoelastic behavior, will play a significant role in defining the minimum D in order to minimize area overlapping in lattice structures. One may also visually assess the ideal printing circumstance by looking at the shape of the filament material issued from the nozzle. The ideal gelation is defined by a uniformly sized, smooth-surfaced, three-dimensional extruded filament. In under-gelation or low total solids concentration circumstances, the ink will behave like a liquid and produce droplets at the nozzle end. Over-gelation or excessively concentrated ink will result in a stiff filament with breaks across the length of it (Ouyang et al., 2016). According to their research, gel-like inks' printing quality was assessed using Eqs. 10.6 and 10.7. According to a report, $Pr = 1$ is reached when the optimal gelation is accomplished. $Pr < 1$ and $Pr > 1$ are the corresponding criteria for under-gelation and over-gelation.

$$\text{Circularity} = C = \frac{4\pi A}{L^2} \tag{10.6}$$

$$Pr = \frac{\pi}{4} \cdot \frac{1}{C} = \frac{L^2}{16A} \tag{10.7}$$

10.11.2 Mechanical Properties of the 3D-Printed Construct

Numerous studies have used texture profile analysis (TPA) as a large deformation test to assess the mechanical qualities of 3D-printed food constructs made of soft materials (such as gels or purees). The outcomes may be linked to textural characteristics that might appear in macro- or microstructure. Owing to their importance in gummy candies, for instance, owing to their ductile capabilities, textural features like chewiness and gumminess may be obtained by measuring the applied force, force working distance and duration of deformation. The twice penetrating technique, which is often used on food goods, allows for the determination of the two-peak force curve from TPA. For gel-like systems, it is important to establish their hardness, adhesiveness, springiness, gumminess and chewiness (Rosenthal, 2010). These gels, which include gelatin, gellan gum, methylcellulose, guar gum, locus bean gum, hydroxypropyl methylcellulose and xanthan gum, have high levels of fracturability in addition to high levels of hardness, which is detrimental to extrusion (Kim, Bae, and Park, 2017).

The springiness quality in gels may be linked to the behavior of some polymers in "remembering" their prior form. The gels with low springiness and a broad gel strength were categorized as printable by Kim, Bae, and Park (2017). Yang et al. (2018) investigated the influence of starch inclusion on the mechanical characteristics of 3D-built lemon juice gels. As a result of the more compacted network provided by the starch molecules, the increase in starch resulted in a more substantial increase in hardness and gumminess. Azam et al. (2018) discovered a similar pattern in their analysis of starch addition in orange concentrates. The maximum hardness, adhesiveness and gumminess values were recorded for the highest starch content (30% w/w), which was associated to crystallinity increase with the presence of sugar and acid in conjunction with starch granules. Wang et al. (2018) recently found a correlation between gel strength (measured by TPA) and rising NaCl content in surimi gels. The behavior was linked to the enlargement of myofibrillar protein produced by an increase in ionic strength, which encouraged an enhanced protein–protein interaction. A texture analyzer was utilized by Lille et al. (2018) to quantify the hardness (cutting test) of protein-rich 3D printed structures that had been oven-dried and freeze-dried. Mantihal et al. (2017) investigated the snappiness of chocolate structures produced via 3D printing. Using a texture analyzer, their study revealed the impact of supports on the snap force. It was found that the cross-supported structures snapped more readily and had a stronger texture than other sample designs, requiring a larger force to break the sample with >56.00 N. Samples constructed without interior support snapped with less force (16.0 N).

10.12 EDIBLE INKS FOR 3D PRINTING

The availability of food inks with the necessary functional properties is one of the key elements influencing the success of 3DP's use for the production of food items. Researcher is concentrating on extrusion-based 3D printers since they are presently the most applicable to a wide range of food industry applications. When deposited onto a surface, a 3D ink must be able to flow through a nozzle while also hardening to form a strong structure that can sustain the weight of the succeeding layers (Joshi and Sheikh, 2015). Only a small number of food substances display this type of behavior in addition to having the desired organoleptic and nutritional qualities.

Food elements including water, carbohydrates, fats, proteins, minerals, vitamins and additives like tastes, colors and preservatives may be combined to make edible 3D inks that can flow through a nozzle before hardening. For bulk lipids, the triglycerides should flow within the nozzle but then crystallize after printing to create a 3D network of aggregated fat crystals that give the surface a semi-solid feel. This may be done by holding the nozzle at a temperature higher than the triglycerides' melting point while keeping the printing platform at a lower temperature. The size, quantity and interactions of the fat crystals that form after deposition may be crucial to manage since they affect the material's optical and mechanical characteristics (Pérez et al. 2019). Chocolate is mostly 3D printed using the hot-melt extrusion technique (Godoi, Prakash, and Bhandari, 2016; Pérez et al., 2019). Triglycerides may be transformed into an oil-in-water emulsion in the case of emulsified lipids. Using a gelling agent like gelatin in the aqueous phase, solidification may then be accomplished after printing. Above the gelatin's gelling point, this would result in a fluid emulsion, but below that, it would be semi-solid. An alternative would be to use a highly concentrated emulsion that behaves like plastic. Therefore, it may flow through the nozzle when high pressure is used (above the YS), but when the pressure is dropped after printing, it will solidify. Another example of this methodology is the 3D printing of potatoes (Liu et al., 2018).

By changing the solution's chemical makeup or the surrounding environment, other food hydrocolloids, such as proteins and polysaccharides, may gel (Gu et al., 2020). Few examples are alginate gels in the presence of calcium, carrageenan gels in the presence of potassium (Godoi, Prakash, and Bhandari, 2016), and protein gel in the presence of some salts and transglutaminase (Wang et al., 2018). Coaxial nozzles may be used in these circumstances to co-extrude the gelling biopolymer and gelling chemical in order to produce gelation (Ko et al., 2021). A number of researchers have investigated the use of coaxial printing with edible materials (Gu et al., 2020; Ko et al., 2021; Jeon et al., 2021; Kim, Kim, and Park, 2021). Examples include developing structures that mimic muscle fibers utilizing different hydrocolloids and surimi to generate imitation crab flesh (Kim, Kim, and Park, 2021) and to develop structures that simulate muscle fibers using various hydrocolloids (Ko et al., 2021). Ion kinds, concentrations and biopolymer types must all be carefully regulated in order to produce a 3D ink with the necessary flow and gelation characteristics.

10.12.1 Printing Factors for Edible Inks

The capacity of a material to be handled and deposited by the printer while maintaining its structural integrity is referred to as printability, according to Godoi, Prakash, and Bhandari (2016). One of the most important factors determining food inks' ability to be printed is their rheological properties. Viscosity, YS and shear modulus are a few of the important parameters that affect a material's capacity to be printed. Shear modulus, which influences both the final printed material's capacity to keep shape and the feeling of texture and mouthfeel, measures the material's hardness. The YS needs to be sufficient to be able to push the material out of the nozzle without causing it to collapse after printing (Liu et al., 2018).

10.13 OPPORTUNITIES OF 3D FOOD PRINTING

10.13.1 Customization

Personal preferences have a big role in determining what people eat and how much they like to eat (Sun et al., 2015). Other influences on food choices include region, culture, gender, age and how healthy they are. The demand for developing fresh strategies that optimize customer personalization while satisfying their dietary requirements is thus growing. The content and structure of meals may be changed using 3DP (Dankar et al., 2018). The inability to chew and swallow food is said to affect 1 in 4 adults over the age of 50 (Sun et al. 2015; Hamilton, Alici, and in het Panhuis, 2018). These people find it difficult to get the nutrients they require to be healthy from commercially available food alternatives because of this. As an illustration, a study by the Netherlands Organization for Applied Scientific Research (TNO) looked at the printing of tailored pureed meals that satisfy all of the nutritional requirements of the elderly (Dankar et al., 2018). Foods that contain 3DP may also be advantageous for those who suffer from specific illnesses such food allergies, diabetes, heart disease, hypertension and compromised immune systems (Dankar et al., 2018). People can use 3D printers to create food that satisfies their unique nutritional needs at home or in organizations (like care facilities), depending on their state of health. Meals may be labeled with specific ratios of macronutrients (fats, proteins and carbohydrates), micronutrients (vitamins or minerals), or nutraceuticals (carotenoids, curcuminoids or polyphenols), for example. The range of easily accessible food inks and printing software may increase customization and expand its usage.

10.13.2 Sustainability

Every stage of the food supply chain experiences food waste, which poses a serious concern for the sustainability of world food production. According to Papargyropoulou et al. (2014), postharvest losses affect up to 50% of all food produced worldwide. Food that is still edible but thrown away because it is no longer deemed appetizing (Papargyropoulou et al., 2014) is still safe to eat, and leftovers from cooked meals are all examples of avoidable food waste. Food waste is produced in developing and poor nations as a result of ineffective supply chains and subpar distribution networks. These issues can be solved in a number of ways using 3D food printing. With the aid of shelf-stable food inks with long shelf lives, such as powders and pastes, food may be printed precisely and on demand using 3D printing technology (Feng, Zhang, and Bhandari, 2019; Jiang et al., 2020). Using shelf-stable food inks with lengthy shelf lives, such as powders and pastes, 3D printing technology may be used to print the food on demand and in exact amounts. A further benefit of 3DP meals is that they may lessen the quantity of packaging materials needed by the food business (Godoi, Prakash, and Bhandari, 2016; López Galdeano, 2015; Jiang et al., 2020). Because food ink cartridges are reusable, recycling is encouraged and waste is further reduced. The cost and energy required for food storage, distribution and transportation have already been proven to be reduced by 3D-printed foods (López Galdeano, 2015). Producing plant-based substitutes for animal-based goods including meat, fish, eggs and dairy products is another area where the use of 3DP may have positive environmental effects. Therefore, substituting plant-based foods for animal-based ones may have a significant negative impact on the environment. The use of 3DP is investigating the creation of high-quality meat analogs using food inks that comprise alternative proteins including those obtained from plants, micro-organisms, or insects (Portanguen et al., 2019).

10.13.3 Shelf Life and Food Safety

Powders used in 3D printing may have a reasonably extended shelf life because to their low water activity, which prevents microbial development and chemical reactions. However, due to their relatively high water activity, pastes used as food inks may have a shorter shelf life. However, by using different processing methods like thermal processing, pH control, cold storage, or the addition of preservative systems, the shelf life of pastes can be increased. According to one research, the food inks utilized had bacterial levels of 4.28 Log CFU/g, which suggests that the 3DP process needs sanitization techniques (Severini et al., 2018). Therefore, additional research is required to assess the security of 3D-printed foods. In order to create effective methods to enhance their storage and cleaning, it will be crucial to pinpoint the elements affecting the safety of food inks and potential 3D printer contamination (Zhang et al., 2021).

10.13.4 3D Printing of Foods for Space

Space food should be small, light, simple to store and easy to consume. Early space missions, including the Mercury and Gemini missions, put a lot of effort into food research and development to supply calorie-dense, nutrient-rich and appetizing meals (Cahill and Hardiman, 2020; Jiang et al., 2020). Before retort pouches, cans and meal bars were created, the food was first packed in tubes or bite-sized cubes. Although generally healthy and safe, these foods lacked flavor quite frequently. The unappealing quality of the food could also be tolerated since the earliest space missions were very short flights (Perchonok and Bourland, 2002)., but it was later realized that longer voyages would need food that was more sophisticated. Food safety and shelf life provide additional concerns in these more prolonged missions (Cahill and Hardiman 2020; Jiang et al., 2020). Foods sent on a journey to Mars must have a shelf life of at least 5 years. According to a recent research, after 5 years of preservation, only 7 out of 65 (or 11% of all the thermostabilized meals) that were evaluated for their possible use in space missions were still edible (Jiang et al., 2020). A varied variety of space meals with anticipated shelf life needs must thus be created using novel methods. Moreover, enticing food items should provide astronauts with the nutrients they need to stay healthy throughout lengthy missions. However, the quantity of ingredients and equipment for food processing/preparation that can be brought on a mission is constrained by the limited storage space on a spacecraft. The importance of food to astronauts' mental and physical wellbeing cannot be overstated. The availability of high-quality, diverse cuisine has been demonstrated to have a significant influence on astronauts' mental health. According to Sirmons et al. (2020), stress may be exacerbated in particular by meals' lack of variety or taste.

Though little actual research has been reported, several researchers have highlighted the potential uses of 3D food printing in space (Hamilton, Alici, and in het Panhuis, 2018; Liu et al., 2018; Joshi and Sheikh, 2015; Jiang et al., 2020). The benefits of having meals available on space missions, such as nutritional stability, shelf life and acceptability, have been emphasized by researchers who have identified research possibilities to develop 3D food printing for in-space uses (Liu et al., 2018). In future work sections or conclusions, printing food for space travel is often cited as a possible use (Hamilton, Alici, and in het Panhuis, 2018; Liu et al., 2018; Joshi and Sheikh, 2015; Dankar et al., 2018). However, in this area, more thorough research is required. Long space missions need careful attention to maintaining the health and wellbeing of the crew. Depending on the specific dietary needs of each astronaut, 3D printing might tailor the nutritional profiles of the items meant for them. For usage in nutrient fortification of 3D-printed meals, nutrients might be kept as powders. These powders might be produced by spray drying nanoemulsions or nutrient-enriched emulsions that have been specifically created to boost the bioavailability of the vitamins, minerals and nutraceuticals they contain. Foods may be specially formulated to satisfy both each astronaut's individual nutritional needs and the daily nutritional requirements of all the astronauts (Cahill and Hardiman, 2020).

10.14 POST-PROCESSING OF 3D FOOD MATERIALS

To improve the taste of the printed material, additional post-processing steps including baking, boiling, frying, heating, freezing and steaming should be taken while producing 3D food. In this post-processing, the finished products' sensory quality is improved (Dankar et al., 2018). According to L. Zhang, Lou, and Schutyser (2018), baking, frying and steaming reduced the product's moisture content, which increased the product's hardness and lowered its stickiness. Due to baking at high temperatures, cereal-based items often lose their original printed structure and develop some additional voids. Transglutaminase (TGase), gelatin, xanthan gum, methylcellulose, calcium caseinate in baking dough are used to improve the quatity of finish goods (Brunton et al., 2019; Kim et al., 2019; Fan Yang et al., 2019; Dick et al., 2020; Vieira et al., 2020; Kim et al., 2018; Thangalakshmi et al., 2021). Since the majority of foods consumed on a daily basis must undergo post-processing (baking, cooking, frying, etc.), the 3D food structures should ideally be resistant to these procedures. Different soft materials, such as cookie dough, cheese and cake icing, have been deposited using the extrusion-based 3D printing method while using recipe control and the inclusion of additives to maintain the form stability of the objects after post-processing (Lipton et al., 2010). Godoi, Prakash, and Bhandari (2016) assert that it is feasible to produce 3D-printed structures that can endure post-processing by modifying the materials' physical, chemical, rheological, structural and mechanical properties.

10.15 CONSUMER PERCEPTION TOWARD 3D FOOD PRINTING

Few studies have been done on consumers' attitudes, perceptions and acceptances of food produced by 3D printing, despite the benefits and myriad potential uses of this technology already being widely discussed (Izdebska and Zołek-Tryznowska, 2016; Sun et al., 2015). Although not all novel food technologies and foods produce the same reactions in consumers, most consumers are skeptical of them (Popa and Popa, 2012). Both technology (Cardello, 2003; Frewer et al., 2011) and a person's experience and knowledge base affect the appraisal process (Greehy et al., 2013; Olsen, Grunert, and Sonne, 2010). Repeated exposure to various types of novel food and visual representations of them can also help make new techniques and new food more familiar to consumers, which in turn affects consumers' attitudes in a positive way (Bruhn, 2007; Cardello, 2003; Cardello, Schutz, and Lesher, 2007; Frewer et al., 2011; Jaeger et al., 2015; Lyndhurst, 2009). Participants who had a favorable experience with the

"3D-printed" cookie were more likely to say they wanted to try 3D-printed foods, thought it may improve food sustainability and thought it could be used to customize food (Manstan, Chandler, and McSweeney, 2021). A few individuals expressed their want to consume more 3D-printed meals in the future; nonetheless, various obstacles to their acceptability were found. Some participants were repulsed by the concept of 3D printing and said they would never consume 3D-printed food again. Their major worries were to the manufacturing process's product safety. Additionally, they objected to the idea of consuming meat products made in 3D printers.

10.16 SUMMARY AND FUTURE DIRECTIONS

The material characteristics and binding processes have a significant impact on 3D food printing. Large efforts have been made in recent years to produce 3D constructions with end-use qualities that are comparable to or better than those achieved using conventional manufacturing methods. To implement 3DFP in specialized industries that need personalization, there are still a lot of obstacles to be addressed, as a result of the absence of adequate software for configuring the printing settings for multicomponent systems of viscous materials. Researchers from all around the globe are now modifying the factors that are often regulated in the fusion deposition of plastic filaments to get the optimal pressure in extrusion procedures, for instance. In our view, hardware and software will be created as soon as the viability of the 3DFP application is established. Researchers and those working in the food industry should be aware of current trends that have an impact on consumer behavior while also keeping an eye on how 3DP is developing as it moves closer to the development of new 3D structures.

10.16.1 Industrial Trend

The traditional techniques used for food production will not entirely be replaced by 3D printing of food. On the other hand, this adaptable technology fits into niche markets with distinct demands and has the potential to generate new needs for demographic groups, the magnitude of which is now unknown. According to several studies (Boutouchent, 2016; Subic and Gallagher, 2017), 3DP technology is a component of the 4.0 industrial revolution. Decentralized production and distribution architectures will bring more and more manufacturing closer to the final customer. Everyone can simultaneously be a producer and a developer thanks to the 4.0 industry's prosumer trend. In light of this, 3D printing is an ideal match for future production processes in a range of industries, including the manufacture of food (Kocovic, 2017).

10.16.2 Evolution of 3DP Technology

3DP technology is advancing significantly in terms of both materials and equipment. The development of 4D printing was a result of the advancement of materials toward 3DP. Ingenious components that can transition between different configurations in response to environmental signals like heat or moisture are produced via the 4DP process, which combines 3D printing with active material technologies (Ding et al., 2017; Ge, Qi, and Dunn, 2013). The gelatin sheets developed by researchers at the Massachusetts Institute of Technology are a prime example, since they can assume 3D shapes when submerged in water (Jennifer Chu, 2017). Since the finished product may need to be cooked or treated before consumption, food 3D printing is sometimes referred to as 4D printing. Additionally, the product's design must account for the shape and size changes that will take place during post-printing. But the phrase "5D printing" originated with the advancement of technology. The development of 5D printing was motivated by the simplicity with which elaborate artwork's supporting structures could be printed. In contrast to the three-axis printer used in conventional 3D printing, a five-axis 3D printer can produce things in five dimensions. A five-axis 3D printer, according to Ethereal-Halo 2023, builds the product essentially from multiple directions (X, Y and Z with rotational and translational motions), producing stronger components than with standard 3D printing.

10.17 SUMMARY

The majority of bioactive chemicals come from fruits and vegetables. Due to these substances' therapeutic benefits, they have been utilized for ages. Large-scale uses, such as functional meals and nutritional supplements, have just lately been made use of them. Peels and seeds, which are processed waste products, are particularly rich in bioactive substances. These substances include colors, pectin, essential oils, polyphenols, flavonoids and flavonoids. The recovery of these vital components from waste created during the preparation of fruits and vegetables has been the subject of recent research. Despite the fact that there are still certain classic extraction techniques in use, there is a growing need for more environmentally friendly and sustainable alternatives that may both hasten the extraction process and reduce its negative environmental consequences. In general, the waste produced during the production of fruits and vegetables is a possible source of bioactive substances that may be gathered and used for a number of purposes.

The material's characteristics and binding methods have a significant influence on 3D printing of food. In the past several years, there have been numerous attempts to produce 3D structures with end-use characteristics that are on par with or better than those attained using conventional manufacturing methods. Before 3DFP can be applied in specialized enterprises that require personalization, there are still many obstacles to be addressed. There isn't enough software available to set the printing settings for several viscous material systems. Internationally recognized researchers are now changing the factors that are often managed in the fusion deposition of plastic filaments to get the optimal pressure in extrusion operations, for instance. We believe that hardware and software will be created when the 3DFP application's feasibility has been established. In addition to keeping an eye on 3DP's growth as it approaches the production of new 3D structures, researchers and other people working in the food business should be aware of the most current advancements that have an impact on consumer behavior.

REFERENCES

Agbor, V. B., Cicek, N., Sparling, R., Berlin, A., & Levin, D. B. 2011. Biomass pretreatment: Fundamentals toward application. Biotechnology advances, 29(6): 675–685.

Ahn, J. H., Jang, Y. S., & Lee, S. Y. 2016. Production of succinic acid by metabolically engineered microorganisms. Current opinion in biotechnology 42:54–66.

Akhtar, N., Gupta, K., Goyal, D., & Goyal, A. 2016. Recent advances in pretreatment technologies for efficient hydrolysis of lignocellulosic biomass. Environmental Progress & Sustainable Energy 35(2): 489–511.

Alvarez-Rivera, G., Bueno, M., Ballesteros-Vivas, D., Mendiola, J. A., & Ibanez, E. 2020. Pressurized liquid extraction. In Liquid-phase extraction (pp. 375–398). Elsevier, Amsterdam, The Netherlands

Angenent, L. T., Karim, K., Al-Dahhan, M. H., Wrenn, B. A., & Domíguez-Espinosa, R. 2004. Production of bioenergy and biochemicals from industrial and agricultural wastewater. TRENDS in Biotechnology 22(9):477–485.

Arikan, E. B., & Ozsoy, H. D. 2015. A review: Investigation of bioplastics. J. Civ. Eng. Arch 9(1): 188–192.

Azam, S.M.R., M. Zhang, A.S. Mujumdar, and C. Yang. 2018. Study on 3D Printing of orange concentrate and material characteristics. Journal of Food Process Engineering 41(5),:1–10.

Azmir, J., I.S.M. Zaidul, M.M. Rahman, K.M. Sharif, A. Mohamed, F. Sahena, M.H.A. Jahurul, K. Ghafoor, N.A.N. Norulaini, and A.K.M. Omar. 2013. Techniques for extraction of bioactive compounds from plant materials: A Review. Journal of Food Engineering 117: 426–36. https://doi.org/10.1016/j.jfoodeng.2013.01.014.

Bajpai, P. 2013. Biorefinery in the pulp and paper industry. Academic press,London, UK.

Basso, L. C., De Amorim, H. V., De Oliveira, A. J., & Lopes, M. L. 2008. Yeast selection for fuel ethanol production in Brazil. FEMS yeast research 8(7):1155–1163.

Berdy, J. 2005. Bioactive microbial metabolites. The Journal of antibiotics, 58(1):1–26.

Bokkhim, H., N. Bansal, L. Grøndahl, and B. Bhandari. 2016. Characterization of alginate-lactoferrin beads prepared by extrusion gelation method. Food Hydrocolloids 53 (February 1): 270–276.

Bommarius, A. S., & Riebel-Bommarius, B. R. 2004. Biocatalysis: Fundamentals and applications. John Wiley & Sons, UK

Boutouchent, A. 2016. Industry 4.0: The 4 Technologies behind It. https://insights.bridgr.co/manufacturing-trends-industry-4-technologies/.

Bridgwater, A. V. 2012. Review of fast pyrolysis of biomass and product upgrading. Biomass and bioenergy 38:68–94.

Bruhn, C.M. 2007. Enhancing consumer acceptance of new processing technologies. Innovative Food Science and Emerging Technologies 8, no. 4: 555–558.

Brunton, L.A., A.P. Desbois, M. Garza, B. Wieland, C.V. Mohan, B. Häsler, C.C. Tam, et al. 2019. Identifying hotspots for antibiotic resistance emergence and selection, and elucidating pathways to human exposure: Application of a systems-thinking approach to aquaculture systems. Science of The Total Environment 687 (October 15): 1344–1356.

Cahill, T., and G. Hardiman. 2020. Nutritional challenges and countermeasures for space travel. Nutrition Bulletin 45, no. 1: 98–105.

CandyFab. 2017. A DIY 3D Sugar Printer. Https://Candyfab.Org/. https://candyfab.org/.

Carbonell Capella, J. M., Sic ˇ Zlabur, ˇ J., Rimac Brnˇci´c, S., Barba, F. J., Grimi, N., Koubaa, M., & Vorobiev, E. 2017. Electrotechnologies, microwaves, and ultrasounds combined with binary mixtures of ethanol and water to extract steviol glycosides and antioxidant compounds from Stevia rebaudiana leaves. Journal of Food Processing and Preservation 41(5):e13179

Cardello, A. V. 2003. Consumer concerns and expectations about novel food processing technologies: Effects on product liking. Appetite 40, no. 3 (June): 217–33. https://linkinghub.elsevier.com/retrieve/pii/S0195666303000084.

Cardello, A. V., H. G. Schutz, and L.L. Lesher. 2007. Consumer perceptions of foods processed by innovative and emerging technologies: A conjoint analytic study. Innovative Food Science and Emerging Technologies 8, no. 1: 73–83.

Chakka, A. K., Sriraksha, M. S., & Ravishankar, C. N. 2021. Sustainability of emerging green non-thermal technologies in the food industry with food safety perspective: A review. Lwt 151: 112–140.

Chakraborty, S., Aggarwal, V., Mukherjee, D., & Andras, K. 2012. Biomass to biofuel: A review on production technology. Asia-Pacific Journal of Chemical Engineering 7: S254–S262.

Chemat, F., Abert Vian, M., Fabiano-Tixier, A. S., Nutrizio, M., Režek Jambrak, A., Munekata, P. E. S., Lorenzo, J. M., Barba, F. J., Binello, A., & Cravotto, G. 2020. A review of sustainable and intensified techniques for extraction of food and natural products. Green Chemistry 22(8): 2325–2353. https://doi.org/10.1039/c9gc03878g

Cherubini, F., & Strømman, A. H. 2011. Chemicals from lignocellulosic biomass: Opportunities, perspectives, and potential of biorefinery systems. Biofuels, Bioproducts and Biorefining 5(5): 548–561.

Ching, S.H., B. Bhandari, R. Webb, and N. Bansal. 2015. Visualizing the interaction between sodium caseinate and calcium alginate microgel particles. Food Hydrocolloids 43 (January 1): 165–171.

Chon, J. H., & Zarbis-Papastoitsis, G. 2011. Advances in the production and downstream processing of antibodies. New biotechnology 28(5):458–463.

Jennifer Chu. 2017. Researchers Engineer Shape-Shifting Noodles | MIT News | Massachusetts Institute of Technology. MIT News Office. https://news.mit.edu/2017/researchers-engineer-shape-shifting-noodles-0525.

Chuyen, H. V., Nguyen, M. H., Roach, P. D., Golding, J. B., & Parks, S. E. 2018. Microwave-assisted extraction and ultrasound-assisted extraction for recovering carotenoids from Gac peel and their effects on antioxidant capacity of the extracts. Food Science & Nutrition 6(1): 189–196.

Coelho, M., Pereira, R., Rodrigues, A. S., Teixeira, J. A., & Pintado, M. E. 2019. Extraction of tomato by-products' bioactive compounds using ohmic technology. Food and Bioproducts Processing 117:329–339.

Da Rosa, G. S., Martiny, T. R., Dotto, G. L., Vanga, S. K., Parrine, D., Gariepy, Y., ... & Raghavan, V. 2021. Eco-friendly extraction for the recovery of bioactive compounds from Brazilian olive leaves. Sustainable Materials and Technologies 28: e00276.

Da Rosa, G. S., Vanga, S. K., Gariepy, Y., & Raghavan, V. 2019. Comparison of microwave, ultrasonic and conventional techniques for extraction of bioactive compounds from olive leaves (Olea europaea L.). Innovative Food Science & Emerging Technologies 58:102234.

Da Silva, H.R.P., da Silva, C., & Bolanho, B. C. 2018. Ultrasonic-assisted extraction of betalains from red beet (Beta vulgaris L.). Journal of *Food Process Engineering* 41(6):e12833.

Dahiya, A. (Ed.). 2014. Bioenergy: Biomass to biofuels. Academic Press, Amsterdam, The Netherlands

Dankar, I., A. Haddarah, F.E.L. Omar, F. Sepulcre, and M. Pujolà. 2018. 3D printing technology: The new era for food customization and elaboration. Trends in Food Science & Technology 75 (May 1): 231–242.

Dash, D. R., Pathak, S. S., & Pradhan, R. C. 2021. Improvement in novel ultrasoundassisted extraction technology of high value-added components from fruit and vegetable peels. Journal of Food Process Engineering, 44(4): 1–11. https://doi.org/10.1111/jfpe.13658.

Derossi, A., R. Caporizzi, D. Azzollini, and C. Severini. 2018. Application of 3D printing for customized food. A case on the development of a fruit-based snack for children. Journal of Food Engineering 220 (March 1): 65–75.

Diaz, J.V., M.W.-J. Noort, B.K.J.C. Van, and K.J.C. Van Noort, Martijn Willem-Jan Bommel. 2015. Method for the Production of an Edible Object by Powder Bed (3d) Printingno. 12: WO 2015/115897. https://encrypted.google.com/patents/WO2015115897A1?hl=zh-CN&cl=en.

Diaz, J., K. Van Bommel, ... M.N.-U.P., and U. 2018. 2016. Method for the Production of Edible Objects Using Sls and Food Products. Google Patents. https://patents.google.com/patent/US10092030B2/en.

Dick, A., B. Bhandari, X. Dong, and S. Prakash. 2020. Feasibility study of hydrocolloid incorporated 3D printed pork as dysphagia food. Food Hydrocolloids 107 (October 1): 105940. https://linkinghub.elsevier.com/retrieve/pii/S0268005X19329790.

Ding, Z., C. Yuan, X. Peng, T. Wang, H.J. Qi, and M.L. Dunn. 2017. Direct 4D printing via active composite materials. Science Advances 3, no. 4 (April 7): e1602890. https://www.science.org/doi/10.1126/sciadv.1602890.

Drapcho, C. M., Nghim, N. P., & Walker, T. 2008. Biofuels Engineering Process Technology. McGraw-Hill Education, Noida, India.

Ebrahimi, P., & Lante, A. 2022. Environmentally friendly techniques for the recovery of polyphenols from food by-products and their impact on polyphenol oxidase: A critical review. Applied Sciences, 12(4), 1923.

Ethereal-Halo. 2023. Ethereal HALO SIMULTANEOUS 5 AXIS CNC MACHINE. Accessed April 28. https://etherealmachines.com/products-ethereal-halo-5axis-cnc-machine/.

Feng, C., Zhang, M., & Bhandari, B. 2019. Materials properties of printable edible inks and printing parameters optimization during 3D printing: A review. Critical reviews in food science and nutrition 59(19):3074–3081.

Fengying, C., Min, Z., Kai, F., & Arun, S. M. 2020. Non-thermal technology and heating technology for fresh food cooking in the central kitchen processing: A review. Food Reviews International . https://doi.org/10.1080/87559129.2020.1740246.

Fernando, S., Adhikari, S., Chandrapal, C., & Murali, N. 2006. Biorefineries: Current status, challenges, and future direction. Energy & Fuels 20(4):1727–1737.

Flieger, M., Kantorova, M., Prell, A., Řezanka, T., & Votruba, J. 2003. Biodegradable plastics from renewable sources. Folia microbiologica 48:27–44.

Frewer, L.J., K. Bergmann, M. Brennan, R. Lion, R. Meertens, G. Rowe, M. Siegrist, and C. Vereijken. 2011. Consumer response to novel agri-food technologies: Implications for predicting consumer acceptance of emerging food technologies. Trends in Food Science and Technology.

Fröhlich, H., Villian, L., Melzner, D., & Strube, J. 2012. Membrane technology in bioprocess science. Chemie Ingenieur Technik 84(6): 905–917.

Gavahian, M., Chu, Y.-H., & Farahnaky, A. 2019. Effects of ohmic and microwave cooking on textural softening and physical properties of rice. Journal of Food Engineering 243: 114–124.

Ge, Q., H.J. Qi, and M.L. Dunn. 2013. Active materials by four-dimension printing. Applied Physics Letters 103, no. 13 (September 23): 131901. https://pubs.aip.org/aip/apl/article/103/13/131901/130318/Active-materials-by-four-dimension-printing.

Godoi, F.C., S. Prakash, and B.R. Bhandari. 2016. 3d printing technologies applied for food design: Status and prospects. Journal of Food Engineering. Elsevier.

Gomord, V., Chamberlain, P., Jefferis, R., & Faye, L. 2005. Biopharmaceutical production in plants: Problems, solutions and opportunities. TRENDS in Biotechnology 23(11):559–565.

Gray, N. 2013. Looking to the Future: Creating novel foods using 3D Printing. Food Navigator: 1–2. https://www.foodnavigator.com/Article/2010/12/23/Looking-to-the-future-Creating-novel-foods-using-3D-printing.

Greehy, G.M., M.B. McCarthy, M.M. Henchion, E.J. Dillon, and S.N. McCarthy. 2013. Complexity and Conundrums. Citizens' evaluations of potentially contentious novel food technologies using a deliberative discourse approach. Appetite 70: 37–46.

Gu, Z., J. Fu, H. Lin, and Y. He. 2020. Development of 3D Bioprinting: From printing methods to biomedical applications. Asian Journal of Pharmaceutical Sciences.

Haberl, S., Miklavcic, D., Sersa, G., Frey, W., & Rubinsky, B. 2013. Cell membrane electroporation-Part 2: The applications. IEEE Electrical Insulation Magazine 29(1):29–37.

Haenggi, R., and N. Hanspeter. 2011. Method and Device for Micro-Dosing a Liquid. European Patent Application. https://patents.google.com/patent/US8556392B2/en.

Hamilton, C.A., G. Alici, and M. in het Panhuis. 2018. 3D printing vegemite and marmite: Redefining "Breadboards." Journal of Food Engineering 220: 83–88.

Hao, L., S. Mellor, O. Seaman, J. Henderson, N. Sewell, and M. Sloan. 2010. Material characterisation and process development for chocolate additive layer manufacturing. Virtual and Physical Prototyping 5, no. 2 (June): 57–64. https://www.tandfonline.com/doi/abs/10.1080/17452751003753212.

Ibrahim, N. I., Shahar, F. S., Sultan, M. T. H., Shah, A. U. M., Safri, S. N. A., & Mat Yazik, M. H. 2021. Overview of bioplastic introduction and its applications in product packaging. Coatings 11(11): 1423.

Izdebska, J., and Z. Zołek-Tryznowska. 2016. 3D Food Printing - Facts and Future. Agro Food Industry Hi-Tech, Italy.

Jaeger, H., D. Knorr, E. Szabó, J. Hámori, and D. Bánáti. 2015. Impact of terminology on consumer acceptance of emerging technologies through the example of PEF Technology. Innovative Food Science and Emerging Technologies 29: 87–93.

Jan, B., Shams, R., Rizvi, Q. E. H., & Manzoor, A. 2021. Ohmic heating technology for food processing: A review of recent developments. J. Postharvest Technol 9(1):20–34.

Jeon, W.Y., J.Y. Yu, H.W. Kim, and H.J. Park. 2021. Production of customized food through the insertion of a formulated nanoemulsion using coaxial 3D food printing. Journal of Food Engineering 311.

Jiang, J., M. Zhang, B. Bhandari, and P. Cao. 2020. Current processing and packing technology for space foods: A Review. Critical Reviews in Food Science and Nutrition 60, no. 21 (November 29): 3573–3588. https://pubmed.ncbi.nlm.nih.gov/31830802/.

Jönsson, L. J., & Martín, C. 2016. Pretreatment of lignocellulose: Formation of inhibitory by-products and strategies for minimizing their effects. Bioresource technology 199:103–112.

Joshi, S.C., and A.A. Sheikh. 2015. 3D printing in aerospace and its long-term sustainability. Virtual and Physical Prototyping 10, no. 4: 175–185.

Katoh, S., Horiuchi, J. I., & Yoshida, F. 2015. Biochemical Engineering: A Textbook For Engineers, Chemists And Biologists. John Wiley & Sons, Weinheim, Germany.

Khalid, A., Arshad, M., Anjum, M., Mahmood, T., & Dawson, L. 2011. The anaerobic digestion of solid organic waste. Waste management 31(8):1737–1744.

Kim, H.W., I.J. Lee, S.M. Park, J.H. Lee, M.H. Nguyen, and H.J. Park. 2019. Effect of hydrocolloid addition on dimensional stability in post-processing of 3D printable cookie dough. LWT 101 (March 1): 69–75.

Kim, H.W., J.H. Lee, S.M. Park, M.H. Lee, I.W. Lee, H.S. Doh, and H.J. Park. 2018. Effect of hydrocolloids on rheological properties and printability of vegetable inks for 3D food printing. Journal of Food Science 83, no. 12 (December 1): 2923–2932. https://onlinelibrary.wiley.com/doi/full/10.1111/1750-3841.14391.

Kim, H.W., H. Bae, and H.J. Park. 2017. Classification of the printability of selected food for 3D printing: Development of an assessment method using hydrocolloids as reference material. Journal of Food Engineering 215 (December 1): 23–32.

Kim, S.M., H.W. Kim, and H.J. Park. 2021. Preparation and characterization of surimi-based imitation crab meat using coaxial extrusion three-dimensional food printing. Innovative Food Science & Emerging Technologies 71 (July): 102711. https://linkinghub.elsevier.com/retrieve/pii/S1466856421001120.

Klass, D. L. 1998. Biomass for Renewable Energy, Fuels, and Chemicals. Elsevier,Amsterdam, The Netherlands.

Knothe, G. 2009. Improving biodiesel fuel properties by modifying fatty ester composition. Energy & Environmental Science 2(7):759–766.

Ko, H.J., Y. Wen, J.H. Choi, B.R. Park, H.W. Kim, and H.J. Park. 2021. Meat analog production through artificial muscle fiber insertion using coaxial nozzle-assisted three-dimensional food printing. Food Hydrocolloids 120 (November 1): 106898. https://linkinghub.elsevier.com/retrieve/pii/S0268005X21003143.

Kocovic, P. 2017. 3D printing and its impact on the production of fully functional components. Advances in Chemical and Materials Engineering. IGI Global. https://services.igi-global.com/resolvedoi/resolve.aspx?doi=10.4018/978-1-5225-2289-8. Hershey, PA, USA.

Köhler, G., & Milstein, C. 1975. Continuous cultures of fused cells secreting antibody of predefined specificity. Nature 256(5517):495–497.

Kumar, P., Barrett, D. M., Delwiche, M. J., & Stroeve, P. 2009. Methods for pretreatment of lignocellulosic biomass for efficient hydrolysis and biofuel production. Industrial & *Engineering Chemistry Research* 48(8):3713–3729.

Kumar, R., Tabatabaei, M., Karimi, K., & Sárvári Horváth, I. 2016. Recent updates on lignocellulosic biomass derived ethanol-A review. Biofuel Research Journal 3(1): 347–356.

Lam, M. K., Lee, K. T., & Mohamed, A. R. 2010. Homogeneous, heterogeneous and enzymatic catalysis for transesterification of high free fatty acid oil (waste cooking oil) to biodiesel: A review. Biotechnology advances 28(4): 500–518.

Lante, A., & Friso, D. 2013. Oxidative stability and rheological properties of nanoemulsions with ultrasonic extracted green tea infusion. Food *Research International* 54(1):269–276.

Lewis, J.A. 2002. Direct-write assembly of ceramics from colloidal inks. Current Opinion in Solid State and Materials Science 6, no. 3: 245–250.

Li, H., Cann, A. F., & Liao, J. C. 2010. Biofuels: Biomolecular engineering fundamentals and advances. Annual review of chemical and biomolecular engineering 1:19–36.

Lille, M., A. Nurmela, E. Nordlund, S. Metsä-Kortelainen, and N. Sozer. 2018. Applicability of protein and fiber-rich food materials in extrusion-based 3D Printing. Journal of Food Engineering 220 (March 1): 20–27.

Lipton, J.I., M. Cutler, F. Nigl, D. Cohen, and H. Lipson. 2015. Additive Manufacturing for the Food Industry. Trends in Food Science and Technology. Elsevier, Oxford, UK.

Lipton, J., D. Arnold, F. Nigl, N. Lopez, D. Cohen, N. Norén, and H. Lipson. 2010. Mutli-material food printing with complex internal structure suitable for conventional post-processing. In *21st Annual International Solid Freeform Fabrication Symposium - An Additive Manufacturing Conference, SFF* 2010, 809–815. United States.

Liu, Z., M. Zhang, B. Bhandari, and C. Yang. 2018. Impact of rheological properties of mashed potatoes on 3D printing. Journal of Food Engineering 220: 76–82.

López Galdeano, J.A. 2015. 3D Printing Food : The Sustainable Future.

Lynd, L. R., Weimer, P. J., Van Zyl, W. H., & Pretorius, I. S. 2002. Microbial cellulose utilization: Fundamentals and biotechnology. Microbiology and *Molecular Biology Reviews* 66(3): 506–577.

Lyndhurst, B. 2009. An evidence review of public attitudes to emerging food technologies. Social Science Research Unit Food Standards Agency 89, no. March: 89. https://www.esr.cri.nz/SiteCollectionDocuments/ESR/PDF/SocialScience/emergingfoodtechsummary.pdf.

Manstan, T., S.L. Chandler, and M.B. McSweeney. 2021. Consumers' Attitudes towards 3D printed foods after a positive experience: An exploratory study. Journal of Sensory Studies 36, no. 1 (February 8). https://onlinelibrary.wiley.com/doi/10.1111/joss.12619.

Mantihal, S., Kobun, R., & Lee, B. B. 2020. 3D food printing of as the new way of preparing food: A review. International Journal of Gastronomy and Food Science 22: 100260.

Mantihal, S., S. Prakash, F.C. Godoi, and B. Bhandari. 2017. Optimization of Chocolate 3D printing by correlating thermal and flow properties with 3D structure modeling. Innovative Food Science & Emerging Technologies 44 (December 1): 21–29. https://linkinghub.elsevier.com/retrieve/pii/S146685641630844X.

Mazumder, M. A. R., Rana, J., Jubayer, M. F., Ranganathan, T. V., & Ansari, M. J. 2023. Sonication microwave synergistic extraction of bioactive compounds from plant source. In Ultrasound and Microwave for Food Processing, ed. G. A. Nayik, M. Ranjha, X. A. Zeng, S. Irfan, & S. M. Zahra, 239–267. Academic Press Singapore .

Mihaylova, D. S., Lante, A., Tinello, F., & Krastanov, A. I. 2014. Study on the antioxidant and antimicrobial activities of Allium ursinum L. pressurised-liquid extract. Natural product research 28(22):2000–2005.

Mohanty, A. K., Misra, M., & Drzal, L. T. (Eds.). 2005. Natural Fibers, Biopolymers, and biocomposites. CRC press, Boca Raton, Florida.

Mood, S. H., Golfeshan, A. H., Tabatabaei, M., Jouzani, G. S., Najafi, G. H., Gholami, M., & Ardjmand, M. 2013. Lignocellulosic biomass to bioethanol, a comprehensive review with a focus on pretreatment. Renewable and Sustainable Energy Reviews 27: 77–93.

Moo-Young, M. 2019. Comprehensive biotechnology. Elsevier, London, United Kingdom.

More, P. R., Jambrak, A. R., & Arya, S. S. 2022. Green, environment-friendly and sustainable techniques for extraction of Food Bioactive Compounds and Waste Valorization. Trends in Food Science & Technology, Amsterdam, The Netherlands.

Mosier, N., Wyman, C., Dale, B., Elander, R., Lee, Y. Y., Holtzapple, M., & Ladisch, M. 2005. Features of promising technologies for pretreatment of lignocellulosic biomass. Bioresource technology 96(6):673–686.

Mulder, M. 1996. Basic Principles Of Membrane Technology. Springer science & business media, Dordrecht.

Myerson, A. 2002. Handbook of industrial crystallization. Butterworth-Heinemann, Cambridge University Press.

Nelson, D. L., Lehninger, A. L., & Cox, M. M. 2008. Lehninger Principles Of Biochemistry. Macmillan, New York City, United States. .

Ng, K. S., Hernandez, E. M., & Sadhukhan, J. 2014. Biorefineries and Chemical Processes: Design, Integration And Sustainability Analysis. John Wiley & Sons, Hoboken, New Jersey.

Nizami, A. S., Rehan, M., Waqas, M., Naqvi, M., Ouda, O. K., Shahzad, K., ... & Pant, D. 2017. Waste biorefineries: Enabling circular economies in developing countries. Bioresource technology 241:1101–1117.

Nüchter, M., Ondruschka, B., Bonrath, W., & Gum, A. 2004. Microwave assisted synthesis-a critical technology overview. Green chemistry, 6(3): 128–141.

Olsen, N.V., K.G. Grunert, and A.M. Sonne. 2010. Consumer acceptance of high-pressure processing and pulsed-electric field: A Review. Trends in Food Science and Technology.

Ouyang, L., R. Yao, Y. Zhao, and W. Sun. 2016. Effect of bioink properties on printability and cell viability for 3D bioplotting of embryonic stem cells. Biofabrication 8, no. 3 (September 16): 035020. https://iopscience.iop.org/article/10.1088/1758-5090/8/3/035020.

Pagano, I., Campone, L., Celano, R., Piccinelli, A. L., & Rastrelli, L. 2021. Green non-conventional techniques for the extraction of polyphenols from agricultural food by-products: A review. Journal of Chromatography A 1651:462295.

Papargyropoulou, E., R. Lozano, J. K. Steinberger, N. Wright, and Z. Bin Ujang. 2014. The food waste hierarchy as a framework for the management of food surplus and food waste. Journal of Cleaner Production 76: 106–115.

Pavlou, A. K., & Reichert, J. M. 2004. Recombinant protein therapeutics-success rates, market trends and values to 2010. Nature biotechnology 22(12):1513–1519.

Peltola, S.M., F.P.W. Melchels, D.W. Grijpma, and M. Kellomäki. 2008. A review of rapid prototyping techniques for tissue engineering purposes. Annals of Medicine. 40: 268280.

Perchonok, M., and C. Bourland. 2002. NASA food systems: Past, present, and future. Nutrition 18, no. 10: 913–920.

Pérez, B., H. Nykvist, A.F. Brøgger, M.B. Larsen, and M.F. Falkeborg. 2019. Impact of macronutrients printability and 3D-printer parameters on 3D-food printing: A review. Food Chemistry 287 (July 30): 249–257. https://linkinghub.elsevier.com/retrieve/pii/S0308814619304133.

Perlack, R. D., Wright, L. L., Turhollow, A. F., Graham, R. L., Stokes, B. J., Erbach, D. C. 2005. Biomass as Feedstock for a Bioenergy And Bioproducts Industry: The Technical Feasibility of a Billion-Ton Annual Supply. Oak Ridge: U. S. Department of Energy.

Polanco-Lugo, E., Martínez-Castillo, J. I., Cuevas-Bernardino, J. C., González-Flores, T., Valdez-Ojeda, R., Pacheco, N., & Ayora-Talavera, T. 2019. Citrus pectin obtained by ultrasound-assisted extraction: Physicochemical, structural, rheological and functional properties. CyTA-Journal of Food, 17(1): 463–471.

Popa, M.E., and A. Popa. 2012. Consumer Behavior: Determinants and Trends in Novel Food Choice. In Novel Technologies in Food Science: Their Impact on Products, Consumer Trends and the Environment, 137–156. Springer, New York City.

Portanguen, S., P. Tournayre, J. Sicard, T. Astruc, and P.S. Mirade. 2019. Toward the design of functional foods and biobased products by 3D Printing: A Review. Trends in Food Science and Technology.

Rifna, E. J., Misra, N. N., & Dwivedi, M. 2023. Recent advances in extraction technologies for recovery of bioactive compounds derived from fruit and vegetable waste peels: A review. Critical Reviews in Food Science and Nutrition, 63(6): 719–752.

Roda, A., & Lambri, M. 2019. Food uses of pineapple waste and by-products: A review. International *Journal Of Food Science* & technology, 54(4):1009–1017.

Rodríguez De Luna, S. L., Ramírez-Garza, R. E., & Serna Saldívar, S. O. 2020. Environmentally friendly methods for flavonoid extraction from plant material: Impact of their operating conditions on yield and antioxidant properties. The Scientific World Journal

Rosenthal, A.J. 2010. Texture profile analysis - How important are the parameters? Journal of Texture Studies 41, no. 5 (October 1): 672–684. https://onlinelibrary.wiley.com/doi/full/10.1111/j.1745-4603.2010.00248.x.

Routray, W., & Orsat, V. 2014. MAE of phenolic compounds from blueberry leaves and comparison with other extraction methods. Industrial Crops and Products, 58:36–45.

Saberian, Hamidi-Esfahani, H., Zohrehb Ahmadi Gavlighi, H., Banakar, A., & Barzegar, M. 2018. The potential of ohmic heating for pectin extraction from orange waste. Journal of Food Processing and Preservation,42(2):e13458. https://doi.org/10.1111/jfpp.13458.

Sagar, N. A., Pareek, S., Sharma, S., Yahia, E. M., & Lobo, M. G. 2018. Fruit and vegetable waste: Bioactive compounds, their extraction, and possible utilization. Comprehensive Reviews in Food Science and Food Safety 17(3):512–531. https://doi.org/10.1111/1541-4337.12330.

Saini, A., Panesar, P. S., &Bera, M. B. 2019. Valorization of fruits and vegetables waste through green extraction of bioactive compounds and their nanoemulsions-based delivery system. Bioresources Bioprocessing, 6(1):1-12 https://doi.org/10.1186/s40643-019-0261-9.

Sarker, A., Ghosh, M. K., Islam, T., Bilal, M., Nandi, R., Raihan, M. L., ...& Kim, J. E. 2022. Sustainable food waste recycling for the circular economy in developing countries, with special reference to bangladesh. Sustainability 14(19): 12035. https://doi.org/10.3390/su141912035.

Sarker, A., Mondal, S. C., Ahmmed, R., Rana, J., Ansary, M. W. R., & Bilal, M. 2023. Prospects and challenges of polymer nanocomposites for innovative food packaging. In Smart Polymer Nanocomposites, ed. N. Ali, M. Bilal, A. Khan, T. A. Nguyen, & R. K. Gupta, 355–377. Elsevier. https://doi.org/10.1016/B978-0-323-91611-0.00021-9, Amsterdam, The Netherlands.

Scragg, A. H. 2009. Biofuels: Production, Application And Development. CABI, UK .

Seerden, K.A.M., N. Reis, J.R.G. Evans, P.S. Grant, J.W. Halloran, and B. Derby. 2001. Ink-jet printing of wax-based alumina suspensions. Journal of the American Ceramic Society 84, no. 11: 2514–2520.

Sekhon, B. S. (2010). Biopharmaceuticals: An overview. Thai J. Pharm. Sc, 34(34):1–19.

Severini, C., D. Azzollini, M. Albenzio, and A. Derossi. 2018. On printability, quality and nutritional properties of 3D printed cereal based snacks enriched with edible insects. Food Research International 106: 666–676.

Sirmons, T.A., P.G. Roma, A.M. Whitmire, S.M. Smith, S.R. Zwart, M. Young, and G.L. Douglas. 2020. Meal replacement in isolated and confined mission environments: Consumption, Acceptability, and Implications for Physical and Behavioral Health. Physiology & Behavior 219 (May): 112829. https://linkinghub.elsevier.com/retrieve/pii/S0031938420301463.

Smay, J.E., J. Cesarano, and J.A. Lewis. 2002. Colloidal inks for directed assembly of 3-D periodic structures. Langmuir 18, no. 14: 5429–5437.

Sridhar, A., Ponnuchamy, M., Kumar, P. S., Kapoor, A., Vo, D. V. N., & Prabhakar, S. 2021. Techniques and modeling of polyphenol extraction from food: A review. Environmental Chemistry Letters 19:3409–3443.

Subic, A., and S. Gallagher. 2017. Industry 4.0 Testlabs in Australia Preparing for the Future. Department of Industry, Innovation and Science and Swinburne University of Technology. https://www.industry.gov.au/sites/g/files/net3906/f/July 2018/document/pdf/industry-4.0-testlabs-report.pdf.

Sun, J., W. Zhou, D. Huang, J.Y.H. Fuh, and G.S. Hong. 2015. An Overview of 3D printing technologies for food fabrication. Food and Bioprocess Technology 8, no. 8: 1605–1615.

Thangalakshmi, S., V.K. Arora, B.P. Kaur, and S. Malakar. 2021. Investigation on rice flour and jaggery paste as food material for extrusion-based 3D Printing. Journal of Food Processing and Preservation 45, no. 4 (April 6): e15375. https://onlinelibrary.wiley.com/doi/full/10.1111/jfpp.15375.

Tipu, M. M. H., Baroi, A., Rana, J., Islam, S., Jahan, R., Miah, M. S., & Asaduzzaman, M. 2022. Potential applications of nanotechnology in agriculture: A smart tool for sustainable agriculture. In Agricultural Development in Asia-Potential Use of Nano-Materials and Nano-Technology, ed. M. Asaduzzaman and M. Afroz, 1–28. IntechOpen, UK.

Ukaga, O., Maser, C., & Reichenbach, M. (Eds.). 2010. Sustainable Development: Principles, Frameworks, and Case Studies. CRC Press, Boca Raton, Florida.

Vancauwenberghe, V., L. Katalagarianakis, Z. Wang, M. Meerts, M. Hertog, P. Verboven, P. Moldenaers, M.E. Hendrickx, J. Lammertyn, and B. Nicolaï. 2017. Pectin based food-ink formulations for 3-D printing of customizable porous food simulants. Innovative Food Science and Emerging Technologies 42 (August 1): 138–150.

Vancauwenberghe, V., P. Verboven, J. Lammertyn, and B. Nicolaï. 2018. Development of a coaxial extrusion deposition for 3D printing of customizable pectin-based food simulant. Journal of Food Engineering 225 (May 1): 42–52.

Vieira, M. V., S.M. Oliveira, I.R. Amado, L.H. Fasolin, A.A. Vicente, L.M. Pastrana, and P. Fuciños. 2020. 3D printed functional cookies fortified with arthrospira platensis: Evaluation of its antioxidant potential and physical-chemical characterization. Food Hydrocolloids 107 (October 1): 105893.

Vinatoru, M., T. J. Mason, and I. Calinescu. 2017. Ultrasonically assisted extraction (UAE) and microwave assisted extraction (MAE) of functional compounds from plant materials. TrAC Trends in Analytical Chemistry 97:159–78. https://doi.org/10.1016/j.trac.2017.09.002.

Wang, L., M. Zhang, B. Bhandari, and C. Yang. 2018. Investigation on fish surimi gel as promising food material for 3D Printing. Journal of Food Engineering 220 (March 1): 101–108.

Wang, X., Q. Ao, X. Tian, J. Fan, H. Tong, W. Hou, and S. Bai. 2017. Gelatin-based hydrogels for organ 3D Bioprinting. Polymers 2017, Vol. 9, Page 401 9, no. 9 (August 30): 401. https://www.mdpi.com/2073-4360/9/9/401/htm.

Wegrzyn, T.F., M. Golding, and R.H. Archer. 2012. Food layered manufacture: A new process for constructing solid foods. Trends in Food Science and Technology.

Wu, D., Forghani, F., Daliri, E. B. M., Li, J., Liao, X., Liu, D., ... & Ding, T. 2020. Microbial response to some nonthermal physical technologies. Trends in Food Science & Technology 95:107–117.

Xaar. 2012. A Guide to Industrial Inkjet. https://www.xaar.com/media/1312/xaar-guide-to-ink-jet.pdf.

Yang, Fan, M. Zhang, Z. Fang, and Y. Liu. 2019. Impact of processing parameters and post-treatment on the shape accuracy of 3D-printed baking dough. International Journal of Food Science & Technology 54, no. 1 (January 1): 68–74. https://onlinelibrary.wiley.com/doi/full/10.1111/ijfs.13904.

Yang, Fan, M. Zhang, S. Prakash, and Y. Liu. 2018. Physical Properties of 3D printed baking dough as affected by different compositions. Innovative Food Science & Emerging Technologies 49 (October 1): 202–210.

Yang, Fanli, M. Zhang, B. Bhandari, and Y. Liu. 2018. Investigation on lemon juice gel as food material for 3D printing and optimization of printing parameters. LWT 87 (January 1): 67–76.

Yang, S. T., & Yu, M. 2013. Integrated biorefinery for sustainable production of fuels, chemicals, and polymers. In Bioprocessing technologies in biorefinery for sustainable production of fuels, chemicals, and polymers, ed. S.- T. Yang, H. A. El-Enshasy and N. Thongchul, 1–26. Wiley, Hoboken, New Jersey .

Zakzeski, J., Bruijnincx, P. C., Jongerius, A. L., & Weckhuysen, B. M. 2010. The catalytic valorization of lignin for the production of renewable chemicals. Chemical reviews 110(6): 3552–3599.

Zhang, J.Y., J.K. Pandya, D.J. McClements, J. Lu, and A.J. Kinchla. 2021. Advancements in 3D food printing: A comprehensive overview of properties and opportunities. https://doi.org/10.1080/10408398.2021.1878103 62, no. 17: 4752–4768. https://www.tandfonline.com/doi/abs/10.1080/10408398.2021.1878103.

Zhang, L., Y. Lou, and M.A.I. Schutyser. 2018. 3D printing of cereal-based food structures containing probiotics. Food Structure 18 (October 1): 14–22.

Zhang, Y. H. P. 2015. Production of biofuels and biochemicals by in vitro synthetic biosystems: Opportunities and challenges. Biotechnology advances 33(7):1467–1483.

11 Biotechnological Approaches to Increase Food Production Quality and Safety

Zahra Maqbool
Government College University Faisalabad

Waseem Khalid
University of Castilla La Mancha
The University of Lahore

Mahum and Maryam Maqsood
Muhammad Nawaz Shareef University of Agriculture

Shakeela Khalid
University of Sargodha

Andres Moreno
University of Castilla La Mancha

11.1 INTRODUCTION

The food industry's positive impact on the economy, employment, and well-being of the populace has led to its widespread recognition as an effective sector. Food security plans emphasize on the amount and quality of food. The organic agricultural methods produced greater nutritive crops than traditional methods Organic agricultural techniques enhance human health, food safety, and food quality (Rempelos et al., 2021).

In general, sensory, mechanical, and functional qualities, as well as nutritional values, form a product's quality. Proteins, minerals, and vitamins are examples of nutritional values. For consumers, food protection is just as vital as food value, and more environmentally friendly organic agriculture goes a long way toward guaranteeing food safety (Navarro et al., 2011). The various sensory qualities, such as nutritional value and safety of the product, are observed to be highly dependent on

DOI: 10.1201/9781003412489-11

factors like the cultivar, growing conditions and environmental effects, fertilization type, harvest time, harvest method, storage, transportation, and processing methods (Hadayat et al., 2018).

This also involved a critical evaluation of the following topics: the definitions and history of plant biotechnology; the effective use of biotechnology in food varieties like vegetables and fruits; dynamics of ripening and perishability; the knowledge of the basic methodologies that affect fruit set, ripening, and maturation; impact of advanced procedures on the shelf life and nutritional worth of food; difficulties involved in the marketing of biotechnologically produced vegetables and fruits; the necessity of modern technologies involved in the food production in the 21st century; and the fresh patterns required in improving the shelf life and nutritive quality of genetically processed food (Abano and Buah, 2014).

While there are many environmental issues that could have an impact on human health, we concentrated on recent technical developments in this analysis that significantly affected the global influence of the quality of food and water. In fact, providing enough water of an acceptable quality has an impact on many elements of human life, especially in light of the population growth and climate change of the past century (Schwarzenbach et al., 2010). Public health faced significant issues as a result of the emergence of contaminants in marine and aquatic environments, including heavy metals, pesticide residues, medicines, and waterborne infections, as well as changes in the physical characteristics of water.

Consequently, substantial progress has been made to create cutting-edge technology for wastewater treatment, sanitation, and water resource monitoring. The World Summit on Food Security's declaration states that all people should have simple access to enough safe and nourishing food in healthy societies (Soozanipour et al., 2023).

The primary goal of the current chapter is to present a thorough summary of the use of methods based upon biotechnology to extend the value and safety of the food production.

11.2 HISTORICAL PERSPECTIVE OF FOOD PRODUCTION

The history of food inspection and hygiene backdates to the Roman era, when inspectors would patrol marketplaces to check on meat transactions and remove any rotting flesh. Scientific study has only been able to clarify the function of zoonotic disease in foodborne pathogens considering from the 20th century (O'Connor et al., 2017). Fresh produce consumption has increased significantly during the past years due to changes in consumer preferences that dietitians have helped to foster. More people are now exposed to viruses present in fresh food as a result of more intake of raw foodstuffs (Summa et al., 2018).

According to the U.S. Food and Drug Administration, this law mandates training, observance of protected and sanitary food care processes, and recording of techniques and remedial measures for harvesting, soil amendments, water quality, and animal intrusions. It has proven challenging to implement Food Safety Modernization Act Produce Safety Rule (FSMA-PSR), and it is unclear whether these regulations would increase food safety. It is evident that further study is required to understand how food diseases interact with plants, water, and soil, as well as how both people and animals can contaminate produce fields. With the growing human population come several problems to global food safety and security. The main focus of problems in the 21st century will remain on providing safe and nutritious food to increase global food security. While there are many important topics in this category, the present review concentrates on three issues: transboundary diseases, natural disasters, and produce contaminated by microbes. These issues are critical to food security and safety and require ongoing attention and discussion (Garcia et al., 2020).

11.3 BIOTECHNOLOGICAL APPROACHES TO INCREASE FOOD PRODUCTION

11.3.1 Genetic Engineering

Crop genetic engineering is seen as a potential remedy for the current situation. For crop enhancement techniques, the identification of superior alleles or haplotypes is essential. Around 3400 gardens worldwide are home to over 80,000 plant species that are conserved, and over 17,000 institutes worldwide focus on the maintenance and viable use of genetic resources in plant cultivations.

Crop improvement greatly benefits from the genetic engineering employed between plant species, which is made possible by rDNA (recombinant DNA) methodology. In this technique, genes are extracted from similar or dissimilar organisms and transmitted to the desired yield. These genes are responsible for beneficial features, including high yield, resistance to illnesses, and harsh conditions (Krenek et al., 2015).

One of the fundamental methods in plant biotechnology is plant transformation, which is used to add desired features into a current genetic setup and maintaining the genetic uniqueness of the seedstocks. This is mostly accomplished by electroporation, gene cannon (particle bombardment), or *Agrobacterium*-mediated gene transformation (Kado, 2014).

Crops resistant to insect pests need less insecticides and pesticides overall. The main benefits of transgenic crops are their increased productivity, bigger net gain per hectare, simplicity of yield supervision, low manual labor requirements, and capacity to employ nontoxic managing approaches. These benefits are achieved through a variety of techniques. The tediousness of the process, problems with trait introgression, deregulation of traits, time commitment, and high expense are some of the disadvantages of this method (Sood et al., 2019).

The rice plant's brassinosteroid levels have been successfully manipulated with this technology, increasing crop output. Improvements in crop production and progress with decreased iron level have also been shown in rice following the insertion of a gene providing the capacity to boost the production of an iron chelator. In a similar vein, genetic transformation has been used to increase plant phosphorus absorbance. By improving a plant's capacity to absorb nutrients from the soil, hazardous agrochemicals can be used less frequently, which lowers pollution in the environment (Tian et al., 2021).

Crop advances have made use of the transformation system in conjunction with a number of other methods, including gene transference applying RNA interference and clustered regularly interspaced short palindromic repeats (CRISPR). Using improved tissue culture procedures, successful genetic transformations on wheat, maize, rice, barley (*Hordeum vulgare*), and sorghum (*Sorghum bicolor*) have been reported, with gains in crop yield (Hayta et al., 2021).

Crop advances have made use of the transformation system in conjunction with a number of other methods, including genome editing using CRISPR and RNA interference. Using optimal tissue culture procedures, successful genetic transformations on rice, maize, wheat, *Hordeum vulgare* (barley), and *Sorghum bicolor* (sorghum) have been documented (Che et al., 2018).

11.3.2 Microbial Interventions

Human nutrition and health are positively impacted by prebiotics and probiotics. Prebiotics are substances that probiotics can use and ferment, enabling them to increase intestinal probiotics' metabolism and reproduction for the body's overall health. A number of dairy and nondairy products now contain probiotics because of their possible health advantages. Probiotic bacteria and yeast microencapsulation is crucial for overcoming the obstacles brought on by the beneficial microorganism's poor viability. Probiotics are more viable when microencapsulated for use in a variety of processing methods and gastrointestinal environments (Sehrawat et al., 2022).

Microbes are essential to many different culinary applications. They are the primary ingredient in the manufacture of a number of food varieties and a cause of food spoiling, which has significant financial consequences. Since ancient times, different strains of fungus and bacteria are involved in making various fermented food yields and enzymes. The normal gut flora following medication treatments has been enhanced nowadays by the utilization of prebiotics and probiotics. However, the microorganisms may also cause unfavorable alterations in food quality that compromise food safety. Thus, preserving food quality and safety requires prompt recognition or real-time checking of the microscopic organisms in the foodstuffs. Various internal (biological structures, water activity, food components, redox potential, pH, nutrient density) and external factors (relative humidity, presence of competing microorganisms, and temperature) affect the development and persistence of microbes in foodstuff. It is possible to monitor these bacteria quickly or using more conventional techniques. Thus, preserving food worth and safety requires prompt exposure or in-time controlling of the microscopic organisms in the foodstuffs (Kumari et al., 2022).

11.3.3 Precision Agriculture

Using Internet of Things (IoT) sensors, near- and far-reaching sensing techniques are combined in precision agriculture (PA) to monitor harvest settings at many growth stages. PA demands gathering and treating a substantial quantity of crop fitness data. The health of plant is affected by temperature and water (Abbasi et al., 2014).

With PA, a farmer may determine exactly which characteristics are desired for a vigorous product, how much of them are required, and where they are required, at a certain time. This forces collecting a massive sum of statistics from various areas of the field, including soil nutrients, insect and weed prevalence, plant chlorophyll level, and certain meteorological conditions (Wang et al., 2006).

The majority of current research suggests that PA-related follows have a significant impact on productivity and sustainability. The purpose of PA is to offer assessment provision methods that take into account several agricultural factors, such as soil nutrients, soil water content, wind speed, sunshine intensity, temperature, humidity, and chlorophyll content. Nevertheless, the process of developing and implementing these systems involves a number of difficulties (Mumtaz et al., 2017).

Prescription maps are essential for resource optimization because they allow agriculturalists to work out the properties needed for strong yields at any particular growth levels. This is because the main objective of PA is to produce surplus harvest by elevating properties like pesticides, fertilizer, water, etc. Many other projects conducted in the arena of agriculture emphasize on using distant detecting stages to gather images that solely display flora indicators. Prescription maps may not be created solely with the use of flora indicators; rather, a number of additional variables, including soil characteristics, soil moisture content, and weather patterns, must be taken into account (Shafi et al., 2019).

From harvesting to land preparation, virtually every facet of PA may benefit from the use of remote sensing technology. The face of PA has altered because of the availability of great three-dimensional perseverance multi-temporal satellite data and availability of ground-based proximity detectors on commercial basis. Numerous sophisticated methods, including as empirical, regression, and skill learning methods, are employed to investigate the possible uses of distant recognition in Pennsylvania. In a similar vein, numerous plantation guides, such as those for variable irrigation scheduling, weed mapping, fertilizer management, yield forecasting, and disease control, have been created and evaluated for their potential to support PA operations. But before remote sensing technology may possibly be widely used in both profitable and non-profitable agriculture, a number of issues must be resolved (Sishodia et al., 2020).

11.4 CASE STUDIES

In recent decades, the field of science and engineering has witnessed notable progress, leading to the emergence of numerous technologies within the foodstuff and agricultural sector. These advancements have significantly enhanced our comprehension of the characteristics of food and its treating methods. In contemporary society, there exists a prevailing belief among consumers that food consumption not only serves the purpose of satiating hunger but likewise shows a substantial part in promoting vigor and mitigating the risk of diseases. Hence, it is imperative to engage in ongoing technological advancements in order to effectively respond to the dynamic demands of customers (Bigliardi and Galati, 2013). This is because technologies can occasionally prove inadequate, necessitating the constant pursuit of innovative solutions that support the specific desires of customers. Moreover, given the significant influence of the foodstuff business on the worldwide low cost, these emerging technologies, including genomic engineering, assume even greater importance (Ouyang et al., 2017). The food industry has experienced notable benefits from recent advancements in genetic engineering, specifically in relation to enhancing crucial agronomic traits such as plant height and weight. Additionally, genetic engineering has also contributed to the improvement of non-agronomic traits, including physical and nutritious properties of crops. As a result of these advancements, it will become feasible to develop food items with improved attributes on a bigger production scale. The utilization of recombinant genetic technology in the agricultural sector has had a significant influence on other domains, such as food science, medicine (where plants are utilized to produce natural medications), agriculture, and animal breeding (Kaboli and Babazada, 2018). Following advancements significantly enhanced the design of food products. In recent times, the utilization of RNA interference (RNAi) to suppress gene expression has gained prominence in plant biotechnology. This methodology involves the usage of tiny siRNA (interfering RNA) particles, which are double-stranded and typically consist of 21–27 nucleotides in length. The application of RNAi-mediated gene suppression in plants has facilitated the incorporation of desirable agronomic features into biotechnology-derived products (Punjabi et al., 2018). This process of RNA interference involving gene expression is suppressed by mechanisms involving translational repression and mRNA degradation (Iwakawa and Tomari, 2015). Although many are persisting welfare distresses associated with crops that endure interfering RNA-mediated gene suppression, a groundbreaking genome-transmitting tool known as CRISPR-Cas9 has been identified. This tool has the ability to achieve the same objective as RNAi, which is the improvement in the expression of gene, but with more efficiency (Stern et al., 2010). This breakthrough has opened up numerous possibilities for submissions in various industrialized sectors (Cong et al., 2013; Doudna and Charpentier, 2014). In summary, CRISPR-Cas9 is a cutting-edge machinery used for genome editing, enabling the precise manipulation of genetic information by introducing, removing, or modifying specific loci within the genome (Mali et al., 2013).

The sgRNA, consisting of a sequence of 20 bases, functions as a protospacer system that governs particularly the CRISPR-Cas9 system. It confirms the precise targeting of the desired region of DNA. On the other hand, the CRISPR-Cas9 machinery works as "molecular scissors," unwinding and subsequently cleaving the target DNA at definite locations (Shalem et al., 2014). CRISPR-Cas9, a potent tool with varied applicability, is founded upon the principles of Watson–Crick base pairing (Doudna and Charpentier, 2014). The utilization of CRISPR-Cas9 in recent studies has presented promising prospects for exploring gene function and uncovering significant biological knowledge. For instance, researchers have employed CRISPR-Cas9 to engineer microbial consortia, thereby enabling the investigation of their collective behavior and interactions (Bittihn et al., 2018). Additionally, the programmable nature and potency of CRISPR-Cas9 systems have been harnessed to develop effective antimicrobials (Bikard and Barrangou, 2017). Furthermore, the use of CRISPR-Cas9 has facilitated the design of vaccination strategies against invasive genetic elements in microorganisms, as well as the regulation of genome expression in a controllable and flexible mode (Di Cristina and Carruthers, 2018). The agricultural and food sciences are among the key domains that stand to gain significant advantages from the prospective applications of CRISPR-Cas9. According to Barrangou and Horvath (2012), the implementation of this innovative approach has the potential to facilitate the creation of nutritious food products that align with the increasing demands of customers. In addition, the application of the CRISPR-Cas9 tool has the potential to boost the resilience of crops against both biotic and abiotic stressors. After the identification of CRISPR-Cas9, a multitude of interdisciplinary investigations were conducted with the objective of modifying the CRISPR methodology to various prospective industrial applications. The food and agricultural science sector has shown considerable interest in utilizing this technology in response to the growing need for food resources with enhanced features. The initial work utilizing the CRISPR-Cas9 method was centered on gene management in rice, a prominent staple yield globally (Feng et al., 2013). The present study employed the CaMV 35S and OsU6-2 sponsors to guide the Cas9 gene and sgRNA, respectively, resulting in successful and effective alteration of the rice genomic setup. The initial research showcased the viability of the method and provided impetus for further research endeavors aimed at applying this skill for more prospective nourishment sources. Within the year under consideration, a research led by (Li et al., 2018) assessed the efficacy of mutagenesis in five distinct mark genes of Arabidopsis by employing sgRNA: pcoCas9. This evaluation involved the examination of seven sequences. The significance of defining mutagenesis was determined to be contingent upon the binding affinity of the sgRNA (single guide RNA) and its mark sequences. Viral infections are a significant contributing factor to the degradation of important crops within natural ecosystems. According to Alexander et al. (2017), the existence of these virions has a notable impact on crop productivity as they manifest specific symptoms in plants, resulting in a large economic burden. Genetic engineering has proven to be a valuable strategy in enhancing plant resistance against a wide spectrum of viral diseases. Nevertheless, the procedures that have been previously established exhibit certain limits in their ability to fully or partially eradicate viral infections in agricultural crops. Given the widely held belief that viral infections in agricultural crops result in a reduction in worldwide production by approximately 10%–15%, the utilization of the CRISPR system presents a promising avenue for addressing these constraints. The primary objective of this approach was to mitigate geminivirus proliferation within a transgenic plant model, hence enhancing the overall efficiency of the system.

Ali et al. (2015) showed a parallel investigation to find out the molecular protection of plants against tomato yellow leaf curl virus (TYLCV), which is classified as a DNA virus. The researchers employed CRISPR-Cas9 technology to increase the plants' immune response. Consequently, the interference and cleavage of the viral genome were effectively achieved, leading to a decrease in viral DNA accumulation within the plant. In a similar vein, Tashkandi et al. (2018) devised a CRISPR-Cas9 method for the purpose of genetically modifying *Solanum lycopersicum* and *Nicotiana benthamiana* plants, with the aim of stimulating immunity against TYLCV. The aim of the research was to measure the plants' resistance to viral infection.

11.5 CHALLENGES AND ETHICAL CONSIDERATIONS

The utilization of two guide RNAs (gRNAs) in conjunction with a repair template has the potential to confer the intended resistance against bispyribac sodium herbicide. Following a time gap of 36 days, it was perceived that the wild-type rice plants did not exhibit survival, whereas the plants subjected to CRISPR-Cas9 mutation had normal growth. Li et al. (2015) implemented a comparable approach in the cultivation of soybean, resulting in the development of crops that exhibited notable resistance to chlorsulfuron. Collectively, the findings of these studies indicate that individual plants necessitate a highly tailored approach to genome editing. The majority of agricultural crops exhibit a limited ability to retain their physical characteristics following mechanical damage such as bruising or slicing. The white button mushrooms, scientifically known as *Agaricus bisporus*, undergo a pronounced enzymatic browning reaction as a result of the presence of a distinct polyphenol oxidase (PPO) gene. A team led by Yinong Yang at Pennsylvania State University conducted a genetic modification by selectively deleting a certain gene family responsible for encoding PPO enzymes, resulting in the development of mushrooms that exhibit non-browning characteristics. The scientific committee became interested in this finding due to its status as one of the initial gene-edited crops that fell beyond the purview of regulatory oversight (Waltz, 2016). Potato is widely recognized as a highly significant staple crop in global agriculture due to its substantial nutritional content. Certain wild varieties of potatoes include higher nutritional content. Nonetheless, the process of producing these varieties using conventional genetic procedures can be time-consuming due to sexual incompatibility. In the present context, the utilization of CRISPR-Cas9 technology holds significant importance in the enhancement of potato crop yields within a reduced time frame. In the subsequent year, the same study team employed an innovative DNA-free genome editing methodology to specifically mark the aforementioned gene in potato. This was accomplished through the utilization of ribonucleoproteins (RNPs), as documented by Andersson et al. (2018).

The introduction of CRISPR-Cas9 machinery by RNP (ribonucleoprotein) complexes directed the improvement of commercial lines exhibiting increased yields, without the need for DNA integration. The conducted investigations have significant commercial significance due to the enhanced quality of amylopectin potato starch. SGAs (steroidal glycoalkaloids) are compounds that have a propensity to gather in various tissues of the potato plant. When consumed by humans, these molecules can impart a bitter taste and exhibit harmful effects on a range of organisms. Hence, it is advisable to decrease the consumption of SGA content in order to enhance nutritional quality. In their study, Nakayasu et al. (2018) effectively disrupted the St16DOX gene, which encodes a steroid involved in the synthesis of SGA, through the introduction of gRNAs into a CRISPR-Cas9 vector. Consequently, they were able to generate hairy roots of potato that were free from SGA. Cereal-derived products that exhibit a high amylose content possess the potential to confer advantageous effects on human health. Therefore, the utilization of CRISPR-Cas9 has become vital in the enhancement of amylose and resistant starch levels in cereal crops, particularly rice. In their study, Sun et al. (2017) utilized the CRISPR-Cas9 scheme to induce directed mutation in the SBEI and SBEIIb genes of rice. The rice mutant exhibited a notable rise in amylose and resistant starch content, with respective increments of 25% and 9.8%. This enhancement in starch nutritional qualities inside the rice grain has been seen as a result. Zhang et al. (2018) conducted a comparable investigation wherein they employed CRISPR-Cas9 machinery to mark the Waxy gene. Their objective was to decrease the amylose concentration in rice and transform it into a glutinous variety, while ensuring that other agronomic features and nutritional values remained unaltered. The aforementioned research has provided the proof to support the notion that the use of CRISPR-Cas9 machinery holds significant promise in the commercialization of high-quality crops such as rice. Tomato, a food resource known for its great nutritional value, is also found to contain a substantial quantity of GABA (γ-aminobutyric acid), a primary neurotransmitter (Takayama and Ezura, 2015). The work studied by Cho et al. (2017) demonstrated the successful enhancement of GABA synthesis in *Corynebacterium glutamicum* through the implementation of CRISPRi-mediated gene regulation. Furthermore, a comparable strategy was employed to develop the production of L-glutamate and L-lysine in the same organism. In their study, Wu et al. (2015) employed CRISPRi technology to modulate key metabolic pathways, including fatty acid synthesis, glycolysis, and the TCA (tricarboxylic acid) cycle, in order to enhance the production of flavonoids in *Escherichia coli*. In addition, Wu et al. (2017) conducted genetic modifications on *E. coli* to boost the manufacture of resveratrol. They successfully obtained a 1.7-fold increase in resveratrol production by manipulating the malonylCoA pathway (eno, adhE, fabB, sucC, fumC, fabF) using CRISPRi technology. Furthermore, the optimization of 1,4-butanediol synthesis was achieved through the utilization of CRISPRi and CRISPR-Cas9 techniques targeting specific genes (gabD, ybgC, tesB) resulting in a 1.8 g/L of yield (Wu et al., 2017). In a study conducted by Li et al. (2015), it was shown that by utilizing a CRISPR-Cas9 editing tool to perform gene deletions, *E. coli* was able to achieve significant levels of β-carotene production, with titers reaching 2.0 g/L. Furthermore, Kim et al. successfully enhanced the generation of lycopene and isoprene in *E. coli* through the manipulation of key components within the mevalonate (MVA) pathway, specifically mvaK1, mvaE, and ispA. Significantly, the utilization of CRISPR-Cas technology depicted a crucial role in the manipulation of industrialized field, such as the yeast strains (*Saccharomyces cerevisiae*) employed in the brewing industry. A group of researchers affiliated with the Keasling lab at the University of California, Berkeley, successfully engineered yeast strains with distinct flavors by presenting four novel genes and enhancing the expression of the mevalonate pathway (Denby et al., 2018). Consequently, the researchers successfully eradicated the utilization of hops throughout the fermentation process, which had hitherto served to impart distinct flavor characteristics to the beverage. The consequences of these discoveries are significant in terms of sustainability, given the substantial resource requirements, including water, fertilizers, and energy, associated with hop cultivation. The CRISPR system has demonstrated remarkable versatility as a genome editing tool, leading to a versatile application within a restricted time period. The utilization of CRISPR technology in the realm of agriculture or food systems offers advantages in terms of enhanced speed, simplicity, precision, and perhaps reduced expenses compared to conventional genetic engineering techniques. Previous research has indicated that the achievement of effective genome editing through the CRISPR-Cas9 system necessitates the incorporation of a single guide RNA (sgRNA) that possesses a distinct 20-base sequence that is particular to the target DNA. The advancement in precision of gene targeting systems has the potential to eliminate the requirement for lengthy nucleotide single guide RNA (sgRNA). This, in turn, would streamline the process of designing highly specific sgRNA. In addition to crop enhancement, genetic engineering has also proven advantageous in the realm of fermentation processes, since it has facilitated the development of strains that exhibit enhanced functional qualities. *Lactobacillus* spp. strains are microorganisms that hold significant importance in the nourishment diligence, particularly as psychobiotics ad probiotics (Misra and Mohanty, 2017). Hence, the utilization of CRISPR-Cas9 technology in these bacterial strains has the potential to substantially raise their resilience to various stressors, thereby resulting in heightened efficiency of the fermentation process and ultimately augmenting the overall quality of food commodities.

11.6 FUTURE DIRECTIONS

Food undergoes various physical, chemical, or biological transformations subsequent to its harvest, resulting in its progressive deterioration. Food preservation is a systematic approach aimed at preventing the introduction and proliferation of microorganisms that may lead to contamination. According to Morata et al. (2021), microbes, along with enzymes, play a vital role in driving change and should thus be the focal point of conservation strategies. The following is a compilation of contemporary evolving technologies aimed at food preservation, with the objective of minimizing or eradicating significant pathogenic bacteria in food, as well as facilitating the isolation of biologically active chemicals that hold utility within the food business. The primary application of the high hydrostatic pressure (HHP) methodology is to induce physical and chemical modifications in chemical compounds found in food, with the aim of enhancing their quality. This technique is sometimes referred to as cold pasteurization or pressurization. The usefulness of this method in preserving food has garnered significant attention in the food business. It is considered superior to conventional thermal processes due to its ability to mitigate the loss of nutritional quality and sensory qualities, which are inevitable consequences of the latter (Knorr 2018). It has been shown to be effective in deactivating bacterial spores and enzymes (Meyer et al., 2021). One of the advantages linked with the utilization of HHP treatment, in comparison to other non-thermal technologies, is its ability to prevent food change. This is achieved through the application of uniform and rapid pressure to the system, hence eliminating pressure gradients. In contrast to heat treatments, HHP treatment is not influenced by the volume and form of the sample, resulting in reduced processing time for large quantities of food. According to Kimura et al. (2019), HHP technology, due to its low temperature nature, does not lead to the degradation of thermolabile nutrients like vitamins. The use of HHP to meals preserves its nutritional values and organoleptic characteristics, resulting in minimal loss of these attributes when compared to fresh products. The numerical value provided by the user is six. The conservation of the freshness and quality of HHP processed foods can be attributed to the use of specific temperature-pressure-time parameters, which result in minimal chemical alterations while maintaining the sensory and nutritional qualities intact. Numerous studies have provided evidence to support the notion that items subjected to high pressure processing retain the attributes often associated with fresh, untreated products. Consequently, distinguishing between high pressure processed products and their untreated counterparts can often provide a challenge (Meyer et al., 2021).

11.7 CONCLUSION

Advances in biotechnology present a promising opportunity to transform food production and guarantee elevated standards for food safety and quality. The investigation of data-driven approaches, PA, and genetic engineering highlights the lasting benefits of biotechnology in crop production and disease prevention. This chapter focuses on the use of biotechnology to improve food quality while also addressing global concerns about food security. Taking responsible and ethical advantage of these developments can help create a future in which biotechnological approaches are essential to supplying the growing needs for safe, sustainable, and high-quality food production worldwide.

REFERENCES

Abano, E.E., Buah, J.N., 2014. Biotechnological approaches to improve nutritional quality and shelf life of fruits and vegetables. *International Journal of Engineering and Technology*, *4*(11).

Abbasi, A.Z., Islam, N., Shaikh, Z.A., 2014. A review of wireless sensors and networks' applications in agriculture. *Comput. Stand. Interfaces* 36, 263–270.

Alexander, H.M., Bruns, E., Schebor, H., Malmstrom, C.M., 2017. Crop-associated virus infection in a native perennial grass: reduction in plant fitness and dynamic patterns of virus detection. *J. Ecol.* 105, 1021–1031.

Ali, Z., Abulfaraj, A., Idris, A., Ali, S., Tashkandi, M., Mahfouz, M.M., 2015. CRISPR/Cas9-mediated viral interference in plants. *Genome Biol.* 16, 238.

Andersson, M., Turesson, H., Olsson, N., Fält, A.-S., Ohlsson, P., Gonzalez, M.N., Samuelsson, M., Hofvander, P., 2018. Genome editing in potato via CRISPR-Cas9 ribonucleoprotein delivery. *Physiol. Plant.* 164, 378–384.

Barrangou, R., Horvath, P., 2012. CRISPR: new horizons in phage resistance and strain identification. *Annu. Rev. Food Sci. Technol.* 3, 143–162. BASF, 2018. BASF Licenses CRISPR-Cas9 Genome-Editing Technology from the Broad Institute. https://www.basf.com/us/en/company/news-and-media/news-releases/ 2017/03/P-US-17-026.html (accessed 10.20.18).

Bigliardi, B., Galati, F., 2013. Innovation trends in the food industry: the case of functional foods. *Trends Food Sci. Technol.* 31, 118–129.

Bikard, D., Barrangou, R., 2017. Using CRISPR-Cas systems as antimicrobials. *Curr. Opin. Microbiol.* 37, 155–160.

Bittihn, P., Din, M.O., Tsimring, L.S., Hasty, J., 2018. Rational engineering of synthetic microbial systems: from single cells to consortia. *Curr. Opin. Microbiol.* 45, 92–99.

Che, P., Anand, A., Wu, E., Sander, J.D., Simon, M.K., Zhu, W., Sigmund, A.L., Zastrow-Hayes, G., Miller, M., Liu, D., 2018. Developing a flexible, high-efficiency Agrobacterium-mediated sorghum transformation system with broad application. *Plant Biotechnol. J.* 16, 1388–1395.

Cho, J.S., Choi, K.R., Prabowo, C.P.S., Shin, J.H., Yang, D., Jang, J., Lee, S.Y., 2017. CRISPR/Cas9-coupled recombineering for metabolic engineering of Corynebacterium glutamicum. *Metab. Eng.* 42, 157–167.

Cong, L., Ran, F.A., Cox, D., Lin, S., Barretto, R., Habib, N., Zhang, F., 2013. Multiplex genome engineering using CRISPR/Cas systems. *Science* 339, 1231143.

Denby, C.M., Li, R.A., Vu, V.T., Costello, Z., Lin, W., Chan, L.J.G., Williams, J., Donaldson, B., Bamforth, C.W., Petzold, C.J., Scheller, H.V., Martin, H.G., Keasling, J.D., 2018. Industrial brewing yeast engineered for the production of primary flavor determinants in hopped beer. *Nat. Commun.* 9, 965.

Di Cristina, M., Carruthers, V.B., 2018. New and emerging uses of CRISPR/Cas9 to genetically manipulate apicomplexan parasites. *Parasitology* 145, 1119–1126.

Doudna, J.A., Charpentier, E., 2014. Genome editing. The new frontier of genome engineering with CRISPR-Cas9. *Science* 346, 1258096.

Feng, Z., Zhang, B., Ding, W., Liu, X., Yang, D.-L., Wei, P., Cao, F., Zhu, S., Zhang, F., Mao, Y., Zhu, J.-K., 2013. Efficient genome editing in plants using a CRISPR/Cas system. *Cell Res.* 23, 1229–1232.

Garcia, S.N., Osburn, B.I., Jay-Russell, M.T., 2020. One health for food safety, food security, and sustainable food production. *Front. Sustain. Food Syst.* 4, 1.

Hadayat, N., De Oliveira, L.M. Da Silva, E., Han, L., Hussain, M., Liu, X., Ma, L.Q., 2018. Assessment of trace metals in five most-consumed vegetables in the US: conventional vs. organic. *Environ. Pollut.* 243, 292–300.

Hayta, S., Smedley, M.A., Clarke, M., Forner, M., Harwood, W.A., 2021. An efficient Agrobacterium-mediated transformation protocol for hexaploid and tetraploid wheat. *Current Protocols.* 1, e58.

Iwakawa, H., Tomari, Y., 2015. The functions of microRNAs: mRNA decay and translational repression. *Trends Cell Biol.* 25, 651–665.

Kaboli, S., Babazada, H., 2018. CRISPR mediated genome engineering and its application in industry. *Curr. Issues Mol. Biol.* 26, 81–92.

Kado, C.I., 2014. Historical account on gaining insights on the mechanism of crown gall tumorigenesis induced by Agrobacterium tumefaciens. *Front. Microbiol.* 5, 340.

Kimura K, Ida M, Yosida Y, et al., 2019. Comparison of keeping quality between pressure-processed jam and heat-processed jam: changes in flavor components, hue, and nutrients during storage. *Biosci Biotech Biochem.* 58(8), 1386–1391.

Krenek, P., Samajova, O., Luptovciak, I., Doskocilova, A., Komis, G., Samaj, J., 2015. Transient plant transformation mediated by Agrobacterium tumefaciens: principles, methods and applications. *Biotechnol. Adv.* 33, 1024–1042.

Kumari, R., Jayachandran, L.E., Kumar, S., 2022. Role of microbes in the food industry. *Role of Microbes in Industrial Products and Pro*cesses. Wiley, pp. 59–79.

Li, Rui, Li, Ran, Li, X., Fu, D., Zhu, B., Tian, H., Luo, Y., Zhu, H., 2018. Multiplexed CRISPR/Cas9-mediated metabolic engineering of γ-aminobutyric acid levels in Solanum lycopersicum. *Plant Biotechnol. J.* 16, 415–427.

Li, Y., Lin, Z., Huang, C., Zhang, Y., Wang, Z., Tang, Y.-J., Chen, T., Zhao, X., 2015. Metabolic engineering of *Escherichia coli* using CRISPR-Cas9 meditated genome editing. *Metab. Eng.* 31, 13–21.

Mali, P., Aach, J., Stranges, P.B., Esvelt, K.M., Moosburner, M., Kosuri, S., Yang, L., Church, G.M., 2013a. CAS9 transcriptional activators for target specificity screening and paired nickases for cooperative genome engineering. *Nat. Biotechnol.* 31, 833.

Meyer RS, Cooper KL, Knorr D, et al. 2021. High pressure sterilization of foods. *Food Technol.* 54(11), 67–71.

Misra, S., Mohanty, D., 2017. Psychobiotics: a new approach for treating mental illness? *Crit. Rev. Food Sci.* Nutr. 59(8), 1–7.

Morata Barrado A., 2021. *New Food Consevation Technologies.* Vol. 2. Madrid, España: A. Madrid Vicente Ediciones..

Mumtaz, R., Baig, S., Fatima, I., 2017. Analysis of meteorological variations on wheat yield and its estimation using remotely sensed data. A case study of selected districts of Punjab Province, Pakistan (2001-14). *Ital. J. Agron.* 12. 254-270

Nakayasu, M., Akiyama, R., Lee, H.J., Osakabe, K., Osakabe, Y., Watanabe, B., Sugimoto, Y., Umemoto, N., Saito, K., Muranaka, T., Mizutani, M., 2018. Generation of α-solanine-free hairy roots of potato by CRISPR/Cas9 mediated genome editing of the St16DOX gene. *Plant Physiol. Biochem.* 131, 70–77.

Navarro, P., Perez-Lopez, A., Mercader, M., Carbonell-Barrachina, A.A., Gabaldon, J., 2011. Antioxidant activity, color, carotenoids composition, minerals, vitamin C and sensory quality of organic and conventional mandarin juice, cv. Orogrande. *Food Sci. Technol. Int.* 17, 241–248.

O'connor, E.L., Sims, L., White, K.M. 2017. Ethical food choices: examining people's Fair Trade purchasing decisions. *Food Qual. Pref.* 60, 105–112.

Ouyang, B., Gu, X., Holford, P., 2017. Plant genetic engineering and biotechnology: a sustainable solution for future food security and industry. *Plant Growth Regul.* 83, 171–173.

Punjabi, M., Bharadvaja, N., Jolly, M., Dahuja, A., Sachdev, A., 2018. Development and evaluation of low phytic acid soybean by siRNA triggered seed specific silencing of inositol polyphosphate 6-/3-/5-kinase gene. *Front. Plant Sci.* 9, 310374

Rempelos, L., Baranski, M., Wang, J., Adams, T.N., Adebusuyi, K., Beckman, J.J., Brockbank, C.J., Douglas, B.S., Feng, T., Greenway, J.D. 2021. Integrated soil and crop management in organic agriculture: a logical framework to ensure food quality and human health? *Agronomy* 11, 2494.

Schwarzenbach, R.P., Egli, T., Hofstetter, T.B., Von Gunten, U., Wehrli, B., 2010. Global water pollution and human health. *Annu. Rev. Environ. Res.* 35, 109–136.

Sehrawat, R., Abdullah, S., Khatri, P., Kumar, L., Kumar, A., Mujumdar, A.S., 2022. Role of drying technology in probiotic encapsulation and impact on food safety. *Dry. Technol.* 40, 1562–1581.

Shafi, U., Mumtaz, R., García-Nieto, J., Hassan, S.A., Zaidi, S.a.R., Iqbal, N., 2019. Precision agriculture techniques and practices: from considerations to applications. *Sensors* 19, 3796.

Shalem, O., Sanjana, N.E., Hartenian, E., Shi, X., Scott, D.A., Mikkelson, T., Heckl, D., Ebert, B.L., Root, D.E., Doench, J.G., Zhang, F., 2014. Genome-scale CRISPR-Cas9 knockout screening in human cells. *Science* 343, 84–87.

Sishodia, R.P., Ray, R.L., Singh, S.K., 2020. Applications of remote sensing in precision agriculture: a review. *Remote Sensing* 12, 3136.

Sood, P., Singh, R.K., Prasad, M., 2019. Millets genetic engineering: the progress made and prospects for the future. *Plant Cell Tissue Organ Cult.* 137, 421–439.

Soozanipour, A., Ejeian, F., Boroumand, Y., Rezayat, A., Moradi, S., 2023. Biotechnological advancements towards water, food and medical healthcare: a review. *Chemosphere* 312, 137185.

Stern, A., Keren, L., Wurtzel, O., Amitai, G., Sorek, R., 2010. Self-targeting by CRISPR: gene regulation or autoimmunity? *Trends in Genetics* 26(8), 335–340.

Summa, M., Henttonen, H. Maunula, L., 2018. Human noroviruses in the faeces of wild birds and rodents-new potential transmission routes. *Zoonoses Public Health.* 65, 512–518.

Sun, Y., Jiao, G., Liu, Z., Zhang, X., Li, J., Guo, X., Du, W., Du, J., Francis, F., Zhao, Y., Xia, L., 2017. Generation of high-amylose rice through CRISPR/Cas9-mediated targeted mutagenesis of starch branching enzymes. *Front. Plant Sci.* 8, 298.

Takayama, M., Ezura, H., 2015. How and why does tomato accumulate a large amount of GABA in the fruit? *Front. Plant Sci.* 6, 612

Tashkandi, M., Ali, Z., Aljedaani, F., Shami, A., Mahfouz, M.M., 2018. Engineering resistance against tomato yellow leaf curl virus via the CRISPR/Cas9 system in tomato. *Plant Signal. Behav.* 13, 1–7

Tian, Z., Wang, J.W., Li, J., Han, B., 2021. Designing future crops: challenges and strategies for sustainable agriculture. *Plant J.* 105, 1165–1178.

Waltz, E., 2016. Gene-edited CRISPR mushroom escapes US regulation. *Nature News* 532, 293.

Wang, N., Zhang, N. Wang, M., 2006. Wireless sensors in agriculture and food industry-Recent development and future perspective. *Comp. Electron. Agri.* 50, 1–14.

Wu, J., Du, G., Chen, J., Zhou, J., 2015. Enhancing flavonoid production by systematically tuning the central metabolic pathways based on a CRISPR interference system in *Escherichia coli. Sci. Rep.* 5, 13477.

Wu, J., Zhou, P., Zhang, X., Dong, M., 2017. Efficient de novo synthesis of resveratrol by metabolically engineered *Escherichia coli. J. Ind. Microbiol. Biotechnol.* 44, 1083–1095.

Zhang, J., Zhang, H., Botella, J.R., Zhu, J.K., 2018. Generation of new glutinous rice by CRISPR/Cas9-targeted mutagenesis of the waxy gene in elite rice varieties. *J. Integr. Plant Biol.* 60, 369–375.

Index

Note: **Bold** page numbers refer to tables and *italic* page numbers refer to figures.